W9-CNY-352

Qualitative analysis for social scientists

The teaching of qualitative analysis in the social sciences is rarely undertaken in a structured way. This handbook is designed to remedy that and to present students and researchers with a systematic method for interpreting "qualitative data," whether derived from interviews, field notes, or documentary materials.

The special emphasis of the book is on how to develop theory through qualitative analysis. The reader is provided with the tools for doing qualitative analysis, such as codes, memos, memo sequences, theoretical sampling and comparative analysis, and diagrams, all of which are abundantly illustrated by actual examples drawn from the author's own varied qualitative research and research consultations, as well as from his research seminars. Many of the procedural discussions are concluded with rules of thumb that can usefully guide the researchers' analytic operations. The difficulties that beginners encounter when doing qualitative analysis and the kinds of persistent questions they raise are also discussed, as is the problem of how to integrate analyses. In addition, there is a chapter on the teaching of qualitative analysis and the giving of useful advice during research consultations, and there is a discussion of the preparation of material for publication.

The book has been written not only for sociologists but for all researchers in the social sciences and in such fields as education, public health, nursing, and administration, who employ qualitative methods in their work.

Anselm Strauss is Professor of Sociology in the Department of Social and Behavioral Science at the University of California, San Francisco. Together with Barney Glaser, he was co-developer of the "grounded theory" approach to qualitative analysis, which was published in *Discovery of Grounded Theory*, (Aldine, 1967), and he has undertaken qualitative research in a wide variety of areas, particularly in the fields of health and medicine

Qualitative analysis for social scientists

ANSELM L. STRAUSS

University of California, San Francisco

and

Tremont Research Institute, San Francisco

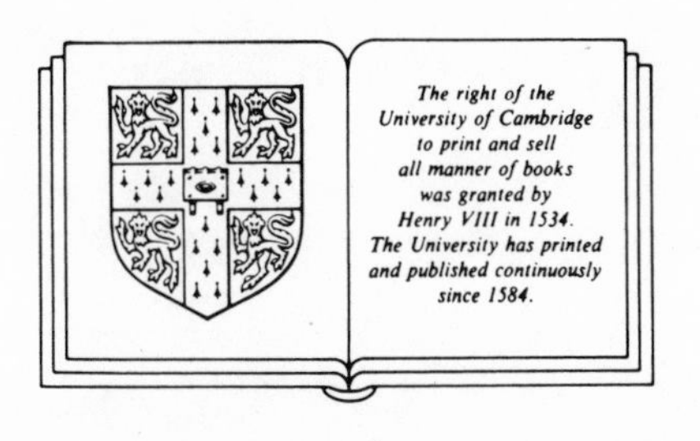

CAMBRIDGE UNIVERSITY PRESS

Cambridge
New York New Rochelle
Melbourne Sydney

Published by the Press Syndicate of the University of Cambridge
The Pitt Building, Trumpington Street, Cambridge CB2 1RP
32 East 57th Street, New York, NY 10022, USA
10 Stamford Road, Oakleigh, Melbourne 3166, Australia

First published 1987
Reprinted 1988, 1989

Printed in the United States of America

Library of Congress Cataloging-in-Publication Data

Strauss, Anselm L.

Qualitative analysis for social scientists.

Includes bibliographies.

1. Social sciences – Methodology. 2. Social sciences – Statistical methods. I. Title.

H61.S8824 1987 300'.72 86-21608

British Library Cataloguing in Publication Data

Strauss, Anselm L.

Qualitative analysis for social scientists.

1. Social sciences – Methodology

I. Title

300'.72 H61

ISBN 0 521 32845 4 hard covers
ISBN 0 521 33806 9 paperback

To Fran

Contents

Preface

This book is a handbook of sorts for the better understanding of social phenomena through a particular style of qualitative analysis of data (*grounded theory*). That mode of doing analysis is only one of many used in qualitative research. It is designed especially for *generating and testing theory*. Although its originators and principal users to date are sociologists, it has been found useful by social scientists from other disciplines, as well as researchers in education, public health, social work, and nursing – found useful because it is a general style of doing analysis that does not depend on particular disciplinary perspectives.

The purpose of this book is to instruct anyone who is interested in learning or improving his or her ability to do qualitative analysis of data. Traditionally, researchers learn such analysis by trial and error, or by working with more experienced people on research projects. Writings on qualitative method, qualitative research, ethnographic method, fieldwork, and interviewing are long on their discussions of data collection and research experiences and short on analysis – how to interpret the data (Miles 1983, pp. 125–6). How often one hears the cry of distress, "What do I do now with all those data I've collected?" Or from more experienced researchers, "I should have done much more with all those data – or at least done it faster."

So in this handbook I have attempted to address the issues of *how* one does theoretically informed interpretations of materials, and does them efficiently and effectively. (I assume experience with or at least knowledge on the part of readers of qualitative data-collecting methods.) To that end, detailed discussions of basic analytic procedures are given, as are rules of thumb for proceeding with them. The illustrative materials are drawn from my research or that of research associates and students, with repeated use of materials from three projects, in order to give a heightened sense of procedural continuity. To these materials, I have added rather specific commentaries to make clearer what is happening analytically in them. Also included are typical problems encountered when learning the grounded theory mode of analysis (and probably any mode of qualitative analysis), as well as how to write up findings and interpretations for publication. One chapter is addressed to the teaching of analysis, and throughout the book one can quite literally see it being taught.

In addition, how qualitative analysis is *actually* done is made vivid by showing through illustration various researchers working together as teachers, learners, and research teammates. The realities of doing analysis – whether one does it as a solo researcher or is fortunate enough to have working colleagues – are particularly difficult to convey, except by showing researchers *at* work. Discussions of how to do analysis, even descriptions of how it should be done, are not enough unless supplemented by visualization of researchers engaged in their work, whether seen in person or shown in the printed form attempted in this book.

A word more about the illustrative materials. In assembling them, a decision had to be made about which ones to use. Originally I had thought of drawing on materials dealing with a relatively wide variety of substantive phenomena. In the end I opted for using those from my own research and teaching, because even with their necessarily restricted scope they would better serve to convey how analysis is taught and learned, as well as to make analytic operations more comprehensible for readers. From considerable experience I have learned that certain operations – particularly the coding, the use of comparative analysis and theoretical sampling, and the integration of findings into a coherent theoretical formulation – are especially difficult to teach and carry out with ease. While my own research interests are reflected in the illustrations (particularly the sociology of health and illness, and the sociology of work/professions), the grounded theory mode of analysis has been used successfully in other substantive areas, including sociological studies of scientists' work (Gerson 1983; Star 1983, 1984, 1985; Star and Gerson, forthcoming; Volberg 1983; Clarke 1984), drug addiction (Biernacki 1986; Rosenbaum 1981), house construction (Glaser 1976), negotiation (Strauss 1977), social support of the elderly (Bowers 1983), the alcoholic policy arena (Wiener 1981), organizational contraction and shifts in the division of labor (Hazan 1985), remarriages of middle-aged divorcees (Cauhape 1983), inheritance (Glaser, forthcoming), biographies (Schuetze 1981; Rieman, forthcoming), abortion (Hoffman-Riem 1984), adult socialization (Broadhead 1983), organization (Gerson 1986), and so forth. In short, grounded theory analysis is a general style of research, not at all restricted in range of analyzable data.

I suggest that readers handle this rather sizable book in the following way. First, it will perhaps be most sensible to read the entire book very quickly, indeed only scanning the illustrative materials. Second, then reread it, carefully. Third, study selected chapters, especially those concerned specifically with techniques: notably those pertaining to coding, memoing, integrative procedures, and the detailed analytic commentaries. In general, think of the book as a smorgasbord: Run your eye down the table of edibles, then move to the essential foods, then return repeatedly to those you especially need – or *still* need. You may find that what you get from this book at one phase of your research (or one stage of your research development) will change as you move from one to another phase or stage. Presumably the book may also function from time to time as a reference volume.

The analytic mode introduced here is perfectly learnable by any competent social researcher who wishes to interpret data using this mode (either without quantitative methods or in conjunction with them). It takes no special genius to do that analysis effectively. True, when students are first learning it, they often listen in awe to their teacher–researcher and mutter about his or her genius at this kind of work, but despair of their own capacities for doing it. They never could! (I shall touch on this psychological problem later in the book.) Inevitably, students get over this phase, if unhappily they have been in it, as they gain increasing competence as well as confidence in that competence. Of course they do not believe they can do it until their first major piece of research – usually a thesis – has actually been completed.

But let us not dwell on students: The point is simply that learning this mode of qualitative analysis is entirely feasible. Like any set of skills, the learning involves hard work, persistence, and some not always entirely pleasurable experience. Furthermore, the latter is requisite to discovering one's own adaptations of any methodology (any technology), a composite of situational context, a personal biography, astuteness, theoretical and social sensitivity, a bit of luck – and courage.

This leads me to a second – though less primary – purpose for writing this book. As mentioned earlier, the literature on qualitative analysis is sparse, and even the ethnographic monographs generally give little clue as to the authors' analytic processes. I would predict, though, that this long era of flying by the seat of one's pants and direct-apprenticeship socialization and relative lack of public communication about analytic techniques, styles, and experiences is about to be supplemented by books like mine or the recent *Source Book of New Methods* by Miles and Huberman (1984). So, my second reason for writing this book was to further the systematic, detailed, and lengthy reporting–illustrating of *analytic* styles: modes, techniques, *and* experiences.

One last comment: My colleague and co-developer of the grounded theory style of qualitative analysis, Barney Glaser, teaches and uses that style in research essentially as I do. There are some differences in his specific teaching tactics and perhaps in his actual carrying out of research, but the differences are minor. I am very much indebted to him, of course, for his crucial part in the evolution of this analytic style, for continued vital discussions over the years about teaching and doing analysis, and in this book for permission to quote extensively in Chapter 1 from his *Theoretical Sensitivity* (1978). Indeed, the second half of that chapter is essentially his except for some amplification. The quotations from his book and some from our co-authored volume, *Discovery of Grounded Theory* (1967), are distinguishable insofar as they are separated from the main text by quotation marks, and occasionally slightly edited or rephrased to suit present purposes. I recommend both books as supplementary reading to this one: *Discovery* for the general background to this approach to qualitative analysis, and *Theoretical Sensitivity* for its greater detail concerning some procedures and further discussion of what lies behind their use (see also Charmaz 1983).

I wish also to express appreciation to many other colleagues for their direct and indirect contributions to this book, for in a genuine sense it is truly a collaborative enterprise. Leigh Star (Tremont Research Institute, San Francisco), Juliet Corbin (University of California, San Francisco), and Joseph Schneider (Drake University) wrote immensely detailed critiques of the initial draft, and I have followed closely many of their suggestions in its revision. Peter Conrad (Brandeis University), Adele Clarke and Nan Chico (Tremont Research Institute and University of California, San Francisco), and Paul Atkinson (University of Cardiff, Wales) also made many useful suggestions. Over the years, I have learned a great deal from students in research seminars and from consulting with them on their research: They will know my indebtedness to them if they read this book, even if they or their materials do not appear in it by name. The same is true of my friends and colleagues, Elihu Gerson (Director, Tremont Research Institute), Leonard Schatzman (University of California, San Francisco), and Fritz Schuetze (University of Kassel, West Germany) with whom I have had countless discussions of methodological issues for many years; and of course there are also my research teammates who appear in these pages – Shizuko Fagerhaugh, Barbara Suczek, Carolyn Wiener, and, again,

Juliet Corbin (University of California, San Francisco) – who furthered and greatly sharpened my teaching and doing of analysis just by working closely with me on research projects. Among the European contributors to my thinking about analysis and its teaching–consultation, I need especially to single out Richard Grathoff (University of Bielefeld, West Germany) and the members of his research teams – particularly Bruno Hildebrand (University of Marburg and University of Frankfurt); Hans-Georg Soeffner and his research teams (Fern Universitaet, Hagen, West Germany); also, for the same reasons, four visiting fellows from overseas – Herman Coenen (University of Tillburg, The Netherlands), Gerhard Rieman (University of Kassel), Christa Hoffman-Riem (University of Hamburg), and Wolfram Fischer (The J-Liebig University). And thanks also to Malcolm Johnson (Open University, England) for suggesting I send this book to Cambridge University Press, a most fortunate suggestion. The prominence in the book of explicit rules of thumb have their source in Leigh Star's insistence that these needed to be spelled out clearly. Also, I have quoted, often extensively, from transcripts and materials in which various colleagues and students have either appeared or which they have written. I am especially grateful to them since their contributions, collectively speaking, form the illustrative heart of this book. These people are Ritch Adison, Barbara Bowers, Nan Chico, Juliet Corbin, Adele Clarke, Shizuko Fagerhaugh, Elihu Gerson, Anna Hazan, Gail Hornstein, Katarin Jurich, F. Raymond Marks, Misty MacCready, Evelyn Peterson, Aaron Smith, Leigh Star, Barbara Suczek, Steve Wallace, and Carolyn Wiener. Most of these people were students in my research seminars, but I wish also to thank their colleagues; after all, it was they who taught me, indirectly but sometimes directly, how to teach qualitative research more effectively. Last in this listing – but not in fact – are two other collaborators. They typed portions of the manuscript and subjected it to most helpful editorial comments: my secretary, Sally Maeth, and my wife, Frances Strauss.

1 Introduction

PART 1

Some assumptions

A set of assumptions lies behind this approach to qualitative analysis, which first will be listed and then briefly discussed.

1. Very diverse materials (interviews, transcripts of meetings, court proceedings; field observations; other documents, like diaries and letters; questionnaire answers; census statistics; etc.) provide indispensable data for social research.
2. As compared with both the quantitative analysis of data and the actual collection of data by qualitative analysts, the methods for qualitatively *analyzing* materials are rudimentary. They need to be developed and transmitted widely and explicitly throughout the social science community.
3. There is need for effective theory – at various levels of generality – based on the qualitative analysis of data.
4. Without grounding in data, that theory will be speculative, hence ineffective.
5. Social phenomena are complex: Thus, they require complex grounded theory. This means conceptually dense theory that accounts for a great deal of variation in the phenomena studied.
6. While there can be no hard and fast rules governing qualitative analysis – given the diversity of social settings, research projects, individual research styles, and unexpected contingencies that affect the research – it is possible to lay out general guidelines and rules of thumb to effective analysis.
7. Such guidelines can be useful to researchers across a broad spectrum of disciplines (sociology, anthropology, political science, psychology, public health, nursing, and education) and, regardless of "tradition" or "theoretical approach," just as long as they believe their work can be furthered by the qualitative examination of materials. Also, such analytic methods can be useful whether researchers are wedded to the idea of social science per se or to more humanistic versions of social research ("understanding," "enlightenment").
8. Finally, research is basically work – sets of tasks, both physical and conceptual – carried out by researchers. Development, use, and teaching of qualitative analysis can be enhanced by thinking specifically of analysis in terms of the

organization and conduct of that work. Thus, what we know about work (from research on that phenomenon) can be applied to the improvement of research methods.

Materials as data

Among social scientists a distinction is commonly drawn between quantitative and qualitative research. The distinction in part has its origins in the history of some disciplines, especially perhaps sociology and social anthropology – in sociology, because so many disciplinary trends since World War II have fostered questionnaires and other survey methods of collecting data and their statistical treatment; and in anthropology, because qualitative analysis of field data is the primary mode, although quantitative methods have lately been more employed, to the distress of many who steadfastly rely on qualitative methods. "Qualitative methods" has generally been used, also, to refer to the work of researchers who work as differently as ethnographers, clinical and organizational psychologists, grounded-theorist sociologists, or macrohistorians/sociologists. *Qualitative* researchers tend to lay considerable emphasis on situational and often structural contexts, in contrast to many *quantitative* researchers, whose work is multivariate but often weak on context. Qualitative researchers tend, however, to be weak on cross-comparisons because they often study only single situations, organizations, and institutions. (See, however, recent discussions and methods pertinent to cross-site qualitative analysis: Miles and Huberman 1983, pp. 151–209; Miles, p. 1284; and see others who are inventing and testing procedures for merging quantitative and qualitative analysis: Louis 1982; Smith and Robbins 1982; Jick 1983; Sieber 1983; McClintock et al. 1983.)

Quite aside from historical considerations, it is our contention that the genuinely useful distinction (which we will touch on further) is in how data are treated *analytically*. (There is neither logical nor any sensible reason for opposing these two general modes of analysis. I do not discuss in this book their use in conjunction with each other because I have had no recent research or teaching experience in combining the two.) In quantitative research, statistics or some other form of mathematical operations are utilized in analyzing data. In qualitative research, mathematical techniques are eschewed or are of minimal use, although assuredly rudimentary or implicit counting and measuring are usually involved (How many? How often? To what degree?). Qualitative analysis may utilize a variety of specialized nonmathematical techniques, as noted below, or as commonly practiced may use procedures not appreciably different from the pragmatic analytic operations used by everybody in thinking about everyday problems. (Leonard Schatzman terms these *natural analysis*. See Schatzman, forthcoming.) Qualitative researchers, however, when addressing scientific rather than practical or personal problems, are more self-conscious and more "scientifically rigorous" in their use of these common modes of thinking.

In any event, moving to the research materials themselves: They occur in a variety of forms, all of which have been utilized by social scientists – as well as by investigators in fields like history, psychology, education, and law – although different disciplines and their specialties have favored one type of material rather than another. For instance, among those primarily utilizing qualitative methods, ethnographers have relied mainly for data on field observations converted into field notes and on interviews. Historians may interview if their work is on contemporary or relatively recent events, but principally they utilize many different kinds of documents, depending on their specific research aims and on the availability and accessibility of materials: records of various types, memoirs, official and personal letters, diaries, newspapers, maps, photographs, and paintings. Researchers in clinical psychology base conclusions primarily on their clinical observations of patients' nonverbal as well as verbal behavior, and on therapeutic interviews. Many sociologists prefer to analyze written texts rather than engage in field research or interviewing; others generate materials through tape recordings of conversations, transcripts of court trials, and the like. While some materials (data) may be generated by the researcher – as through interviews, field observations, or videotapes – a great deal of it already exists, either in the public domain or in private hands, and can be used by an informed researcher provided that he or she can locate and gain access to the material – or is lucky enough to stumble on it.

These materials, then, are useful for qualitative analysts in all of the social sciences. In some disciplines or their specialties, materials are converted into quantitative data through counting and measuring operations. In others, counting and quantitative measurement are minimal and these operations may even be rejected on reasonable, well-thought-out grounds. Whether qualitative or quantitative analysis predominates is sometimes a matter of ideology (which can be frozen into tradition), but more often is a matter of rational choice. At any rate, qualitative analyses are more than merely useful: They are often indispensable.

Of course in daily life everyone engages in some form of qualitative analysis – much as Moliere's citizen used prose – without thinking twice about the matter since no judgments, no decisions, no actions can be taken in their absence. So, in a genuine sense, both common sense and "researcher" conclusions are based on "qualitative data." Without denigrating the care, self-awareness, and systematic character of a large proportion of everyday, pragmatic analyses (indeed, researchers themselves would be irate if accused of lacking those virtues in their daily thinking), it is clear enough that researchers are expected by their colleagues to adhere to disciplinary practices associated with the "good researcher," and will criticize or ignore as incompetently done any research products judged deficient in careful, scrupulous, systematic treatment of reliable data.

More important for our purposes here is that improved qualitative analysis requires more explicitly formulated, reliable, and valid methods than currently exist. *Analysis* is synonymous with *interpretation* of data. It refers to research activity which, as will be detailed later, involves several different but related elements (or operations). (See Miles and Huberman 1983, p. 214, for slightly different emphases.) Qualitative analysis occurs at various levels of explicitness, abstraction, and systematization. At the beginning of a research project, when the researcher reads a sentence or sees an action, the analysis may be quite implicit; but analysis it surely is insofar as perception is selective, mediated by language and experience. Later in the investigation or even during the first days when an observed scene, interview, or perused document challenges the researcher's analytic sense, the conclusions will be drawn more explicitly and probably more systematically. Depending on the purposes of the investigator, the final conclusions drawn in the course of the research can vary greatly by level of abstraction. At the lowest levels they can be "descriptive," and at the highest levels, the researcher may aim for the most general of theory. But description itself can be "low level" – perhaps only reproducing the informants' own words or recording their actions – or can be reported at a much more complex, systematic, and interpretative level. If social theory is aimed for, it can be formulated with more or less systematic treatment and with varying degrees of abstraction. In addition, the theory at any level can be broader or narrower in scope; and it may be linked with other theory which is more or less developed.

Methods for qualitative analysis of data

Social scientists who engage entirely or primarily in qualitative analysis generally would agree that quantitative methodology is much more explicitly presented in standard manuals and during training. As we noted some years ago in *The Discovery of Grounded Theory* (1967), quantitative analysts since the 1920s have developed relatively rigorous methods for collecting and treating their data, and have written extensively about those methods. By contrast, much of the attention of qualitative researchers is still focused on improving and making explicit their techniques for the collection of data – analytic considerations being at best quite secondary and, such as they are, transmitted on an apprenticeship basis in tacit rather than explicit fashion. However, a number of researchers have developed effective methods for the qualitative analysis of different types of materials. The character of some of these methods is suggested by their respective names: conversational analysis, (qualitative) network analysis, biographical analysis, sociolinguistic analysis, dramaturgical or social drama analysis, textual analysis. These methods, or sets of techniques, have evolved in conjunction with particular lines of research and theoretical interests or commitments.

Grounded theory

The methodological thrust of the grounded theory approach to qualitative data is toward the development of theory, without any particular commitment to specific kinds of data, lines of research, or theoretical interests. So, it is not really a specific method or technique. Rather, it is a style of doing qualitative analysis that includes a number of distinct features, such as theoretical sampling, and certain methodological guidelines, such as the making of constant comparisons and the use of a coding paradigm, to ensure conceptual development and density.

This approach to qualitative analysis was developed by Glaser and Strauss in the early 1960s during a field observational study of hospital staffs' handling of dying patients (1965, 1968). Contributing to its development were two streams of work and thought: first, the general thrust of American Pragmatism (especially the writings of John Dewey, but also those of George H. Mead and Charles Peirce) and including its emphases on action and the problematic situation, and the necessity for conceiving of method in the context of problem solving; second, the tradition in Chicago Sociology at the University of Chicago from the 1920s through the mid-1950s, which extensively utilized field observations and intensive interviews as data-collecting techniques, and furthered much research on the sociology of work. Both the philosophical and the sociological traditions assumed that change is a constant feature of social life but that its specific directions need to be accounted for; they also placed social interaction and social processes at the center of their attention. In addition, Chicago Sociology almost from its inception emphasized the necessity for grasping the actors' viewpoints for understanding interaction, process, and social change. The study of dying by Glaser and Strauss, with its initial use of the grounded theory style of analysis, drew from both of those philosophical and sociological traditions. (For a fuller historical understanding of the background of grounded theory, it would be useful to read John Dewey's *Logic: The Theory of Inquiry*, 1937, and Everett C. Hughes's papers on occupations and work and on fieldwork in *The Sociological Eye*, 1971.[1])

Of course, theory is generated and tested even by researchers whose analytic methods remain relatively implicit, but the grounded theory style of analysis is based on the premise that theory at various levels of generality is indispensable for deeper knowledge of social phenomena (Glaser and Strauss 1967; Glaser 1978). We also argued that such theory ought to be developed in intimate relationship with data, with researchers fully aware of themselves as instruments for developing that grounded theory. This is true whether they generate the data themselves or ground their theoretical work in data collected by others. When we advocated that position in 1967 there was perhaps more need to remind social scientists of that necessity for grounding their theory than now.

Complex theory

One of our deepest convictions is that social phenomena are complex phenomena. Much social research seems to be based on quite the opposite assumption; either that, or researchers working in various research traditions describe or analyze the phenomena they study in relatively uncomplex terms, having given up on the possibility of ordering the "buzzing, blooming confusion" of experience except by

[1] Barney Glaser had studied with Paul Lazarsfeld at Columbia University, and so brought to the development of the grounded theory approach some of Lazarsfeld's emphasis on multivariate analysis. The Chicago tradition similarly emphasizes variation.

ignoring "for a time" its complexity. Their assumption apparently is that later generations will build on current endeavors – a kind of accumulation premise that seems reasonable, since one cannot study everything at the same time. Nevertheless much more complexity can be handled than is often done by quite competent or even gifted researchers. This is why grounded theory methodology emphasizes the need for developing many concepts and their linkages in order to capture a great deal of the variation that characterizes the central phenomena studied during any particular research project. We shall have much to say about this issue of complexity throughout this book.

Guidelines and rules of thumb, not rules

Affected by a mistaken imagery (based on speculative philosophy) of effective scientific research – exact, precise, explicit about its technology – students of social life often assume that is should be possible to lay down rules (later if not right now) for carrying out social investigations. We do not believe this is an accurate characterization of how any kind of work is carried out; and it is not likely ever to be true for researchers who aspire to developing new theory or to extending extant theory. Even in the more precise scientific investigations of physicists or chemists, contingency is inevitable; thus, discretion is advisable and often essential. Moreover, the best opinion among philosophers these days holds that such codification of investigation is impossible anyhow.

We shall not argue the point further except to repeat that several structural conditions mitigate against a neat codification of methodological rules for social research. These include the diversity of social settings and their attendant contingencies which affect not merely the collection of data but how they are to be, and can be, analyzed – quite aside from researchers' often different aims in doing their analyses. Researchers also have quite different investigatory styles, let alone different talents and gifts, so that a standardization of methods (swallowed whole, taken seriously) would only constrain and even stifle social researchers' best efforts.

Hence we take the stand about our own suggested methods that they are by no means to be regarded as hard and fixed rules for converting data into effective theory. They constitute guidelines that should help most researchers in their enterprises. For that – as we shall attempt to show – researchers need to be alive not only to the constraints and challenges of research settings and research aims, but to the nature of their data. They must also be alert to the temporal aspects or phasing of their researches, the open-ended character of the "best research" in any discipline, the immense significance of their own experiences as researchers, and the local contexts in which the researches are conducted.

Our guidelines for developing theory are not merely a kind of laundry list of suggestions, however: they are stronger than that, for they emphasize that certain operations must be carried out. Coding must be done, and generally done early and continually. Analytic memos must be done early and continually in conjunction with the coding. And a few concepts, loosely strung together, cannot satisfy the requirements for formulating social theory. Yet, we emphasize also that personal pacing and experiences can be ignored only to the detriment of effective and analytic work. We do not believe that strict instructions can be given for how to proceed in detail with all kinds of materials, by everyone, holding for all kinds of research, at all phases of the research project. Methods, too, are developed and change in response to changing work contexts. However, we have throughout this book included lists of rules of thumb. These are to be thought of as operational aids, of proven usefulness in our research. *Study* them, *use* them, but *modify* them in accordance with the requirements of your own research. Methods, after all, are developed and changed in response to changing work contexts.

Our guidelines and rules of thumb, then, will be useful to any researcher who shares our concern for achieving better comprehension of social phenomena – through the development of some level of theory – regardless of the substantive character of the materials or of the particular discipline in which he or she has been trained. We believe that the same assertion holds for researchers who are committed to different traditions or theoretical approaches, even within the same discipline; this, provided these traditions and approaches cash in on their strengths – raising important problems or looking at relevant or neglected areas of social life – rather than box their adherents into dogmatic positions which foreclose on the possibility of actually challenging some of what their own traditions currently stand for.

Underlying some contemporary positions are the contrasting assumptions that either a social science is possible or that it is to be eschewed in favor of more humanistic versions of knowledge about human activity. Our own position is somewhere between these extremes, though some practitioners of grounded theory methodology might lean in either direction on that continuum of belief. Nevertheless, we believe that the methodological guidelines and general procedures can be of service to researchers regardless of where they stand on this particularly divisive and long-standing dispute among social scientists.

Research investigation as work

The last assumption that underlies the grounded theory approach is that research should be understood and analyzed as work. Essentially we are advocating a highly self-conscious approach to the work of research: to how it is and can be actually carried out under a variety of circumstances, during its various phases, by researchers who stand in different relationships to the work of getting and examining and interpreting the information that becomes their data. Consequently, this book is not only based on an explicit sociology-of-work perspective, but is designed to help readers think in those terms about their own research endeavors. We should note also that research work consists of more than sets of tasks or a clear formulation of the goals of those tasks. It involves the organization of work – the articulation of tasks (itself a type of work) including the management of physical, social, and personal resources necessary for getting the research work done, whether working alone, with someone else, or in a team.

Perhaps it is also necessary to add that a sociology-of-work perspective emphasizes temporal features, both of the investigatory process itself and of the phenomena being studied. This constitutes our own bias toward reality, of course. For all that, we believe a sociology-of-work perspective on research activity can be useful even if a reader chooses to ignore for the moment or to downplay or deny temporal considerations when doing his or her research work. Admittedly, however, our approach to analysis, which emphasizes complexity of phenomena and the unexpected contingencies affecting both the phenomena under study and the course of the research itself, tends to bring temporality into focus for the analyst.

We should add that while much research involves routine operations and can at times be boring, assuredly also at its most creative it is exciting, fun, challenging, although sometimes extremely disturbing and painful. This means that researchers, as workers, can and should *care* very deeply about their work – not being simply possessive about its products or jealous of their research reputations, but find deep and satisfying meaning in their work. They and it are immensely interactive in exactly the sense used by John Dewey when writing about artists (he did not regard artistic and scientific activity as basically different): An "expression of the self in and through a medium, constituting the work of art, is *itself* a prolonged interaction issuing from the self with objective conditions, a process in which *both of them* [our italics here] acquire a form and order they did not first possess" (Dewey 1934, p. 65). In short, the researcher, if more than merely competent, will be "in the work" – emotionally as well as intellectually – and often will be profoundly affected by experiences engendered by the research process itself.

Qualitative analysis of data: an introduction

Besides those general assumptions that lie behind our approach to the qualitative analysis of materials, some additional remarks will be useful before the more technical details of grounded theory analysis are discussed.

Complexity

The basic question facing us is how to capture the complexity of reality (phenomena) we study, and how to make convincing sense of it. Part of the capturing of course is through extensive data collection. But making sense of complex data means three things. First, it means that both the complex interpretations and the data collection are guided by successively evolving interpretations made during the course of the study. (The final products are analyses done at a relatively high level of abstraction: that is, *theories*.) The second point is that a theory, to avoid simplistic rendering of the phenomena under study, must be conceptually dense – there are many concepts, and many linkages among them. (Even the best monographs often are rather thin in their conceptual treatment, as betrayed by the monograph's index, which lists few if any new concepts.) The third point: It is necessary to do detailed, intensive, microscopic examination of the data in order to bring out the amazing complexity of what lies in, behind, and beyond those data. (Later, we shall say much more about complexity and capturing it through analysis.)

Experiential data

To that analysis, as will be seen, analysts bring experiences of various kinds. If not new to the research game, then they bring research skills and savvy to their analyses. What is in their heads also in the way of social science literature also affects their analyses. This is true, whether in the form of specific hypotheses and concepts or, more diffusely, an informed theoretical sensitivity (ways of thinking about data in theoretical terms) – to nuances in their data that less well-read researchers may lack in some degree. Equally important is the utilization of experiential data, which consists not only of analysts' technical knowledge and experience derived from research, but also their personal experiences (see also the next section, Induction, Deduction, and Verification). These experiential data should not be ignored because of the usual canons governing research (which regard personal experience and data as likely to bias the research), for those canons lead to the squashing of valuable experiential data. We say, rather, "Mine your experience, there is potential gold there!"

Experiential data are essential data, as we shall see, because they not only give added theoretical sensitivity but provide a wealth of provisional suggestions for making comparisons, finding variations, and sampling widely on theoretical grounds (Schatzman, forthcoming). All of that helps the researcher eventually to formulate a conceptually dense and carefully ordered theory. The researcher's will not be the only possible interpretation of the data (only God's interpretations can make the claim of "full completeness"), but it will be plausible, useful, and allow its own further elaboration and verification.

We should add that the mandate to use experiential data gives the researcher a satisfying sense of freedom, linked with the understanding that this is not license to run wild but is held within bounds by controls exerted through a carefully managed triad of data collection/coding and memoing (to be discussed shortly). This triad serves as a genuinely explicit control over the researcher's biases.

Induction, deduction, and verification

The grounded theory of analysis involves – as does all scientific theory which is not purely speculative – a grounding in data. Scientific theories require first of all that they be conceived, then elaborated, then checked out. Everyone agrees on that. What they do not always agree on are the exact terms with which to refer to those three aspects of inquiry. The terms which we prefer are induction, deduction, and verification. Induction refers to the actions that lead to discovery of an hypothesis – that is, having a hunch or an idea, then converting it into an hypothesis and assessing whether it might provisionally work as at least a partial condition for a type of event, act, relationship, strategy, etc. Hypotheses are both provisional and conditional. Deduction consists of the drawing of implications from hypotheses or larger systems of them for purposes of verification. The latter term refers to the procedures of verifying, whether that turns out to be total or a partial qualification or negation. All three processes go on throughout the life of the research project. Probably few working scientists would make the mistake of believing these stood in simple sequential relationship.

Because of our earlier writing in *Discovery* (1967) where we attacked speculative theory – quite ungrounded in bodies of data – many people mistakenly refer to grounded theory as "inductive theory" in order to contrast it with, say, the theories of Parsons or Blau. But as we have indicated, all three aspects of inquiry (induction, deduction, and verification) are absolutely essential. Of course, deduction without verification or qualification or even negation of an hypothesis or set of hypotheses is truncated inquiry. Obviously, too, verification cannot occur without deduction: Hypotheses for data collection without reference to implications of theoretical hypotheses are useless. And how can there be hypotheses without either thinking through the implications of data or through "data in the head" (whether experiential or from previous studies) that eventuates in so-called hunches, insights, and very provisional formulations of hypotheses?

In fact, it is important to understand that various kinds of experience are central to all these modes of activity – induction, deduction, and verification – that enter into inquiry. Consider induction first: Where do the insights, hunches, generative questions which constitute it come from? Answer: They come from experience with this kind of phenomenon before – whether the experience is personal, or derives more "professionally" from actual exploratory research into the phenomenon or from a previous research program, or from theoretical sensitivity because of the researcher's knowledge of technical literature.[2] As for deduction: Success at it rests not merely on the ability to think logically but with experience in thinking about the particular kind of data under scrutiny. The researcher is able to think effectively – and propositionally – because he or she has experiences to draw upon in thinking about those data, including the making of comparisons that help measurably in furthering the lines of deduction. Further, a special kind of prepa-

[2] See the writings of Charles Peirce, the American Pragmatist, whose concept of *abduction* strongly emphasized the crucial role of experience in this first phase of research operations (Fann 1980; Hartshorne et al. 1958).

ration underlies this deductive ability: experience not only with deductive procedures but with those used specifically in research endeavors. And verification: Quite clearly, this is not primarily a matter of activity or ability. It involves knowledge about sites, events, actions, actors, also procedures and techniques (and learned skills in thinking about them). Again that knowledge is based on personal and professional experience.

The crucial role of experience has been underplayed by philosophers of science, probably because they do not actually have a working knowledge of research, and by positively minded if methodologically reflective social scientists, who wish to rule out of court anything that smacks of "subjectivity" and who wish to minimize soft data in favor of hard (or "real") data.

If, then, experience and associated learned skills at verification, deduction, and induction are central to successful inquiry, do not talent–gifts–genius contribute to that success? Obviously the answer is yes; but not so obviously it is a qualified and complex yes. Why? Because different abilities are relevant for each of these central investigative modes. Some people are better at generative questions, intuitive flashes, hunches, etc. Some are better "theorists" – better at developing hypotheses and drawing out implications. And some are best at doing the verifying work: the laboratory whizzes, the gifted interviewers, the sensitive field observers, the highly skilled questionnaire designers. Some people can do two of these central modes of inquiry well, and some all three. Moreover, doing each well or not so well implies a continuum for each mode (verification, deduction, induction). The "real geniuses" do them all, and brilliantly. Yet as should be evident (we shall say more about this later in the book), analytic capacities can be developed, and competent if not brilliant accomplishments at one or more modes of inquiry achieved. Good research analysis can be taught and learned: It is not at all merely an innate skill.

We should add that in the event an extant grounded theory is used at the beginning or early in the research project, then deductions are made from it in the form of theoretical questions, hypotheses, suggested theoretical sampling, possible categories, and so on. They lead directly into the initial phase of collecting and analyzing data. Thus the role of deduction is the same as if the researcher began without using such a grounded theory. (See the Appendix, Discovering New Theory from Previous Theory.) This is in marked contrast to a very frequent mode of using previous theory – usually drawn from a well-known theorist, like Goffman, whose theory may be well grounded – but this theory is misused because it is not really checked out in the further inquiry. It is only applied like a label to one's data. This practice almost totally relieves the researcher of three very important responsibilities: of (1) genuinely checking or qualifying the original data; (2) interacting deeply with his or her own data; and (3) developing new theory on the basis of a true transaction between the previous and newly evolving theory. While this practice and its citations may flatter the theorist, and may give the illusion of adding to "knowledge," it really does not advance the collective scientific enterprise. In this regard, effective social science research must follow the example of physical science research in its intertwining of the formulating of provisional hypotheses, making deductions, and checking them out – all with the use of data.

An example

Here is an example of the beginnings of a complex analysis, based on field observational data but certainly supplemented by experiential data. It will serve as a brief introduction to the grounded theory style of analysis and introduce a couple of important terms for analytic processes. This example is taken from an actual study, and the field workers did make the observations and go through some of the analytic operations described. (See Strauss et al. 1985. Other materials from this project are given in chapters on coding and memoing.)

Imagine that in a study of whether and how the use of machines in hospitals affects the interaction between staff and patients, we observe that many machines are connected to the sick persons. We can formulate a category – machine–body connections – to refer to this phenomenon. Our observations also lead us to make a provisional distinction (which may or may not turn out to be significant after further research) between those machines where the connection is external to the skin of the patient, and those where the connection is internal (through various orifices: nose, mouth, anus, vagina). This distinction involves two dimensions of the machine–body category: internal and external connections. The basic operation of making those distinctions is *dimensionalizing.*[3] But since further distinctions can be made – either by thinking about previous observations or making new ones – the process of dimensionalizing will continue. That is termed subdimensionalizing. Subdimensions may also be generated analytically by questions that

[3] This discussion of dimensionalizing was much furthered by a working session with Leonard Schatzman, a colleague, who has been thinking through the details and implications of dimensions and dimensionalizing (see Schatzman, forthcoming).

sooner or later will occur to us about some of those distinctions. Thus, about the internal connections: Don't they – or at least some of them – hurt? Are they safe? Are they uncomfortable? Are they frightening? We can think of these subdimensions (hurt, safety, discomfort, fear) dichotomously – as yes or no – or as continua running from very much to not at all. Or we can slice up a continuum roughly into "more or less" subcategories, as for instance, terribly uncomfortable, very uncomfortable, somewhat uncomfortable, a bit uncomfortable, not at all uncomfortable. (In quantitative analysis, continua can be given "values," running from 0 to 100.) All of these subdimensions, subcategories, and questions come not only from inspection of field/interview data but, understandably from our experiential data (those orifices are sensitive, so that connection probably hurts; or, that tube looks horrible coming out of his belly, so is it really safe?).

Those last questions refer to consequences: "If it looks like that, then it may have the consequences of endangering life." This may be amended by specific conditions through adding: "It may endanger his life, especially if he moves too quickly or turns over in his sleep or it falls out and then he gets an infection." Or there may be questions raised which involve the staff's strategy: "Why did they connect it up that way rather than another?" Or the patient's strategy: "Did he try to bargain to get it done another way?" Questions about interactions will also arise: "What went on between the personnel and the patient when he was being hooked up? Did they tell him beforehand and warn him? Did they just do it and so he got frightened?" (That last is a question also about consequences of interaction.)

Those questions are given provisional answers – that is, they have the status of hypotheses. Some may be checked out by further observations or interviews. But now the researcher can be more directed than previously in making observations and doing interviews. He or she is likely to realize (recognizing when observing) that a nasal connection is likely to be uncomfortable but perfectly safe, and so will interview around that hypothesis. Or thinking about unsafe conditions, the researcher may either ask staff for examples of when those connections proved unsafe for the patient – thus eliciting relevant data – or be on the lookout for unsafe nasal connections in terms of further conditions, like: because the connection got disconnected, or because of the way the connection was made.

This line of reasoning can lead to further subdimensionalizing and further questions and provisional hypotheses. Thus, for connections that become disconnected more or less easily: How do they become disconnected? By accident, carelessness, purposefulness (as on the part of the annoyed or uncomfortable or fearful patient)? What tactics or techniques are used by the personnel to minimize or prevent disconnection: Special care? Warning the patient about moving? Emphasizing that one's safety depends on staying immobile or in not loosening the connection no matter how it hurts? Or by eliciting "cooperation," promising that the connection will remain only for several hours or be removed periodically to give relief? These questions and hypotheses and distinctions may not turn out to be "realistic"; but if they are, then further *directed inquiry* will tell the researcher: yes–no–maybe; as well as, *why*. Understandably, the researcher is likely to raise questions eventually (or observed events will occur that raise and partly answer the questions) about many more conditions; also about consequences not only for the patient but for kin, staff, different types of personnel, for the ward's functioning, and probably also for the redesign of particular models of machinery.

The directed inquiry will also very naturally and easily lead the researcher to ask: Where can I find instances of "*x*" or "*y*"? The technical term for this is *theoretical sampling* – for the researcher, after previous analysis, is seeking samples of population, events, activities guided by his or her emerging (if still primitive) theory. This sampling is harnessed at least implicitly (explicitly by the experienced researchers) to *making comparisons* according to various subdimensions. Thus, the researchers may compare, either "in imagination" or through their own experiential data, certain machine connections that are uncomfortable with those that are not. The researchers have already thought about the discomfort or the anxiety engendered by various connections. But they can go further afield and, say, make (or discover) comparisons between what happens when a dangerous disconnection occurs versus a nondangerous one: For example, once when there was actually a power blackout in the hospital, the researchers rushed around observing what was happening because various pieces of equipment had got disconnected electrically. They discovered much variation, one of the most interesting being the manual emergency motoring, done for about two hours, of dialysis machines in an associated clinical building that had no backup for the dangerously malfunctioning electrical system.

Directed by his or her theorizing, the researcher can sample even more widely by thinking about safety or discomfort with respect to other machines – whether body-connected or not – like x-ray equipment, airplanes, toasters, lawnmowers, or the body-shaking power tools manipulated by men who are employed to break up cement on street surfaces. The purpose of thinking about those comparisons is not to pursue a more encompassing theory about machines in general or safe/dangerous machines in general, but to stimulate theoretical sensitivity in the service of generating theory about medical machinery in hospital work. Out-sampling then links with in-sampling.

Understandably, too, some ideas and thinking about those comparisons can come from personal experiences with the machines, from watching others use the machines, from reading novels or autobiographies or nonfictional literature about them.

Several points about work processes

Next, several things are especially worth noting about the basic research work processes – thinking, going to the field, observing, interviewing, note taking, analyzing. *First*, the raising of *generative questions* is essential to making distinctions and comparisons; thinking about possible hypotheses, concepts, and their relationships; sampling, and the like. These come from examination and thinking about the data, often in conjunction with experiential data. The original generative question may come from insight, which actually sparks interest in an aspect of some phenomenon and thus challenges the researcher to study "it." But these insights occur along the course of a study (although perhaps especially in the earlier phases), and open up questions about other phenomena or other aspects of the same phenomena.

Second, the researcher will be making a number of interesting, if at first quite provisional, linkages among the "discovered" (created) concepts. The coding is beginning to yield *conceptually dense theory* which will of course become much more dense as additional linkages are suggested and formulated.

Third, the theory is not just discovered but *verified*, because the provisional character of the linkages – of answers and hypotheses concerning them – gets checked out during the succeeding phases of inquiry, with new data and new coding.

Fourth, the *relevance of the coding* to the real world of data is a central issue. Of course, "there is no end to the logical elaboration of dimensions, the drawing of distinctions, the making of linkages, but to run riot with logical elaboration is dangerous – if fun. This thought process *must* be linked with, tied in tightly with, the examination and collection of new data" in order to be of service to the research itself. (We shall discuss this point later, under the heading of Deduction and Induction.)

Fifth, there is the issue of *integration*: Which dimensions, distinctions, categories, linkages are "most important," most salient – which, in short, are the *core* of the evolving theory? This issue becomes solved during the course of the inquiry. Conveying how integration happens is not easy, and we shall discuss and illustrate this work later. Suffice it here to say that integration actually begins primitively and provisionally with the first linking up of dimensions, categories, etc. Integration becomes increasingly more certain and "tighter" as the research continues. The *core category or categories* that will best hold together (link up with) all the other categories – as they related to it and to each other – will take hard work and perhaps special techniques to put together in a convincing fashion: convincing both to the researcher and to those who will read his or her resultant publications.

Sixth, theoretical ideas are kept track of, and continuously linked and built up by means of *theoretical memos*. From time to time they are taken out of the file and examined and sorted, which results in new ideas, thus new memos. As research proceeds to later phases, memo writing becomes more intense, more focused, and memos are even more frequently sparked by previous memos or sum up and add to previous ones. *Sorting* of memos (and codes) may occur at any phase of the research. Both examination and sorting produces memos of greater scope and conceptual density. The systematic operation of sorting is especially important in later phases, as the analyst moves into planning the writing up of materials for publication.

Seventh, it is vitally important to recognize the *temporal* as well as *relational* aspect of the triad of analytic operation: data collecting, coding, memoing. Grounded theory practitioners need to understand how very different their perspective on that triad is from that of most other styles of analysis. Figure 1, a simplified diagram of a coding paradigm will illustrate some of the main features of this triad. Note that data collecting leads quickly to coding, which in turn may lead equally quickly – or at least soon – to memoing. Either will then guide the searches for new data. *Or* – and this is important to understand – they may lead directly to additional coding or memoing. Or – please note! – they may lead to inspecting and coding of *already* gathered (and perhaps already analyzed) data. The *return to the old data* can occur at any phase of the research, right down to writing the last page of the final report of the theory. Furthermore, as the diagram indicates, at any phase of the research coding can lead to more coding; or memoing, directly to further and more integrated memos, helped out of course by the sorting of codes and memos.

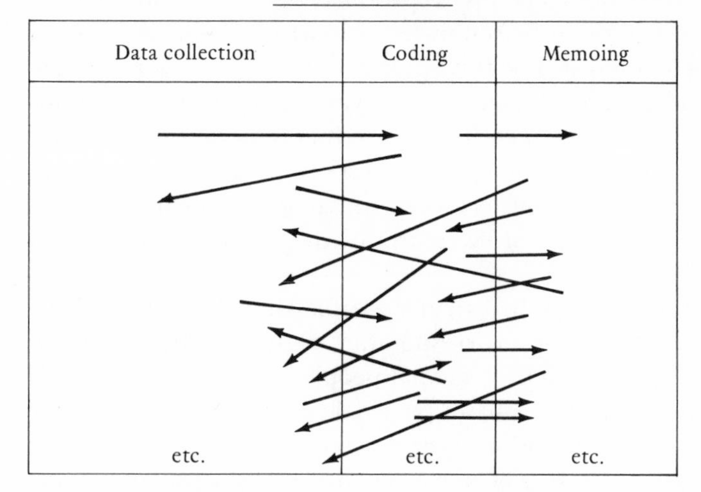

Figure 1. Coding paradigm.

This reexamination of all data throughout the life of the research project is a procedure probably engaged in by most qualitative researchers. But they do not usually double back-and-forth between collecting data, coding them, memoing in terms of data collection, coding, and memoing. The more positivistic research traditions proscribe the use of old data for verifying hypotheses, and so drive the researcher forward in a more linear direction, thereby cutting out the potential dividends of this recommended doubling back-and-forth procedure.

Eighth, during the *writing*, need for additional integration will often be recognized, the researcher sometimes then going back to the data, collecting some new data, or thinking through the sorted memos and codes, to "fill in," thus achieving the necessary integration. However, there is much variation concerning how much those operations will be relied upon during the writing period. How much depends on the degree of thoroughness with which the coding and memoing has been carried out; also on what the researcher realizes ought to be emphasized for particular audiences for whom he or she is writing; also on the writer's previous research/writing experience. Also, in team research it happens that so much data will accumulate, so fast, that although much coding is done and many theoretical memos are written, when the researchers sit down to write their various papers and monographs, they discover substantial holes in the previous analyses. This is especially so when some decisions about what to write, and for whom, evolve fairly late in the study.

The writing then does not just reproduce what is in the memos, although memos can often be rephrased or parts of them can be used pretty much as written in the final publication. The writing is, then, both analytic and creative. It can result in various types of publications (papers, monographs) and speeches, depending both on the substance of the research and the researcher's perceptions of the audiences. But the main point is that all the technical operations touched on in this section go on *continually*, from the outset of the research project until its close.

In the reception to published theory of this kind there is, we have found, a double-edged irony corresponding to two contrasting audiences. When lay people, or professional people of the population who have been studied – such as nurses or physicians – read the paper or monograph, they do not read it as theory, but either as a more or less accurate description of what's been happening to themselves and others of their acquaintance, or as "a new way of seeing what we all know that's very useful" – even an eye-opener. Then there is the audience of social scientists, who may read the publication, recognizing its "solid sociology," to quote an admirer of one of our publications, but without recognizing that the bright and even "brilliant ideas" in the publication arose not from personal gifts but from the hard work of research. The first irony should very much please the grounded theorist. The second will on occasion drive him or her wild with annoyance; but so be it: More-informed social science colleagues will know better.

A glossary of major terms

A number of important terms pertaining to qualitative analysis have appeared in the preceding section. They will be further discussed in the next chapter and then used throughout the book. We shall give capsule definitions of them now, since it is essential to have a firm grasp of them.

Data collection. the finding and gathering – or generating – of materials that the researcher will then analyze

Experiential data. data "in the head," drawn from the researcher's personal, research, and literature-reading experiences

Coding. the general term for conceptualizing data; thus, coding includes raising questions and giving provisional answers (hypotheses) about categories and about their relations. A code is the term for any product of this analysis (whether category or a relation among two or more categories).

Dimensionalizing. a basic operation of making distinctions, whose products are *dimensions and subdimensions.*

Category. since any distinction comes from dimensionalizing, those distinctions will lead to categories. (Thus, *Machine–body connection* is a category.)

Property. the most concrete feature of something (idea, thing, person, event, activity, relation) that can be conceptualized, which will allow the order of specificity required by the analyst for purposes of his or her research

Hypotheses (used exactly as in the usual scientific lexicon). a provisional answer to a question about conceptual relationships

Core category. a category that is central to the integration of the theory

Theoretical sampling. sampling directed by the evolving theory; it is a sampling of incidents, events, activities, populations, etc. It is harnessed to the making of *comparisons* between and among those samples of activities, populations, etc.

Theoretical saturation. when additional analysis no longer contributes to discovering anything new about a category

Conceptual density. the multiplicity of categories and properties *and* their relationships

Integration. the ever-increasing organization (or articulation) of the components of the theory

Variation. product of comparisons; grounded theory analysis rests on a multitude of comparisons – directed by theoretical sampling – and so grounded theory is multivariate. Making comparisons among categories and properties involves connecting (*crosscutting*) them.

Theoretical sensitivity. sensitive to thinking about data in theoretical terms

Theoretical memos. writing in which the researcher puts down theoretical questions, hypotheses, summary of codes, etc. – a method of keeping track of coding results and stimulating further coding, and also a major means for integrating the theory

Theoretical sorting. sorting of the theoretical memos in the service of integration: Codes are also sorted, toward the same end.

Integrative diagrams. a visual device which also furthers cumulative integration along the full course of the research

Generative questions. questions that stimulate the line of investigation in profitable directions; they lead to hypotheses, useful comparisons, the collection of certain classes of data, even to general lines of attack on potentially important problems.

PART 2

Grounded theory analysis: main elements

In this portion of the introductory chapter, a number of essential research operations are presented. Some of the discussion cannot be completely understood, at least in detail, until the illustrative materials in later pages help to provide visualization for the points made here. So, you might wish to read this chapter quickly to get an overview, then return to it, or parts of it, for reading or study later.

Our approach to the qualitative analysis of data is termed grounded theory "because of its emphasis on the generation of *theory* and the *data* in which that theory is *grounded*."[4]

Grounded theory "is a *detailed* grounding by systematically" and intensively "analyzing data, often sentence by sentence, or phrase by phrase of the field note, interview, or other document; by 'constant comparison,' data are extensively collected and coded," using the operations touched on in the previous section, thus producing a well-constructed theory. The focus of analysis is *not* merely on collecting or

[4] As noted in the preface, this part of Chapter 1 is reproduced almost wholly from Barney Glaser's *Theoretical Sensitivity*, 1978, with some editing and supplementation. The quoted sentences and paragraphs are identifiable by the relevant quotation marks. For more detailed statement of these technical aspects of the grounded theory mode of analysis, readers are advised to consult *Theoretical Sensitivity*.

ordering "a mass of data, but on *organizing many ideas* which have emerged from analysis of the data."

We have already seen the basic ingredients in producing complex, conceptually woven, integrated theory; theory which is discovered and formulated developmentally in close conjunction with intensive analysis of data. These procedures vary during the course of a research project. So, that issue will be discussed first, then we shall turn to a more detailed discussion of elements of the main procedures touched on previously. They are:

1. the concept-indicator model which directs the coding
2. data collection
3. coding
4. core categories
5. theoretical sampling
6. comparisons
7. theoretical saturation
8. integration of the theory
9. theoretical memos
10. theoretical sorting

Research phases and the operations

We shall now discuss the essential procedures for discovering, verifying, and formulating a grounded theory. These are in operation all through the research project and, as the case illustrations later will show, go on in close relationship to each other, in quick sequence and often simultaneously. But what about their relations to different phases of the entire research project? More will be said in answer to that question, but a few words should help in reading the concrete materials to be presented throughout this book.

As we shall see, the earliest phases of the research are more "open" than later ones are. There is no attempt to foreclose quickly on one or more categories. Many months may pass before the researcher is more or less certain of them and very many more before those core categories are saturated, and linked in a multiplicity of ways with other categories. In the earliest phases, a number of categories probably will be generated which later will be dropped as not very useful, or as unrelated to the core categories. Likewise a number of hypotheses will fall by the wayside, but are freely if provisionally generated by the enthusiastic researcher. Yet, from the earliest days, theoretical sampling directs the data collection and comparative analysis is done from the word go. The first memos are far less integrative than later they will be, and they too may poke up blind alleys or be focused very closely on the early microscopic analysis of data.

Once the core category or categories have been committed to, then the researcher will be seeking to relate other categories to them, thereby gradually densifying the theory. Also, more confidence will be placed in any new categories that "emerge" from further coding. Further highly directed theoretical sampling will function to generate additional relevant categories and properties. There is likely to be some sorting too, both of codes and memos, during this later phase (presumably by the middle of the project). Memos are likely to become increasingly elaborate, summarizing the previous ones; or focused closely on closing gaps in the theory. Earlier integrative diagrams will be made more elaborate, covering both more concepts and more connections among them. All of that continues until the last phases of the project.

Near the end, achieving integration will be a major focus. Also, considerable thought will be directed at which audiences to write for or speak to, and about what topics; also, what published papers to begin aiming for. Finally, there is the task of pulling the entire theory together for its presentation in a monograph. If a team is involved in this research, then there will be conferencing over who will write which papers, give which talks, write which chapters of the monograph. Or if they decide to publish more than one monograph, there is the question of: Who will write which monographs or portions of them?

Having said all that, we should emphasize that no sequential mini-steps can firmly be laid out in advance of the evolving phases of a given research project. Each enterprise will have its own detailed sequences, depending on: the circumstances of what kind of data are available, accessible and required; the nature of the data and the interpretations that the researcher will make of them; the experience of the researcher or researchers; the many contingencies that affect both the researcher personally (and interactionally, if a team also); the character of the audiences for whom they decide to write their publications; and the scope and generality of the theory for which the researchers aim. Only the general lineaments of the unfolding project can be anticipated in advance. The major differences between the grounded style of qualitative analysis and other qualitative analysis modes, however, is not in the relative unpredictability of project phases, but the differences per stage in the combinations and permutations of the operations (theoretical sampling, comparative analysis, theoretical saturation, memo

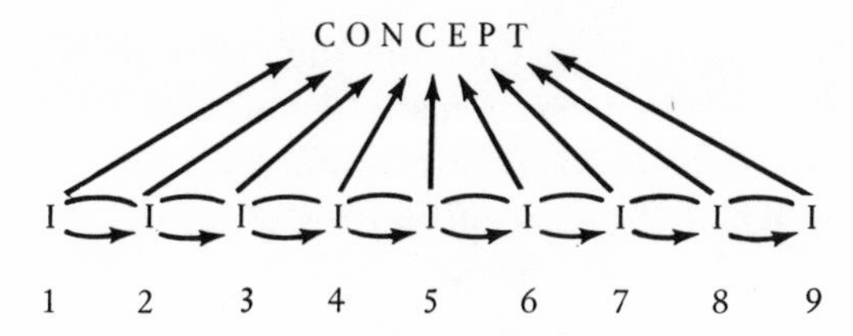

Figure 2. Indicator-concept model.

sorting, and so forth). These operations are essential to the development of densely woven and tightly integrated theory.

Basic operations

Concepts and indicators

"Grounded theory is based on a *concept-indicator* model, which directs the *conceptual* coding of a set of *empirical indicators*. The latter are actual data, such as behavioral actions and events, observed or described in documents and in the words of interviewees and informants. These data are indicators of a concept the analyst derives from them, at first provisionally but later with more certainty." (See the chapters on codes and memos and the chapter illustrating the research seminar analyses, where many illustrations of this indicator-concept model are given and sometimes pointed out explicitly for the reader.)

The concept-indicator model in Figure 2 is based first of all on the constant comparison of indicator to indicator. That is: Many indicators (behavioral actions/events) are examined comparatively by the analyst who then "codes" them, naming them as indicators of a class of events/behavioral actions. He or she may give this class a name, thinking of it then as a coded category. By making "comparisons of indicator to indicator the analyst is forced into confronting similarities, differences, and degrees of consistency of meaning among indicators. This generates an underlying uniformity, which in turn results in a coded" *category*. A second procedural step is that after "a conceptual code is generated, then indicators are compared to the emergent concept . . . . From the comparisons of additional indicators to the conceptual codes, the codes are sharpened to achieve their best fits to data." Meanwhile "further properties of categories are generated, until the codes are verified and saturated," yielding nothing much new.

In this model of concept indicators, "concepts and their dimensions have *earned* their way into the theory by systematic generation from data . . . *Conceptual specification* is at the focus of grounded theory . . . because the operational meaning of the concept derives from the use of its earned distinctions in the grounded theory."

"Changing indicators, thereby generating new properties of a code, will proceed only so far before the analyst discovers saturation of ideas through the *interchangeability of indicators*." That is, the events/behavioral actions which are converted analytically into indicators may vary in detail or in fact just be repetitious – but anyhow the indicators seem to "add up to the same thing" analytically. So the more the researcher "finds indicators that work similarly regarding their meaning for the concept, the more the analyst *saturates* the properties of the concept for the emerging theory. Nothing new happens as he or she reviews the data. The category and its properties exhaust the data. Meanwhile the analyst continues to saturate other categories by use of the constant comparative method."

Data collection

There is some ambiguity associated with the term *data collection*. Many social scientists do generate their data, through field observation, interviewing, producing videotapes, taping proceedings of meetings, and so on. But, as noted earlier, there are other sources of data: published documents of all kinds and private documents like letters and diaries. Use of those latter sources involves work too – searching for the data, getting access to them, taking notes on them, and nowadays xeroxing those data. In some kinds of library research, the researcher will even use the library much like an ethnographer, deciding upon which shelves to find the data sources (books, periodicals), and like the ethnographer happily coming upon fortuitously useful data, too (see Glaser and Strauss 1967).

The initial data collected may seem confusing, the researcher flooded by their richness and their often puzzling and challenging nature. It should not remain *that* confusing (only challenging) for very long because the analysis of these data begins (in our style of research) with the very first, second, or third interview or after the first day or two of fieldwork if at all feasible. It follows also that the next interviews and observations become informed by analytic questions and hypotheses about categories and their relationships. This guidance becomes increasingly explicit as the analysis of new data continues.

Data collection never entirely ceases because coding and memoing continue to raise fresh questions that can only be addressed by the gathering of new data or the examining of previous data. Theory-guided data collection often leads to the search for – or quick recognition of – valuable additional sources of data: for example a series of directed interviews to supplement the more casual interviews done during the daily fieldwork; or the use of published biographies to supplement a series of interviews. We call these "slices of data," for different kinds of data give different views or vantage points, allowing for further coding, including the discovery of relationships among the various categories that are entering into the emergent theory.

Coding

Coding, as noted in a previous section, is an essential procedure. Any researcher who wishes to become proficient at doing qualitative analysis must learn to code well and easily. The excellence of the research rests in large part on the excellence of the coding. (See Chapter 3 for illustrations and further discussion.)

Coding paradigm. One important point about coding that is sometimes misunderstood is this: While coding involves the discovery and naming of categories, it must *also* tell the researcher much more than that. It is not enough, for instance, to code an event qua indicator as an instance of a category – say, as "machine breakdown" – by writing the name of the category in the margins of the page next to the indicating lines of print. Also, the researcher needs to code the associated subcategories which are reflected either in the same lines or which will be reflected in other lines within the same or different interview, fieldnote, or document. (See especially Chapters 3–5.)

So we suggest the following *coding paradigm.* It is central to the coding procedures. Although especially helpful to beginning analysts, in a short time this paradigm quite literally becomes part and parcel of the analyst's thought processes. Whether explicit or implicit, it functions as a reminder to code data for relevance to whatever phenomena are referenced by a given category, for the following:

conditions
interaction among the actors

strategies and tactics
consequences

Because beginning researchers sometimes seem to experience difficulty in discovering "conditions" when inspecting their data, we shall note the following. Conditions are often easy to discover – indeed sometimes the interviewees or actors will point to them specifically – but if not, then look for cues like the use of words such as "because," "since," "as," or phrases like "on account of." Likewise, consequences of actions can be pointed to by phrases like "as a result," "because of that," "the result was," "the consequence was," and "in consequence." Strategies and the more specific tactics associated with strategies seem to present no difficulties for inexperienced analysts. *Interactions* are also easy to discover: They are those interactions occurring between and among actors, other than their straightforward use of tactics and strategies. Exemplifications of how the coding paradigm works will be found throughout this book. Remember that without inclusion of the paradigm items, coding is not coding.

Open coding. The *initial* type of coding done during a research project is termed open coding. This is unrestricted coding of the data. This open coding is done (as some of the case illustrations will show) by scrutinizing the fieldnote, interview, or other document very closely: line by line, or even word by word. The aim is to produce concepts that seem to fit the data. These concepts and their dimensions are as yet entirely provisional; but thinking about these results in a host of questions and equally provisional answers, which immediately leads to further issues pertaining to conditions, strategies, interactions, and consequences. As the analyst moves to the next words, next lines, the process snowballs, with the quick surfacing of information bearing on the questions and hypotheses, and sometimes even possible crosscutting of dimensions. A single session with a single document can often astonish even the experienced researcher, especially when the document at first glance seemed not to promise much in the way of leads. The point is really that the potential is not so much in the document as in the relationship between it and the inquiring mind and training of a researcher who vigorously and imaginatively engages in the open coding.

Novices at this type of coding characteristically get hung up, will argue intensely, about the "true" meaning of a line – or about the "real" motives of the interviewee lying behind the scrutinized line. In terms of open coding, their concern is entirely irrelevant. Why? Because the aim of the coding is to *open up* the inquiry. Every interpretation at this point is tentative. In a genuine sense, the analyst is not primarily concerned with this particular document, but for what it can do to further the next steps of the inquiry. Whatever is wrong in interpreting those lines and words will eventually be cancelled out through later steps of the inquiry. Concepts will then work or not work, distinctions will be useful or not useful – or modified, and so forth. So the experienced analyst learns to play the game of believing everything and believing nothing – at this point – leaving himself or herself as open as the coding itself. For all that, the coding is grounded in data on the page as well as on the conjunctive experiential data, including the knowledge of technical literature which the analyst brings into the inquiry.

This grounding in both sources of data gets researchers away from too literal an immersion in the materials (documents, fieldnotes, interviews, etc.) and quickly gets them to thinking in terms of explicit concepts and their relationships. This stepping away into conceptualization is especially difficult for even experienced researchers who may, in a particular study, either have gone a bit native through personally participating in the field of study, or who know too much experientially and descriptively about the phenomena they are studying and so are literally flooded with their materials. Yet the conceptual stepping back must occur if one is to develop theoretical understanding and theories about the phenomena reflected in the materials. Open coding quickly forces the analyst to fracture, break the data apart analytically, and leads directly to excitement and the inevitable payoff of grounded conceptualization. In research seminars, open coding is additionally valuable since students often find it much easier to code someone else's data, being more emotionally distant from them, and so learn through the open-coding procedures how more quickly to fracture their own data.

A word should be said here, however, about the difficulties novices often have in generating genuine categories. The common tendency is simply to take a bit of the data (a phrase or sentence or paragraph) and translate that into a precis of it. For instance: The interviewee is expressing grief or joy or aggression since he or she has declared "I was full of grief" or "I was mad as hops and so slugged him." The novitiate analyst is merely writing shorthand translation notions on the side of the interview page rather than generating theoretical categories. (In effect they are, as are many researchers who use other methods of analysis, remaining totally or mostly on a descriptive level, not much different from that of the actors themselves.) However, when a nurse tells the researcher that "I tried to keep my composure when the patient was yelling, by leaving the room" then that phrase can be converted analytically into "professional composure," plus notations about the structural condition threatening her composure and the tactic she uses for maintaining her composure. This can lead the researcher to write a memo in which questions are raised immediately about other pertinent conditions and tactics, as well as about situations where the nurse's tactic failed, or she had no chance to use one, and so lost her composure.

In our teaching experience, the most difficult step (other than integrating the total analysis) for beginners at this style of analysis is actually to get off the ground with genuine coding. Until they have learned this, they are frustrated. Yet it is essential that they learn this skill, since everything that follows rests on it. Other than the general guidelines given directly below (and in Chapter 3, on coding), we find in teaching students that the following *rules of thumb* are useful:

1. Look for in-vivo codes, terms used by the people who are being studied. The nurse's "tried to keep my composure" is an instance.
2. Give a provisional name to each code, in-vivo or constructed. Do not be concerned initially about the aptness of the term – just be sure to name the code.
3. Ask a whole battery of specific questions about words, phrases, sentences, actions in your line-by-line analysis.
4. Move quickly to dimensions that seem relevant to given words, phrases, etc.
5. These dimensions should quickly call up comparative cases, if not then concentrate on finding them.
6. Pay attention to the items in the coding paradigm, as previously listed.

There are several additional guidelines for open coding that tend to ensure its proper use and success. "The *first* is to ask of the data a set of questions. These must be kept in mind from the very beginning. The most general question is, *What study are these data pertinent to?* This question keeps reminding the researcher than an original idea of what the study was may not turn out to be that at all – in our experience often it is not. [The case illustrations drawn from the research seminars will show how that can happen.] Another question to ask continually when studying the data is, *What category does this incident indicate?* This is the short form. The long form is, What category or property of a category, or what part of the emerging theory, does this incident indicate? As the theory becomes increasingly well formulated this question becomes easier to answer. The continual asking of this question helps to keep the analyst from getting lost in the rich data her/himself, by forcing the generation of codes that relate to other codes. Lastly, the analyst continually asks: *What is actually happening in the data?* What is the basic problem(s) faced by the participants? What accounts for their basic problem or problems? [Another way to phrase all of this is, What's the main story here, and why?] All of these questions tend to force the generation of a core category or categories which will be at the center of the theory and its eventual write-up."

The *second* guideline for open coding – remember, this is primarily an initial coding procedure – is to *analyze the data minutely*. As noted several times earlier, this means frequently coding minutely. This effort is entirely necessary "for achieving an extensive theoretical coverage which is also thoroughly grounded." A contrasting "approach to open coding (the *overview approach*) is to read the data over rather quickly, which yields then an impressionistic cluster of categories. We do not recommend this approach by itself because it yields only a few ideas and does not force the evolution of conceptual density. It does not, either, give any idea of what has been missed. To continue in that vein gives conceptually thin and often poorly integrated theory."

The more-microscopic approach to open coding "minimizes the overlooking of important categories, leads to a conceptually dense theory, gives the feeling – to the reader as well as to the analyst – that probably nothing of great importance has been left out" of the theory, and forces both verification and qualification of the theory. We should note, however, that when a code seems relatively saturated – "nothing new is happening" – then the analyst will find himself or herself moving quickly through the data, finding repetitions in the line-by-line examination, and so will scan pages until something new catches the eye. Then the minute examination begins again. Indeed, additional data gathering, especially when guided by careful and imaginative theoretical sampling, is very likely to call again for microscopic analysis. (The seminar cases in this book will illustrate very clearly this intense scrutiny, as the students linger for many minutes over particular words, phrases, and sentences, doing their line-by-line analyses.)

So this kind of intensive analysis may be done from time to time. The rule of thumb here is to do this if you sense that some portions of the total analysis are not satisfying or important relationships among categories might be nailed down by additional open coding. Of course, given the usual masses of data, you cannot continue to do open coding more than occasionally – but then there would be no point in doing that anyhow. However, once you sense the usefulness of again engaging in open coding, do not delay the work. The sooner, the better, since that may lead quickly to useful theoretical sampling and slightly redirect your new data collecting.

A *third* important guideline for open coding is: "frequently, to interrupt the coding in order to write a theoretical memo. This leads quickly to accumulated memos as well as moves the analyst further from the data and into a more analytic realm." A *fourth* guideline is: "The analyst should not assume the analytic relevance of any 'face sheet' or traditional variable such as age, sex, social class, race, until it emerges as relevant. Those, too, must earn their way into the grounded theory."

It is important to understand that "open coding both verifies and saturates individual codes." Initially they are likely to be crude, so they will need much modification. Anyhow they are provisional so will end up considerably modified, elaborated, and so on. Hence, the analyst must not become too committed to the first codes, must not become "selective too quickly, tempting as that is, since initial codes can seem highly relevant when they are actually not. Open coding proliferates codes quickly, but the process later begins to slow down through the continual verifying that each code really does fit . . . . Eventually the code gets saturated and is placed in relationship to other codes, including its relation to the core category or categories – if, indeed, they or it are not actually the core."

Axial coding. Axial coding is an essential aspect of the open coding. It consists of intense analysis done around one category at a time, in terms of the paradigm items (conditions, consequences, and so forth). This results in cumulative knowledge about relationships between that category and other categories and subcategories. A convenient term for this is *axial coding*, because the analyzing revolves around the "axis" of one category at a time. It is unlikely to take place during the early days or even weeks when the initial data are collected and analyzed. However, axial coding becomes increasingly prominent during the normally lengthy period of open coding, before the analyst becomes committed to a core category or categories and so moves determinedly into selective coding (to be discussed next). During the open-coding period, however, the very directed axial coding alternates with looser kinds of open coding, especially as the analyst examines new aspects of the phenomena under study. It also runs parallel to the increasing number of relationships becoming specified among the many categories, whether this part of the coding is done as intensively as the axial coding or not. Of course, within this increasingly dense texture of conceptualization, linkages are also being made with the category, or categories, that eventually will be chosen as "core."

Selective coding. Selective coding pertains to coding *systematically* and concertedly for the core category. "The other codes become subservient to the key code under focus. To code selectively, then, means that the analyst delimits coding to only those codes that relate to the core codes in sufficiently significant ways as to be used in a parsimonious theory." The core code becomes a guide to further theoretical sampling and data collection. The analyst looks for the conditions, consequences, and so forth, that relate to the core category, coding for them. Selective coding then, is different from open coding but occurs within the context developed while doing open coding. During selective coding, understandably, the analytic memos become more focused and aid in achieving the theory's integration. Selective coding can begin relatively early, but becomes increasingly dominant, since it is more self-consciously systematic than is open coding.

Sociologically constructed codes and in vivo codes

"The categories are of two types" – sociological constructs and in vivo codes. The latter "are taken from or derived directly from the language of the substantive field: essentially the terms used by actors in that field themselves." Often while doing open coding, the researcher will hear the actors using these terms, and will incorporate them into his or her analysis. "In vivo codes tend to be the behaviors or processes which will explain to the analyst how the basic problem of the actors is resolved or processed. These codes fracture the data directly because they represent analytic categories, as used by the researcher." They can also lead to associated theoretical codes: "for example, 'monitoring' a patient's clinical conditions implies – and the actors often say this explicitly – various conditions under which the monitoring is done, the consequences of the monitoring, and so on."

In vivo codes "have two characteristics: analytic usefulness and imagery. Their analytic usefulness relates the given category to others, with specified meaning, and carries it forward easily in formulation of the theory. Imagery is useful insofar as the analyst does not have to keep illustrating the code in order to give it meaning. Its imagery implies data that have sufficient meaning so that the analyst does not clutter his or her writing with too many illustrations. In vivo terms have a very vivid imagery, inclusive of much local interpretative meaning: they have 'grab' for the participants. And they are seldom forgotten by readers because their terms are colorful. They also have much analytic force since the actors do use them with ease and with sufficiently precise meaning."

"Sociological constructs, on the other hand, are codes formulated by the sociologist ('awareness context,' 'illness trajectory')." (The constructs of course need not be sociological but psychological or anthropological, and so forth, depending on the disciplinary theory that is being formulated.) These constructs "are based on a combination of the researcher's scholarly knowledge and knowledge of the substantive field under study. As a result, they can add more sociological (social science) meaning to the analysis than in vivo codes. They add scope by going beyond local meanings to broader social science concerns. They have much analytic utility because they are constructed clearly and systematically. They may have little imagery (some analysts think that the flatter they are, the more scientific and less impressionistic they are; but others prefer them to resonate with more imagery)."

In the illustrations given later, readers will see the analysts generating many in vivo and sociological codes. As mentioned earlier, this generation is a provisional matter and so is the *labeling* of codes, which is easily changed if better terms are invented later. It is important that researchers should feel free to invent and change those terms. "There is little point in struggling to find exactly the right term, especially when one first notices the phenomenon which leads to the labeling – the important activity is first to notice and then invent or apply a term resonant enough so that the category can be referenced, focused on, and remembered." Analysts can learn to coin these terms with some facility after some experience in doing this style of qualitative analysis. Of course, that facility is not just a linguistic matter but a matter of improving one's theoretical sensitivity *and* associated analytic ability.

Core categories

"The goal of grounded theory is to generate a theory that accounts for a pattern of behavior which is relevant and problematic for those involved. The generation of theory occurs around a *core* category (and sometimes more)." "Since a core category accounts for most of the variation in a pattern of behavior," its different kinds of appearances under different conditions, "the core category has several important functions for generating theory. It is relevant and works. Most other categories and their properties are related to it, which makes it subject to much qualification and modification. In addition, through these relations among categories and their properties, it has the prime function of *integrating* the theory and rendering it *dense* and *saturated* as the relationships are discovered. These functions then lead to theoretical *completeness* – accounting for as much variation in a pattern of behavior with as few concepts as possible, thereby maximizing parsimony and scope."

"The analyst should consciously look for a core variable when coding data. While constantly comparing incidents and concepts, he or she will generate many codes, being alert to the one or two that might be the core. The analyst constantly looks for the 'main theme,' for what appears to be the main concern of or problem for the people in the setting, for what sums up in a pattern of behavior the substance of what is going on in the data, for what is the essence of relevance reflected in the data." (As noted earlier, What's the main story here? is a kind of motto– question that the analyst asks repeatedly, to remind himself or herself to keep trying to answer the above questions.)

"As the analyst asks those questions, while analyzing, he or she becomes sensitized to their potential answers." "Possible core categories should be given a 'best fit' label as soon as possible, so that there is a handle for thinking about them. The researcher may have a feel for what is the core, but be unable to formulate it to his or her satisfaction, so must use a provisional label until a better one can be formulated."

"After several workable coded categories develop, the analyst attempts to *theoretically saturate* as much as possible those which seem to have explanatory power." Thus, relations among categories and their properties become apparent and conceptually dense. Theoretical sampling is done to further the saturation of categories because they are related to many others and recur often in the data. With qualitative analysis, "these relationships must be kept track of in memos, which get spread out or filed until sorted," and get built into integrative memos. "The core category must be proven over and over again by its prevalent relationship to other categories."

"The more data, the more certain one can become of the eventually chosen core category. Time and data can be expensive; in smaller studies the researcher often has to take a chance: and certainly deciding on a core category can test skill and ability. If the analyst decides too rapidly, using a relatively small amount of data, there is a risk that he or she might end up with an undeveloped theory which has little integration and little explanatory power."

There are several criteria for judging which category should serve as the core category.

1. "It must be *central*, that is, related to as many other categories and their properties as is possible, and more than other candidates for the position of core category. This criterion of centrality is a necessary condition for putting a category at the heart of the analysis: It indicates that the category accounts for a large portion of the variation in a pattern of behavior."
2. "The core category must appear *frequently* in the data. (More precisely: The indicators pointing to the phenomena represented by the core category must appear frequently.) By frequent recurrence it comes to be seen as a stable pattern, and consequently becomes increasingly related by the analyst to other categories. If it does not appear frequently, that does not mean that it is uninteresting, only that it is not the core category."
3. "The core category *relates easily* to other categories. These connections need not be forced; rather they come quickly and abundantly. But because the core category is related to many other categories and recurs frequently, it takes more time to saturate the core categories than the others."
4. "A core category in a substantive study has *clear implications for a more general theory*. (See Chapter 11, on generating a formal theory.) Thus, an analyst looking at hospital shifts sooner or later may realize the implications of shifts as a basic structural condition for any twenty-four hours a day work operation, and begin to conceive of generating a theory about work shifts in organizations. The various analytic operations which follow, however, have to utilize data bearing on work shifts from many different substantive areas. Intensive scrutiny of these data is necessary, of course, before the core category or categories for this general theory can be determined."
5. "As the details of a core category are worked out analytically, the theory moves forward appreciably."
6. "The core category allows for building in the *maximum variation* to the analysis, since the researcher is coding in terms of its dimensions, properties, conditions, consequences, strategies and so on." All of these are related to different subpatterns of the phenomenon referenced by the core category. Such variation (also called variance) is, as a colleague once expressed to us, emphasized more usually in quantitative analysis than in discussions of qualitative analysis. He spoke accurately, since many qualitative analysts do not seek for variance, but for very general patterns. It is one of the hallmarks of the grounded theory mode, however, to seek variation. (See additional remarks on this topic, a little further on.)

Who should code?

When it is a matter of an individual researcher embarked on his or her project, the answer to that question is obvious. But what if a team is working together on a project? Should all its members code, or only the most experienced, the most efficient, the most brilliant coders; or the professor rather than student assistants; or, on a large project, the top echelon and not the mere data collectors? Some years ago, a qualitative researcher, Julius Roth (1963) severely criticized the principle investigators of survey researches for their exploitation of the "hired hands," who did nothing but the dirty work of data collecting – contrasting this situation with the deep commitment and involvement of the typical fieldworker, who of course did all the research work, including the brainy-work of coding. Those are the two extreme answers to the issue of who should do the coding.

However, the reasonable answer to this issue takes its cues from structural and organizational conditions bearing on the project, on its aims and its audiences. For instance, a large cross-site qualitative project with, say, two professors back home directing it, and concerned with producing "good results," and fast results (for career reasons), might handle the who-should-code issue quite differently than might – and probably do – smaller and more collaborative teams consisting of peers or virtual peers (cf. Miles 1983, especially pp. 131–2). In these terms, then, think of organizational conditions like amount of funding, numbers of data sites, amount of data to be collected, number of team participants, the degree of homogeneity of team composition. The team structures can correspondingly look different: some are hierarchical, some quite collaborative, and so on. But also, the aims of the project might include – in their various combinations of salience – reaching fast results, or the "best" results, or the most effective results for a given expected audience. Or they might include furthering the creativity of each team member, or of the total team which is expected to do further research together. And the product of all this productive research activity can take various forms during a given project: a collectively written monograph, or two or more monographs written by different members or combinations of members, individual or joint or collective papers – or all of these.

So the answer to the coding issue is going to be inevitably and profoundly affected by such considerations. Each person on a team may code his or her materials, because of greater familiarity with the data – and because there is *so* much of it cumulating for the total project. But, at a team meeting, they may together begin coding someone's presented material, or throwing in individually collected data during the analytic discussion. (See Chapter 6 for an instance of this.) Or one may code some of another's data after reading a memo by the other. Or two teammates may meet in a session to do (or end up doing) joint coding. And they may do that after a team meeting, too, or reading a memo, etc.

My own research projects over the years have involved small teams, composed of more or less experienced people, all trained initially by me, and ended up doing highly collaborative work. And work designed to produce both "findings" and theory, but also designed to increase the creative potentials of each member. So every team participant engages in all the research procedures outlined in this chapter. Sometimes each does that separately, sometimes in twos, or as an entire team, depending on circumstance or ad hoc design. In large part, they tend to code their own data: That is understandable, but perhaps we have failed a bit in not forcing more intercoding of each other's materials, leaving that mainly to joint and team sessions.

Anyhow, to summarize with these *guidelines* pertaining to non-solo projects devoted to doing really creative research – I believe:

1. Each data collector should code much of his or her own data, but
2. code some of the others' data, separately as well as jointly and as a total team,
3. and this should be done from the onset of the initial data collecting to the very end of the project;
4. meanwhile, all should be engaged in theoretical sampling, making comparative analyses, conceptually densifying, integrating, etc.

I should add that there sometimes is one especially difficult problem encountered by students taught in our research seminars. When they attempt to code their own materials alone, without the support of the seminar's analytic discussion, then they may find this not nearly as easy or "deep" and may not have sufficient self-confidence in their coding. For this reason they are urged to meet occasionally without the instructor, as well as to work jointly with another student, between the only occasional opportunities to present their materials in class or to confer individually with the instructor.

Theoretical sampling

Theoretical sampling is a means "whereby the analyst decides *on analytic grounds* what data to collect next and where to find them." "The basic question in theoretical sampling is: *What* groups or subgroups of populations, events, activities (to find varying dimensions, strategies, etc.)" does one turn to *next* in data collection. And for *what* theoretical purpose? "So, this process of data collection is *controlled* by the emerging theory." It involves, of course, much calculation and imagination on the part of the analyst. When done well, this analytic operation pays very high dividends because it moves the theory along quickly and efficiently. This type of sampling, so essential to the grounded theory mode of analysis, is of course neither the same as is utilized in quantitative research nor subject to the same canons (see Glaser and Strauss 1967).

Neither is it what Leonard Schatzman has aptly termed *selective sampling* (Schatzman and Strauss 1973), a frequently used sampling method in qualitative analysis. "Selective sampling refers to the calculated decision to sample a specific locale or type of interviewee according to a preconceived but reasonable initial set of dimensions (such as time, space, identity) which are worked out in advance for a study."

2 Two illustrations

After that long introductory discussion, a "methods" book would ordinarily move directly to presenting in concrete detail the initial steps of research procedure – in this instance, the coding of data. We shall not do that yet. Readers who are eager to get quickly to the procedures can skip the present chapter, returning to it later. But it is placed here for those readers who need some overall visualization of the spate of terms discussed rather abstractly in the preceding pages.

There are two reasons for placing the material given below at this precise point in the book. The first is to give some sense of how a grounded theorist operates with data, since that style of analysis is somewhat different than other modes. Thus the analyst–teacher will be seen developing theory by using both "real" and experiential data, making constant comparisons, discovering and naming categories, suggesting possible theoretical samples to be examined later, emphasizing all of the elements in the coding paradigm, and raising a host of theoretically informed questions. The second reason for presenting this material here is to give a more concrete sense of how grounded theory is taught in research seminars, providing thereby useful imagery before readers are plunged into the technicalities of coding, memoing, and so forth. Indeed the teaching of grounded theory rests on collaborative work by the seminar participants (although sometimes it is strongly guided by the instructor), and is designed to facilitate thinking about and analysis of data in free if disciplined ways. (See also Chapters 4 and 14.) This chapter, in other words, sets the stage for the "how to do, and how it is done, and what the products look like" demonstrations of the next chapters.

Each of the two cases presented illustrates features of the general style of analysis outlined in the opening chapter. It consists of the transcript of a session of a seminar on qualitative analysis, participated in by graduate students in sociology (two of whom were also trained nurses). These students were about one year into their graduate training and two months along in their training in qualitative analysis. The second case will further show the instructor at work, this time doing a microscopic analysis of a short interview.

A class session: pain management

The seminar discussion was focused on the phenomenon of pain management, for the instructor was then doing research on this topic (Fagerhaugh and Strauss 1977). Using it as a springboard for teaching about theoretical sampling, comparative analysis, the generating of categories and the labeling of them, the analysis of dimensions, tactics, etc., he is drawing on his research knowledge as well as on the experiential knowledge of the students. As a teacher, he is purposely very active, since he wishes, as speedily as possible (and since this is only the second session of the seminar): (1) to demonstrate how rapidly initial data – even *experiential data* – can begin to generate theory (about pain management) through coding, theoretical sampling, and comparative analysis; (2) to illustrate *how* these operations are carried out; also (3) he is not concerned with exploring at great depth any given comparison in this particular session, but rather with showing the variety of analytic strategies that are available for carrying out generation of theory.

This case illustration, unlike most of the others in this book, will be presented without detailed commentary of what is transpiring, analytically speaking, throughout the session as it proceeds phase by phase. Here, though, is an overview of some notable things that occurred. There was open coding, which resulted in a number of categories and terms for them (assessing, balancing of priorities, pain expectations, inflicted pain). In relation to those categories, the teacher showed how to find numerous comparison groups, suggesting theoretical sampling. Some exploration was made of conditions, tactics, dimensions, interactions, consequences. There was much emphasis on variation, including how to potentially qualify an interesting hypothesis. So the open coding was already leading to some conceptual density, through exploring possible relationships between categories.

As for pedagogy: The teacher explicitly emphasized and elicited the use of his own and students' experiential (research, personal, professional) knowledge qua data. He converted data into questions, categories, hypotheses, dimensions, consequences, etc., and had the students beginning to do that also – by giving examples, by requesting, by raising questions, by stimulating the students. When the students made comments or offered data which combined two or more issues, he then sometimes clarified by making analytic distinctions. He suggested also when memos could be written and why. He set "next directions" for discussions, or shifted the discussion into new paths (not wanting to go into much analytic depth for any point). He gave summaries and recapitulations occasionally, along with small lectures (lecturettes) on various elements in the grounded theory style (e.g., a little data and then analysis). The teaching style itself emphasizes the provisional nature of open codes, with the corollary that the researcher can correspondingly be relaxed, even if the analytic session is intense and intellectually demanding. A key feature of this kind of teaching of analysis is the raising of generative questions for discussion: expecting that students will be able to respond intelligently, not merely because they are intelligent and can think logically, but because they have experiential data in which to ground their answers – and need not be afraid to do so, once they get over shyness in front of the teacher or their classmates.

The instructor began by saying that the seminar could use informants to begin building its theory, since two members of the seminar were nurses. "They have data in their heads. We can regard what they say as the equivalent of initial field observations." He urged the nurses to say something based on their own experiences. One said that hospital nurses have a problem in "assessing pain," although it is necessary to do so. Why? "Because it is related to the nurses' actions" with regard to patients.

Instructor: "So, we have to ask how they go about assessing pain; that is, what *assessing tactics* are used. But first, let's ask *what* they assess." The class quickly came up with such items as: the kind of pain, its intensity, its bodily location, its duration, its progression over time. The instructor called these *dimensions*. He added: "Next time you go into the hospital, if you were studying this, you would keep an eye open for other kinds of dimensions. But let us suppose that for the time being we just write a memo, noting down what dimensions we have thought of. Later we are going to see if it works out or not. Meanwhile, there is no point in getting anxious about whether they will or not. But the memo tells us that now we have an inquiry: What other dimensions are there, and will they be relevant to building a theory?"

Then he turned to the tactics of assessment, asking: How do nurses go about assessing pain? Jan, please give examples." Jan replied that the assessing nurse "will compare this patient with past cases." Instructor: "Let us theoretically sample by asking what happens in two contrasting cases – the nurse is quite experienced (say she has been on the ward a long time) versus a nurse not very experienced (say she is new, at least to this kind of ward). Now you *can look* for each kind of situation and observe what happens. Or, if you don't want to immediately, then write a brief memo about it and do it later."

Jan offered, as a second assessing tactic: getting a description from the patient of what it is like, using a question-and-answer technique. Instructor: "How many dimensions does she hit?" Also, we can theoretically sample: "What happens when you get a patient who isn't articulate, and can't give her good answers?" Seminar members: "How about semicomatose patients? Little children?" Instructor: "We have to ask what nurses do under those conditions. All these questions lead us to inquire about (as potential conditions) variations in persons, styles, tactics, etc., as they relate to *pain assessment.* We can see that assessment is an important category for studying pain management, and its properties will guide our next steps in searching for data."

The instructor then asked one of the nurses about what happens when a patient continually complains without seeming to have much cause to – the hospital nurses discounting the complaints, but then the patient insisting that the pain is getting more intense. The nurse answers, "It may depend on the patient's reputation among the nurses." The professor comments: "Then, it is not just pain assessment done only 'in the present,' but it can be done in the context of *mutual experiential careers* (that is, these persons have been around each other for a while)." He continued: "Let us theoretically sample then. Suppose they have no mutual careers whatever! Suppose they have a very long one. Suppose they have had mutual careers, and there was no pain – but now there is? Or there was always a lot of pain, through the whole trajectory? (Compare this with a patient they like or don't like.) Well, we could go on with this kind of sampling, along the notion of mutual careers, but that is enough to make the point now."

A student excitedly says: "Look, when sometimes someone comes into a clinic with an overdose of barbiturates, sometimes there won't be as great an effort to relieve the pain, because they assume somehow that then the pain will show him he shouldn't do this. And once, my mother, having a child at a hospital, was told that it was *necessary* for her to suffer during childbirth!"

Another student, stimulated by the foregoing, notes that when people come in for gonorrhea treatments, the staff may stretch out the treatment for twenty days, giving the patients a painful shot each day, as a form of punishment. They are supposed to learn via the pain.

Instructor: "I hear you saying two kinds of things. We've moved to – When is pain *not* relieved? The second is: When is pain actually administered? You now have another set of memos to write, which have to do with, Under what conditions, when pain is discovered by the professionals, do they manage the situation so that the pain will go on? You can break that into two: when pain is inflicted by the professionals and when it is not. There are, for example, many treatments and tests where it is necessary to give pain to get the job done. Professionals can also try to maximize or minimize the pain: with the gonorrhea treatment and a disliked patient they are maximizing it. And the nurses will tell us that there are situations when pain is desirable because it makes good diagnosis possible. So now we have a series of distinctions: pain for *diagnosis*, pain through *negligence*, pain for *punishment*. There also is the situation where a cancer patient's pain is not relieved fully because the staff withholds medication, believing they need those drugs in reserve for 'near the end.' So now I have introduced the idea of *pain trajectory* as a condition for not minimizing pain." (We had touched on trajectory before.)

The instructor then summarizes the major points touched on thus far, adding that, "We have not done much yet. We have done a little coding, sampling, and a little comparative analysis; and have memos that can either cover a couple of sentences or some pages. We could follow through where we have just brushed by: thus, the question of negligence as a condition for giving pain or for not relieving it much. We could explore that whole area. We could look further into the giving of pain through diagnostic tests: You can look for situations where the inflicted pain is unexpected by the patients, or where they are warned or understand there will be pain; or where the hospital personnel themselves are surprised. So you can lengthen particular memos, adding to them, as you either think through these matters from data in your heads or as you latch onto them in the field. But concerning what's in the memos, for each point, you can theoretically sample by asking: Who do I ask about this (interview), or go and observe? Then you can actually interview or observe these people. You can also theoretically sample by turning the initial query upside down. We have been looking at pain assessment. What about where assessment is virtually nonexistent? For instance, I trust the person very much, so when she tells me of her pain, then I don't doubt it. ('I have a headache today,' my wife says.) But where there is little or no trust, then canny assessment tactics prevail, as in the hospital. So you now write a memo on this type of assessment tactics. If you haven't enough experience with that (i.e., data) or you want more immediately, then observe or interview."

"For instance," the instructor says, "let us ask one of the nurses in the class." From one of the nurse students, he gets three canny tactics in short order: the staff gives placebos; the staff asks trapping questions; they pop their heads in the patients' doorways to catch them unawares. Then the instructor added: "Let us stay for one more moment with pain assessment, but do a flip-flop. Suppose a patient doesn't admit to as much pain as he has! We could ask our nurses about that, too."

Rather than pursuing that path, he suggested that next they pick a specific ward, say a surgical unit. He sketched out something of the typical passage of patients through such a ward: pre-op, surgery, then post-op. If one looks at the post-op phase and walks around a surgical ward, he or she would observe that some patients were temporarily closer to leaving the hospital and some were closer to their surgical operations: "like trains traveling along a railroad track." However, since the patients may have gone through different types of operations, their post-op phasing might be comparably different: hence, they are on multiple tracks. Say there are ten tracks and thirty patients: What would that suggest?

The class grasps his point and answers, "Different *expectations* concerning the steps of recovery." The instructor asks them to disregard the patients' expectations and to focus on the nurses' expectations. From the class discussion there arises the idea of *patterns of recovery*, with possible associated pain. Thus, operation #1 might give much early pain but then a quick drop in it; whereas #2 might give minor pain, but long lasting. The instructor: "Look now at the comparisons you are making that are useful for fieldwork. You can look at them as they appear on the ward. Or you can look at any one patient immediately after an operation, and watch a pattern emerge through the next days. It follows that if expectations of nurses match the recovery course (i.e., it is fairly routine), then that pain management will be fairly routine in the sense that they know what to do to relieve it (however much work is involved). Let us say the pattern was: at first, terrible pain; then, mild dropoff; then, no pain at all, but with the big concern being an increase of the patient's weight and strength. Now, how do you theoretically sample that pattern of recovery?" The class discusses doing it by phase, though it might take several weeks to pin down the pattern by following several patients. The instructor: "As you did that, you would begin to get the crisscross of things like energy, weight, liveliness. What else?" he asks. Student: "You could sample by nurses and patients; how they handle each phase. You could find the range of the nurses' tactics and how the patients responded to the tactics." Instructor: "And you could watch the variety of tactics also; some nurses might use predominantly one or a few tactics, and so on. Most likely you would find nobody who would do the same thing all the time."

Instructor: "Any other kinds of sampling?" Student: "You could ask if the priorities are on weight as over against pain? If there is a priority on relieving pain, then they would sedate more – which would maybe cut down on appetite." The instructor remarked that it was likely that this would be relatively a minor issue with most surgery patients, but it is a major consideration on the cancer ward (which he had studied). Let us call this *priority balancing*.

A student remarks that one surgical patient might look at another and say, "That's where I was yesterday." This comment leads to discussion about patients' comparing notes; also to nurses telling patients, "See, that's where you will be in three days," as they point to other patients. But there are other wards where patients can't so easily make such comparisons. Instructor: "Now you can theoretically sample – you can look for (or stumble on), in your fieldwork, patients who are in single rooms and so can't easily compare themselves to other patients."

The instructor says that they ought to think of another theoretical sample where a patient "breaks the pattern" of surgical recovery that they have been talking about. "Now we will see the crosscuts of intensity and duration: Think of a patient who appears to have no pain at all after the first day of the post-op period, but she keeps saying she has pain, and goes on saying that for another seven days." The class agrees that the nurses' reaction would be that, "There's something wrong." The instructor adds: "Yes, and now the suspicion game would scarcely take place, however much they suspected her complaints before." A nurse adds: "They might think something else was going on (physiologically)." Instructor: "*Now* suppose that she shows pain visibly on the day of alarm. Then comes great concern, diagnostic work, calling in consultants, and so on. So inadvertently we have theoretically sampled again. First, it was to break the pattern. Now we have sampled again – her pain is also visible; as over against her continued complaints, but it is invisible."

He then lectures: "So, at every step you're asking about opposites, variations, and continua. Sometimes in actual research you don't follow all of these leads – it is just too exhausting. But sometimes a phenomenon just forces itself on you from the nature of what you are seeing or hearing, day in and day out. Or you see something on one day and on another something fits in with it. But at every step *this* is what you are doing. That is why you don't want to rush out and get a lot of data, because you would get submerged. You get a *little* data, then you stop and *think*! At virtually every point in your initial fieldnotes or interviews, you must do this kind of thing."

The instructor then began to recapitulate the ground they had covered in the last minutes and mentioned multiple patterns of recovery. Then he remarked that we might imagine what would happen if on a given ward all the patterns were evinced simultaneously, on a given day, all patients being in great pain but one patient was not – what would happen to him? Clearly, he would get rather little attention unless he was judged critically ill. "So we can vary the picture further with a little imagination." Suppose one imagined a patient who was recovering "on schedule," but then she began to evince considerable pain when she oughtn't to? Probably the staff would discover that the pain comes from a source other than lack of recovery, like an infection ("or a pair of scissors," someone quipped).

Another student remarked that before we had been talking about nurses' expectations about the duration of pain. What would happen if the two didn't match, as when a patient started with extreme pain but then it just kept on? It wouldn't be just breaking the pattern but "going into another category, for there they are no longer uncertain about the duration. The nurses then probably would go into another whole set of feelings and actions toward that patient." The instructor said: Now you are in the situation of "pain certain, etiology unknown." The student followed up with: "And for that the nurses have been trained, so there will be a momentary shifting of gears but then they will go into another set pattern." "Good," the instructor said, "now we have a word for that: the *shifting of gears*. Now we can raise other kinds of shifting of gears; like shifting upwards or downwards in seriousness of pain. So you write a memo on that and maybe that is all you do with the item for a while, until you see more incidents of that kind, or see it occur on another ward. Realistically, if incidents and memos keep occurring, then you begin to think it is important as a phenomenon." The student who originally raised the issue: "You ask, Under what conditions do the nurses shift gears – like if the patient calls attention to this continued serious pain."

A student suggests the idea that, "From what we've been saying today, what is really important is the assessment; once the assessment is made, the treatment somewhat naturally follows." The instructor answers that this is an hypothesis, "but suppose that I say 'yes but' – and pick a theoretical sample where you get a situation in which an assessment of pain is made and treatment follows, but then we look for a situation where an assessment is made but you aren't sure what the treatment for the pain should be? Under what conditions would we expect to find they don't know what to do despite assessment? Because we might just qualify your hypothesis that way." He added that now we could turn to the nurses in the class for an answer, since they were so full of experiential data. A nurse answered with the illustration that a patient was assessed by a staff nurse as having a sudden pain that needed relieving, but for which there was no prescribed medication or physician's order.

The student who had spoken previously objected, saying that the staff nurse actually knew what to do, she wasn't uncertain in the sense we had spoken of before. The instructor broke in, saying that the *legitimacy* issue was being raised: "That is, who is allowed to kill off this pain legitimately; and the nurses are not allowed to do this, except on a doctor's instruction – that is, delegation." (Another student gave an anecdote about being in the hospital, and suddenly needing medication, but the nurse refused to give her any without a doctor's order.) After an animated confusion of tongues, the instructor said: "Look, we have at least two things going. Let us be clear analytically. First, there is the patient's tactic when no legitimate agent is available. Second, the nurse cannot act because he or she has not been legitimated. They both pertain to the same category of legitimacy. So we now have a new memo about who can give pain killers and under what conditions. For instance, I can give myself aspirin for pain at home, but if I go into the hospital, they take it away – I'm not even allowed to use my own aspirin. The only persons who are allowed to give drugs there are legitimated or delegated agents. So here is an entirely new dimension, and it is entirely understandable why it hasn't come up until now, because probably on surgical wards the phenomenon doesn't strike the eye very easily. But to return to the question raised before and my responding query: Could there be conditions under which the staff could make the pain assessment but not know what the proper treatment was? You, Larry, are assuming they always would, on the analogy that if you have a headache, then you take aspirin."

Larry answers that he believes assessment also entails treatment, and that if the nurse has faith in the treatment she will assume she won't be punished for using it with a patient in dire pain, even though she has not been given orders to use the medication. "Ok," the instructor replies, "pain management always involves where the pain is, how long it has been going on, its intensity, and it involves calculations about how much medication, what kind, how it is acting or how fast acting, etc. But the game I wanted to play with you now is that of possibilities – based on your own experiential data. One works out possibilities even if you can't find immediate examples of them, provided you *do* have that data. (Usually you eventually do find them in the real data.) You say, 'Under what conditions would you expect to find . . . ?' Our example was one where they can assess the pain adequately, but don't know how to treat. What we had previously been talking about was *standardized* pain, and now here is *unstandardized* pain." Larry responds with the situation where the staff feels they need more tests to determine how to treat. The professor says, "Yes, but there is still another possibility. That is where they can assess pain to their own satisfaction, but they don't know how to manage it because the appropriate treatment would disturb some important physiological function in this particular patient. So they are hamstrung. They are balancing priorities (now I happen to know it is true, because you see it with people who are dying) . . . . Anyhow, that is a possibility. You always try possibilities even if they seem not too likely; and you write memos to that effect. When you actually hit them in the field or in interviewing, then you will recognize them!" (A nurse speaks up and gives several varied examples of that particular logical possibility from her own experience as a nurse.)

There was a long silence. The instructor then pointed out that at least two paths could be followed now. (He warned them never to be too anxious, since it didn't much matter which one they followed – they could come back to the other eventually, assuming this was a genuine inquiry. They had the whole hospital available plus their own aches and pains.) "First, given the dimensions we have now got memos written about these; we can follow any and all of them up, and anywhere in the hospital – via the kinds of sampling questions we have been asking." (For example, it could be asked under what conditions someone doesn't have to legitimate, and what would happen; or when one is one's own legitimator and then comes into the hospital; and vice versa, when one is suddenly on one's own after a spell in the hospital?) The second path is to just pick another locale where somewhat different things are likely to happen: thus on another ward the patterns of pain, the pain trajectories, etc., will be somewhat different. You can go to another kind of ward, to a nursing home, you can turn your attention rather to what goes on at your own home, or you can take some pattern like that associated with arthritis and interview many patients with arthritis. This kind of nonrigorous, or related, theoretical sampling is analogous to going to the library and deciding to go to the reference section and just start skimming through the books in that section, just to see what happens, because you feel you are bound to find something interesting that you didn't necessarily anticipate.

"It does make a difference whether you follow the locational path or not, so we will do it both ways to show what happens. Let us first take the locational path, perhaps visiting a pediatrics ward, and there looking for pain and pain management. You have in mind all the categories we have already discussed today. But you aren't going to restrict yourself just to them – you are going to leave yourself wide open for anything new. One new phenomenon that you will find, I know, for instance, is that on pediatrics the staff must very often take into account the child's parent. The parent, here, can intervene for the child patient in matters of pain relief. (For an adult patient, someone might intervene also, but that is not so likely as on a pediatrics ward.) Let us keep that in mind. But now, Fran, can you describe the type of event where pain is a problem that one sees on pediatrics?" Fran describes a young child with leukemia who has lesions in his mouth, so that feeding becomes a painful process and there is danger of infection; but he must be fed. Instructor: "So here is *pain inflicted* by staff or parent on a very young child; but it is *necessary pain* in the sense that feeding must go on for survival. OK – tactics?" The two nurses and rest of the class begin to list the tactics (aside from the obvious avoidance of pain through intravenous feeding): Give bland foods, give him foods he likes, give him constant encouragement, the mother does the feeding, give him rewards (like ice cream) for suffering the inflicted pain.

Instructor: "So now we have a new, important category: *pain inflicted for the good of the patient.* Here it happens to be by food, but it could be by tests or treatment. So let us try each one of these now. How do they handle pain that accompanies treatment?" The class offered: distracting, having her participate in it as an event, saying to her, "Soon it will be over." The instructor: "You see how *dense* your analysis is becoming. But analytically, not anecdotally. You can say, for instance, 'It will be over soon,' because the pain duration is short, or because you can tell the child, 'It is necessary,' or because you promise her it will only happen once, or because it is not really too frightening. If you are going to do it every day, then you have another kind of problem. Or if it is the kind of treatment where the pain begins mildly and then increases, then you have another kind of problem that probably will be handled by different interactional tactics. Every one of those possibilities you might like to be on the alert for, or you might just happen to come across during a day's observation."

"Now," he continued, "you can see why parents are so important on Pediatrics – just in this matter of pain management and relief. And that runs into the issue of trust. That can be theoretically sampled: What happens, for instance, when a child has been trusting, and then you give him a shot that hurts him terribly? And what about a painful treatment that he can see works out right away against one that he can't see work immediately? Or what about something like an application of iodine which he is familiar with, so he knows it is a short pain if only he can get through it, as against one whose efficacy is invisible to him? And how do you explain to a young child about a test, that it is not a treatment but that it is necessary to find out what is wrong with him so they can help him? That is one additional step upward in abstractness. So you watch the tactics associated with giving that test, for they probably will be a little different than with a treatment per se."

"Then you flip-flop your sampling: You look now for treatments and tests that *don't* cause pain, in order to make necessary comparisons – to see the similarities and differences." A student suggests: "A parent may not be necessary during painless treatment or tests." The instructor continues: "Or, you look at situations where the kid might be frightened but actually no pain will be inflicted. So you are theoretically sampling in terms of what is absent. For a dimension like fright with pain – now you take pain away and see what happens in an actual case of that. All that kind of *microscopic sampling* you can do. Or you don't bother with finding those situations, but you run across them and observe them. Then you realize that something different is going on that is well worth noting and thinking about. In short, you can't just study pain, but must study all the penumbra of events around it. You will see it, anyhow, but you must build it into your analysis." Student: "This is probably what happens when people think, anyhow; it makes ordinary thinking explicit." Instructor: "Yes, explicit but theoretical sampling drives this kind of thinking to its limits. You realize rather quickly where your holes are."

He added that while he had been trying to show the class how to do theoretical sampling and find comparison groups, he would also suggest that it doesn't take a genius to do this kind of work. Some people do it better, more efficiently, and can operate on more abstract levels. And of course, one learns to do it faster and better. A student hazards that the genius part is, "How far to take it and in what direction. You can't check out everything!" Instructor: "I also feel that it is temperamental. You cannot allow yourself to be too compulsive – following every last lead, dotting all 'Is,' and crossing all 'Ts.' Also, when you see you are not getting anywhere with a given line, you drop it and move on to something else. If, for example, you can't figure out a legitimacy angle, leave it in your notes, and move to something else. If you are totally blocked, go off to the park or somewhere else where it's fun to be, and then come back refreshed!"

Student: "So, it's not a question of finding out the truth, but *which* truth. Whatever you find out, it will be true and it will be valid. What you leave out may not be interesting or important to you, but later may be important to somebody else." Instructor: "And if your theory is sufficiently comprehensive and dense – as we say in the *Discovery* book – their work then can fit right into it. It is just a matter of systematically integrating your theory. And that is how theories are built up. If anybody argues that, 'I didn't see what you saw,' then you say – 'of course not, you went in with a different frame of reference, or went to observe pain management, say, on a different ward. But if you, on the other hand, follow me step by step, you will certainly see what I saw.'"

Coding through detailed analysis

The next set of materials foreshadows the extended discussion of codes and coding in the following chapter. Again, there will be no commentary on the materials, but here too one can see the same experienced instructor–researcher carrying out an analysis. This time it is based on a detailed scrutiny of data, done line by line and also paragraph by paragraph. He had been given a one-page interview, also reproduced below, by a graduate student in sociology, some time after a conference between them (see the précised version of the conference, Chapter 7, Case 5). The central theme of the student's research seemed then to be this: crucial contributions made by parents to the physical survival of their babies and young children, who had been born with severe congenital heart conditions. The student had had little expereince with coding, for he had been unable to attend the research seminars because of the constraints of his position, working as a social worker with the parents at a medical center where the babies were born and given medical treatment. After scrutinizing the interview, the instructor conveyed his coding results (and associated queries) on the telephone, while the student took careful notes.

Probably it is not necessary for readers to understand the details of the reported analysis, except to understand that the instructor's focus is on parental monitoring and assessment of danger signs – something he was much sensitized to by his own research on clinical safety work done in hospital settings – along with considerable work done on the *biographical-time* conceptions of chronically ill persons, derived from a study of those patients and their spouses (see Chapter 9). Although he brought that knowledge and sensitivity to the interview data, he did not bring the concepts of monitoring and assessing to the data beforehand. Rather, he examined the interview, word by word and line by line, coding as he went. Note again the theoretically informed questions, the suggested hypotheses, the potentially useful categories, the dimensions, possible conditions, consequences, and so on. All these are provisional, offered by the instructor to the student–researcher as guides to focus his analysis and further data collection.

The interview (by Aaron Smith)

The parents installed a very sensitive, high-frequency intercom in the baby's and their room. It was kept on at all times. They wanted to hear her breathing and to know that she was OK.

M.: We did what we had to do, there was no other choice. When we first took her home, we seldom slept, at least, not too soundly. They (the doctors) told us to watch her and not let her get excited; so we played a lot with her until she fell asleep. They told us to look for reactions, so the only way we could do that was to stay up with her. Lucky for us, she slept a lot, but it didn't help because we were afraid that she would die and we would be asleep. We took turns sleeping and then sitting with her. This lasted for two months. I don't know how we did it, but we did.
F.: It was hard to work all day, sleep for three hours, and stay up the rest of the night and go to work the next day. We did what we had to do – no other choice. We got to know her very well and at times we would wake her up, just to see if she was alive. That might sound silly to you, but we understood that she had a serious heart condition, and needed watching for any signs and symptoms of changes that would mean trouble. We called the doctors at the slightest change in anything. Our doctor was kind and didn't seem to get upset by our frequent calls. I don't care if he had, we would rather be wrong than sorry.
M.: I got the idea about the monitor from the smoke detection device. I talked to a friend about it, and we went in search of a sensitive intercom system that could hear it. She scared us once or twice when we didn't hear her and both of us ran into her room and she was turning her head over. That scared us.
F.: She's been back to her own doctor three times in the first two weeks we had her home. We also brought her to the emergency room once in that period. She was cranky and irritable and we just wanted her checked. She was fine each time, but as my husband said, we'd rather be wrong than sorry.
M.: She's three months old now. We've turned the intercom off. We're a little calmer, too. We don't watch her as much or as close as we did. We know more about her heart now than we did when we first went home. I'm sure we didn't learn anything that we hadn't been told, we are just better able to hear it now. Her heart hasn't been fixed – we're still waiting and hoping that those damn symptoms won't show up. I guess we do still worry, that sledgehammer is still over our heads.
F.: I guess it will always be. If things don't change one way or the other, we'll have to wait it out and see what happens. I guess we've had it easy compared to other families we've heard about, but it really hasn't been easy, overall, for us. I guess each situation is by itself in terms of what it means.

Analysis and queries (by the Instructor)

M.: "*We did what we had to do, . . . no other choice. . . .* The doctors told us to watch her and *not let her get excited. . . .* They told us to *look for reactions*, so the only way we could do that was to stay up with her. . . . We were afraid that she would die and we would be asleep. We took turns sleeping and then sitting with her. This lasted for two months."
Instructor's response: The parents shouldered responsibility for the child's survival. They assumed an active role (saw themselves as taking on this role); saw themselves as having no choice – what is choice about? Moral obligation and/or the child's survival, the child is theirs and no one else's.

When the doctors said not to let the child get excited, what did they mean? It's dangerous, it's serious? The doctor's comment, "not let her get excited," suggests it's an action under the voluntary control of the parents; they could stop it if they liked, it was up to them. There were three possible responses to that: (1) prevent it, get rid of it altogether; (2) look for reactions suggesting its occurrence; (3) stop her as she becomes excited.

The doctors gave the parents instructions. What did they tell them as the process for keeping the child alive? Watch her. What's involved in that watching? All day, part of the day, when she's asleep? When she's awake?

Specificity – What did they tell them? Intensity – how did they tell them?

sternly?
eagerly?
casually?
nicely?
excitedly?

Do they tell them how to do it and what to *assess* for?

"The only way we could do it was to stay up." An active role of the parents, their job; they couldn't delegate it to someone else (e.g., grandparents, teenagers if any, older children, etc.).

"Lucky for us, she slept a lot" – Was she in less danger when she slept, yet they were afraid she would die in her sleep? Double whammy. Is there an element of ambivalence here? Stay awake to get cues to be alert or could they take a chance, get some sleep themselves but run the risk of her dying while they were asleep?

For how many months was this their life? Was anything else in their lives? – this emergency, crisis division of labor, this ceaseless burdensome existence, where their negligence could have been fateful as far as their child's survival.

Central issue. Monitoring, labels for categories of monitoring.

F.: "It was hard to work all day . . . and stay up the rest of the night and go to work the next day. We did what we had to do – *no other choice.*"
Instructor's response: An interminable two months, scheduling living activities. Parents accumulated tiredness and fatigue. How did they stay alert and be tired at the same time? Where does this lead? How do you juggle this? Does some cheating on monitoring take place?

There's a great deal of anxiety going on – being familiar with the cues. They continued to wake her up to see if she was alive. There is a discrepancy here.

What do they understand about their child's condition – What's their imagery? That she has a serious illness?

What do they know as a result of their monitoring activities? If she doesn't improve or change, is she getting worse? If she seems as though she's getting better, is she getting better?

First-order assessment. Parents do their own assessing, evaluate the situation based on their awareness of the child's condition. Partial, sometimes, half-time, all the time – When does monitoring take place? How do they know what to monitor?

There is a discrepancy between lay (parents') and professional (physicians') knowledge, between the message and information given and the message and information understood.

However, what happens when their anxiety gets too much or too great for them?

Second-order assessment. Apparently, after some deliberation they call the doctor. They would rather be wrong than sorry. They do juggle their awareness activities.

The doctors are back-up assessing agents. Parents are careful that they do not disturb, upset, or scare off the doctor – cautious in their approach to him.

Their use of the intercom as a second-level assessment agent – they used one machine no one told them about. It is not a medical device; they devised it; the doctors did not tell them to use it. Why not?

The telephone was also used as part of the backstopping activities.

Third-order assessment. Taking the child into the doctor's office, the parents give information to the doctor, maybe in answer to his inquiries; he may have found the information significant or may not.

After 2 months, the crisis passed – monitoring slowed down. Their assessing was less frequent. How long? How much space between assessments?

They turned off the intercom.

Dimensions. How often did they monitor her? How closely did they monitor her?

They said that they've learned more technical things. What? Are they saying if they knew then what they know now, they would be less tense? Don't know. Would they have been calmer? Would they have monitored less? We don't know. They still worry, though calmer. It is still a fateful situation.

Time:

Work time (relationships):

a. monitoring and assessing
b. turn taking
c. marital relationship

Survival time:

We'd better get it in time or she will die.

Intervention time:

Stop problems before they begin.

Appropriate act and appropriate time:

If you don't act right now or at the right moment.

Time to act:

1. parents, directly – emergency room
2. parents, indirectly – telephone
3. doctor, indirectly – telephone
4. doctor, directly – sees child

Time – fateful:

Future is always part of the present. Time is very important.

Time in relation to the last 3 months – biographical. In the proximate future, it will always be part of the present.

Parents – wait it out; it may be cured or improved.

Monitoring and *assessing*:

Quite complex – What is the relationship between the two?

Words and labels (gives categories).

Crisis division of labor as done by parents in their monitoring and assessing activities.

Illness time and *activity time*.

Patient work in high relief; parents enter into work field with/for child.

First-order doctor work: Doctor's diagnosis and warnings, instructions, etc.; division of labor, lay (parents') role and professional (doctor's) role.

Second-order parent work: Assessing the child's condition and state, monitoring responses, reactions, overall awareness of child's present state.

Third-order doctor and parent work: Parents bring child to clinic; child examined, doctors ask parents questions; parents answer to best of their knowledge and awareness; solicited and unsolicited information.

Fourth-order parent work: Back stopping; physician called if cues suggest something the doctor should be aware of; contacts primarily via (indirect) telephone.

Fifth-order parent work: Will bring child to emergency room or take to local doctor or into outpatient clinic; contact is direct.

3 Codes and coding

Coding is the most difficult operation for inexperienced researchers to understand and to master, as noted earlier. Even when understood theoretically, the actual procedures are still baffling for some people, despite watching an instructor or some other experienced researcher do the coding. What is needed, apparently, are examples of coding steps, and visualizations of actual codes. Finally, considerable practice at coding is requisite. The materials in this chapter are designed to help that learning process.[1]

But first recollect that coding: (1) both follows upon and leads to generative questions; (2) fractures the data, thus freeing the researcher from description and forcing interpretation to higher levels of abstraction; (3) is the pivotal operation for moving toward the discovery of a core category or categories; and so (4) moves toward ultimate integration of the entire analysis; as well as (5) yields the desired conceptual density

[1] There are several misconceptions of the grounded theory approach to qualitative analysis which hopefully will be partly, at least, laid to rest by a reading of this book, and especially this chapter. I say "misconceptions" rather than criticisms, for the former rest on inaccurate readings of previous writing on grounded theory. These misconceptions include that the approach: (1) is totally inductive and (2) does not verify findings. In Miles and Huberman (1984, p. 57) there is also a misunderstanding about grounded theory technology. The materials in my book, written before their publication appeared, run directly counter to some of their remarks, that: the grounded theory approach "has a lot going for it. Data get well molded to the codes that represent them, and we get more of a code-in-use flavor than the generic-code-for-many-uses generated by a prefabricated start list . . . . The tradeoff here is that earlier segments may have different codes than later ones. [They may, in part, of course.] Or to avoid this *everything* may have to be recoded once a more empirically sculpted scheme emerges. [No.] This means more overall coding time, and longer uncertainty about the coherence of the coding frame. [Probably, but deliberate, in part.] And there is another risk: The danger of finding *too* much coherence in the data during recoding of earlier segments – retrospective hindsight is at work. [Not at all. The technology attempts to maximize true coherence, not spurious coherence.]" Miles and Huberman also maintain (pp. 63–64) that coding is not much fun, contrasting it with other aspects of the research, like data collection or memo writing. This simply does not apply to the grounded theory style of coding – hard work, yes; boring, no. However, they offer some excellent rules of thumb for coders (pp. 64–9), some of which pertain even to our style of coding. I recommend also looking at their entire discussion of coding procedures.

(i.e., relationships among the codes and the development of each) (Glaser 1978, pp. 55–82).

To supplement that summary statement, readers should examine again the sections on codes and coding in Chapter 1, and this should be done *before* studying the materials given below. These consist of several illustrations. The first will illustrate getting off the ground with *open coding*, by presenting what a research seminar of beginning students did with a fragment of interview data. Next, there is an instance of open coding done with a section of a fieldnote, showing how the initial open coding is done step by step by an experienced analyst. This is followed by a discussion of *axial* coding illustrated by a set of coding notes done by the same analyst on the same fieldnote. Next, a coding session on other data from the same research project is reproduced, with the associated lines from the fieldnote included. To the coding items specific commentaries have been added, to underline *how* such coding proceeds and *what* the codes look like when written. Next there is a discussion of *selective* coding, using the materials of the first illustration. This is followed by a set of coding notes, and commentaries on them, drawn from another research project. The next section elaborates on coding procedures for linking structural and interactional aspects in one's theory. This is not really a separate species of coding, but analysts sometimes do not learn how to code specifically for those aspects of their phenomena. The chapter closes with a few useful rules of thumb for coding.

Besides these materials, there are throughout this book many instances of the coding features exemplified and discussed here. You will see there, and here also, many instances of generative questions leading to coding; of line-by-line or paragraph-by-paragraph eliciting of categories, and queries about them; of discovering in vivo categories as well as the provisional labeling of sociological categories; of relationships drawn between categories; and of relationships between a category and its conditions, consequences, and the strategies and interactions associated with it. (Later chapters address the issue of how to integrate those codes, including into sucessive integrative diagrams.) Particular instances of coding can place emphasis on various of those items.

Initial steps in open coding: a seminal session

Here is an instance taken from the first coding session of a research seminar. The students were told to scan the following lines, taken from an interview with a disabled young man.

Once I'm in the shower, I'm pretty much on my own. I've got a wire chair that I use in the shower and a grab bar over the shower and I stand up, hanging onto the bar. I sort of walk over to the shower and sit down in the chair. I have a chair just on the edge of the shower, like on the lip, so when I sit down and get straightened around in the seat, then my attendant will sort of lift up my legs and push me all the way back against the wall. Once I'm there, the knobs for the hot and cold water are right in front of me. When I'm going to take a shower, I usually have my attendant get the towels and stuff ready before I even get my clothes off, and warm up the water a little bit. But once it gets warm enough for me to stand it, then I have the shower turned off. That way when I get in, it's at a point where I can adjust it to where I want it, rather than going in and having it just be totally cold (Lifchez and Winslow 1979).

The students were next asked what they saw in these data. They came up with some quite good observations – which are the equivalent of *themes* seen in the materials. In fact, many researchers do this kind of theme analysis; for instance, finding themes reflected in interviews with unemployed workers. Theme analysis tends to remain at the level of very careful journalism. Not surprisingly, many researchers read the morning newspapers with this kind of sharp, theme-oriented eye, but because they are only reading newspapers, they do not bother to carry their analyses further.

However, to continue with the seminar example: The instructor's next tactic was to have the students read the first sentence of the interview again, but then concentrate on the first word in it. "What can 'once' mean?" A student answered: "The man felt independent once he was finally in the shower." Then where else would he feel independent once he was there? The answers included: in bed and in a wheelchair. "Where would he then *not* feel independent?" The students gave answers to that, too. Then: "What *else* could 'once' mean?" Someone pointed out that "once" was a condition for the next step in the man's activity. Another followed up, suggesting this meant the end of one phase and the beginning of a new one. The instructor agreed, then asked the class to consider a "far-out" comparison – such as a sprinter entering competition for the first time, who might recollect that, "Once I was set to go off with the gun, I forgot all about the months of grueling training." So the class explored the differences and similarities between the two cases concerning their respective pre-once phases. They talked about rates of movement through the phase, amount of pre-planning, and other dimensions. "Once" also suggested to someone that it might refer to the amount of effort put into getting to the "once." The instructor remarked that comparisons to other instances might suggest readiness to begin a new phase, as when coaches read cues in the people whom they are coaching of "now ready to move on." He asked what this might suggest about preparations for and readiness to get into the shower by the disabled man.

Only then was the class allowed to move along to thinking minutely about, raising questions about, giving provisional answers to, and making comparisons around the next word in the interview. This was the phrase "I'm." An hour went by in this fashion, as the class covered only "Once, I'm . . . ," the first two words in the first sentence of this interview fragment.

So, in this seminar session the participants can be seen focusing their efforts toward quickly stepping up from the data, as they develop analytic abstractions that are nevertheless grounded on close inspection of the data. The analytic operations included word-by-word inspection, the generating of theoretical questions and possible answers to them (hypotheses), the use of stimulating internal and external comparisons, and the exploration of similarities and differences. Before the session was over, the participants had discovered an in vivo code (I'm pretty much *on my own*). And they had rendered explicit some of the conditions, interactions, tactics, and consequences that were implicit in the respondent's words.

If one does not code industriously in some such fashion as this, these paradigm elements will tend to be left implicit or at least unsystematically linked with the phenomena under study. Also, the variations in why consequences differ, strategies differ, interactions differ will be underplayed in the analysis. So, in reading the cases below, it is important to focus not merely on the naming of categories and their supporting data, but on how, in the coding process, these categories are related through an active search for the specific and variable conditions, consequences, etc. (In fact, a good exercise for beginning researchers is to ask, after a coding session, what they have done in their coding, in terms of these paradigm items. And another is to take an in vivo code – like, "I'm pretty much on my own," and ask about its different meanings in different contexts, so as to focus more sharply on what it might mean in the particular context in which it has actually been used.)

Open, axial, and selective coding

A fieldnote taken from a study of medical technology and medical work, discussions of which will appear from time to time in this book, will be used to illustrate the processes of open and axial coding. Afterward, another set of field observational data will be used to illustrate selective coding.

Open coding

The portion of the first fieldnote reproduced below records some observations done on a cardiac recovery unit, where patients are brought immediately after cardiac surgery. They are intensively monitored and cared for on a one-to-one, nurse-to-patient basis. The nurses are abetted by occasional visits of house physicians and, less frequently, by visits of surgeons. During the first postsurgical hours, patients are likely to be unconscious or barely sentient. Each is hooked up to numerous pieces of equipment, vital for survival or for monitoring their bodies. In one fieldnote, a young but highly skilled nurse is described as working on, around, and with a barely sentient patient. The observer was focused on details of her work with the equipment in relation to her patient care, and so reports exclusively on that.

These field observational data have been specially coded for this book in order to illustrate coding procedures as they occur during the first days of a research project. While this coding was actually done long after the project's close, so that the observer and analyst (Anselm Strauss) knew more about those materials than ordinarily a researcher might know, analytically speaking, nevertheless, the coding processes and techniques would be the same as if he had done it early in the research project. It is recommended that you first scan the entire fieldnote, then study the coding discussion and commentary that follow.

I watched Nurse T. working today for about an hour with a patient who was only four hours post-op. In general the work was mixed. She changed the blood transfusion bag. She milked it down, and took out an air bubble. Later she changed it again; later, got the bottle part filled through mechanical motion. She milked the urine tube once. She took a temperature. She put a drug injection into the tube leading to the patient's neck. She added potassium solution to the nonautomated IV. But, all the while, she had in focus (though not necessarily glancing directly at) the TV which registered EKG and blood pressure readings. Once, she punched the computer button to get the fifteen-minute readout on cardiac functioning. And once she milked the infection-purifier tube leading from the patient's belly. And periodically she marked down both readings and some of what she had done. Once the patient stirred, as she was touching his arm: She said quite nicely then that she was about to give him an injection that would relax him. He indicated that he heard. Another time, she noticed him stirring and switched off the light above his head, saying to him, "That's better isn't it?" At one point, she assessed that blood pressure was not dropping rapidly enough, and told the resident, suggesting they should do something.

After the analyst had scanned this fieldnote, he focused on the first five lines, pertaining only to the blood transfusion equipment and what the nurse did in relation to it. His analysis took him several minutes. Then he wrote the following lines, elaborating his brief notes, underlining those words which especially but not exclusively struck his attention. The analysis begins with the first line: "She changed the blood transfusion bag." His explanatory comments in brackets briefly indicate what is happening analytically.

"She *changed* . . . . " This is a *task* [a category, drawn from common experience].

"*She* changed . . . . " She is doing the task by herself. This apparently does not require any immediate division of labor [a category drawn from technical literature]. However, there is a division of labor involved in supplying the blood, an issue I will put aside for later consideration [raising a general query about that category].

" . . . *blood transfusion bag*." "Blood transfusion" tells us that this piece of equipment, the bag and its holder, requires *supplies* [a category]. Again, a fascinating issue, about which I can ask questions in a moment.

Let's look now at the "*changed*," qua task. What are its properties, or what questions can I ask about its properties? It is visible to others [the dimension here being visible–invisible]. It seems like a simple task. So it probably does not take much skill. It's a task that follows another (replace one bag with another). It seems routine. It doesn't take long to do. Is it boring or just routine? It's not a strenuous task either. And it certainly doesn't seem challenging. How often must she do this in her day's work? That is, how often does it take for the blood to get transfused into the patient? Or, perhaps, how much time is allowed to elapse before new blood is actually transfused to the patient between each bag? Or does that depend on her assessment of the patient's condition? What would happen if they temporarily ran out of the bags of blood? [Implication of safety of the patient, which will be looked at later.] I would hypothesize that if there is no immediate danger, then replacing it would have low salience. But if there were potential danger for certain kinds of patients, then there would even be organizational mechanisms for preventing even a temporary lack of blood bags. Well, I could go on with this focus on the task, but enough!

Back to the division-of-labor issue now. Since the patient is virtually nonsentient, the nurse gets no help from him when changing the blood bag. It is a nonworking relationship – she is working on or for but not together with him. This means also that he cannot interfere with her work. He can't complain either, to her or anyone else, that he doesn't like what she's done or how she's done it [implicit comparisons]. I know that some patients object to getting blood transfers, especially nowadays, when they might be anxious about contamination of the blood [explicit comparison, with condition and consequence specified; also touching explicitly again on the supplies issue].

As for that issue: To begin with, there must be supplies for the equipment or it *is* no equipment, of no use whatever. Let's call it *equipment supplies* [category]. But those are very different supplies than for other equipment I have seen around the hospital. Thinking comparatively about those will tell quickly about the special properties of this particular equipment supply, as well as raise questions about those supplies in general. Well, there are machines that use plastic tubing, which when it gets old must be replaced. Blood is a natural supply, not artificial – but the sources of both are somewhere? Where? [I will think about sources later.] Plastic tubing and blood are also replaceable. Blood costs more. Automobiles need gas supplies, but you have to go to a station for it, while blood supplies are brought to the user. What about storage of blood? Where on the ward or in the hospital is it stored? How long is it safe to store? And so on.

Blood transfusion also yields supplies for the patient. It is replacing a loss of blood. But that leads me to see there's a *body–equipment connection* here. This raises various questions. Ease of making the connection? Skill it takes? And more of the dimensions, like, amount of time to do it, duration in the body? Need to monitor it [another category]? Potential hazard to patient? Discomfort to patient in doing it, or keeping the connection to the body? This patient is immobile, but could he walk around with this equipment if need be, as with intravenous liquid drip bags that I have seen? Blood transfusion connections are internal, not like electrocardiograph (EKG) connections, which incidentally won't allow the patient to move an inch – besides, they tend to fall off anyhow. Could that have important implications for our research considerations?

Body invasion is involved in making the transfusion connection, unlike many other body–equipment connections. Incidentally, with ordinary intravenous connections (IVCs) for liquid feeding, I recollect they put in a kind of semipermanent "lock" in the arm so that the IV equipment can be hooked and unhooked through it at will. Is this done with blood transfusions? One hypothesis then might be that the longer the lock remains in, the more likely an infection around the insertion point. Who monitors that possibility? I'd predict that sometimes the monitor overlooks the beginnings of an infection, so if the patient discovers it then there is consequent anger and complaint.

If now I think about the sources of blood supply, then I draw on experiential knowledge, though perhaps I should look further into that issue. For instance, this is a pooled supply contributed to by many people, and stored probably in some sort of central storage place. They screen people as suppliers, which is similar to getting relatively pure oil or gas, I guess. Is this a commercial process or governmental (public health)? In general, I should get data on the source of blood supplies. Well, let's look at the next sentence in this fieldnote.

"She *milked it down,* and *took out an air bubble.*" Milked and took out: The verbs reflect mini-tasks, done in immediate sequence [property]. They follow two others, taking down a depleted bag and replacing it with a full one. So I was watching a sequence of mini-tasks, with more to come. Aren't all the things that she's doing with the transfusion equipment a task cluster [category]? Milking down and taking out a bubble aren't done for aesthetics but surely because of potential danger – a bubble in a vein is bad business. So now we have the issue of clinical safety and danger clearly in focus. This mini-task pertains to something I'll call *clinical safety work*. I need to look into both safety work and other types of work. [Generative questions, since both categories turn out later to be central to the study, and the latter becomes a core category.] Theoretically sample, with safety work done with drugs and other equipment in the hospital. For instance, with some equipment there is no potential danger at all (EKGs), but with others there is. So, observe and interview about this, including some that are potentially very dangerous, which the air bubble in the blood can be. Or where its potential is of low probability if carefully monitored, or of high probability if done too frequently, like x-rays. By looking at such theoretical samples I can begin to open up the issue of clinical safety, including how danger is *prevented, assessed, monitored*, and *rectified* if there has been an error [categories and subtypes of safety work]. Questions can be asked about these, such as how is potential danger assessed, what are the grounds? How is safety actually monitored: by eye, ear, or by equipment, like blood pressure equipment and stethoscopes? What is the relation to monitoring of assessing? How are the mistakes made in monitoring rectified? Who does it? Is there a division of labor in all of this? In the air bubble instance, if she makes a mistake, would she notice? How? By what signs? How soon? Maybe I'd better ask if anyone ever made a mistake and did not stop a bubble, and if so, what happened then.

"*Later it changed again.* Later she got the bottle part *filled through a mechanical motion.*" All that comes to mind with these new sentences is that this task *is* repeated, within the hour. So she is engaged not only in a repeated task but a *series of tasks* [category], repeating the whole series at intervals – *a repeated series* [another category], that seems to occur, in fact, about every twenty minutes. So I won't work on these sentences further, but move on.

"*She milked the urine tube once.*" This is the next task in the series. Well, I will put off questions until later about repeated series of tasks.

"*She took a temperature.*" Now she is *body monitoring*. Not necessarily monitoring for safety, but for state or progression of the illness itself. Some equipment is for monitoring – such as blood pressure equipment, sonoscopes, x-rays. But body monitoring has to do with location on the illness trajectory [this will be the other core catetory], and the properties of monitoring surely include: repeated, varying degrees of skill called for, and clinical hazard.

In the above coding process, one can see categories and subcategories being noted and labeled, and a few connections among them suggested. A variety of questions are asked, some probably truly generative in terms of the future of the study. Also, comparisons are made and thought about that further the more direct itemization of category properties. Theoretical samples are implicitly or explicitly touched on. A few conditions and consequences are touched on also, but not especially pursued. A couple of explicit hypotheses are hazarded but many more are left implicit in terms of discussions of their implied conditions and consequences. Strategies are not noted, but interactions concerning what the nurse "does to" the patient are. Some necessary data are also flagged. Both the observational data and the analyst's experiential data (personal, research, and technical) are used at various junctures. Note also the kinds of choices that the analyst can make: To dimensionalize. To make comparisons. To follow through with a topic – or to put off thinking about that until later. To write memos immediately, or later, on these initial codes and on lines of thought suggested by the coding session.

It cannot be emphasized too much that, at such an early stage of open coding, the analyst has many options which can be followed in the same coding session or in succeeding ones. The inexperienced analyst is likely to be somewhat anxious about what option is "the best." The *rule of thumb* here is: Don't worry, almost any option will yield useful results. Typically, for instance, in a research seminar the class will face the following options after an hour's open coding: (1) to follow through on one or more of several comparisons already touched on, (2) to return to the actual data again to do more microscopic coding, (3) to follow up on something suggested by an operational diagram that has sketched out relationships among the categories discussed so far, (4) to further relate some of those categories, (5) to code specifically any of the categories in terms of the coding paradigm.

One question that might be raised about this kind of fine-grained, microscopic coding is whether it requires collecting data through tape recordings and videotapes. The answer is, definitely not. As we have said before, one can code microscopically on researcher notes from interviews, field observations, and other documents including published material. One uses tape recorders and videotapes when the research aims require very great accuracy of wording and gesturing of and by the people observed. Whether the analysis is microscopic or not on this kind of data depends completely on what the researcher wishes to use this material for.

It is especially important to understand that these initial open-coding sessions have a "springboard" function. The analyst does not remain totally bound within the domain of *these* data, but quickly jumps off to wonder or speculate or hypothesize about data, and phenomena, at least a little removed from the immediate phenomenon. In the example, we see an experienced researcher focusing from the outset on task, division of labor, equipment, supplies – but already suggesting the larger issues of those matters, rather than restricting himself to this particular working situation. He does this by using his technical knowledge and theoretical sensitivity (e.g., division of labor), his experiential knowledge, and his research knowledge (supplies for equipment). He also jumps off from the immediate situation by thinking of comparisons. Some are made within the same domain (medical work, medical equipment) but others are made much further out (see Chapter 1, Part 2 for comparisons). Both types of comparisons help to broaden the scope of analysis. Researchers who are inexperienced in how initial open coding can spring quickly off the data – while yet firmly rooted in it – tend to keep their analyses too limited in scope, intent as they are to crack the shell of this specific set of data so as to get at its analytic kernel.

They also sometimes have considerable anxiety about whether their coding is only reflecting their biases, rather than what is "in" the data. Their anxiety is understandable, but they need have no fear; for the codes are only provisional and later coding sessions either will or will not sustain their usefulness. What is needed is time and a bit of patience.

Axial coding

Among the most important choices to be made during even these early sessions is to code more intensively and concertedly around single categories. By doing this, the analyst begins to build up a dense texture of relationships around the "axis" of the category being focused upon. This is done, first, by laying out properties of the category, mainly by explicitly or implicitly dimensionalizing it (this task is visible to others, requires little skill, seems routine). Second, the analyst hypothesizes about and increasingly can specify varieties of conditions and consequences, interactions, strategies, and consequences (the coding paradigm) that are associated with the appearance of the phenomenon referenced by the category. Third, the latter becomes increasingly related to other categories. For instance, monitoring for clinical safety might be related to other subtypes of safety work: (1) rectification (of monitoring mistakes), and (2) assessment of the degree of potential hazard, under what probable conditions (setting the stage then for monitoring activities). This relating of categories and subcategories is done in terms of monitoring being a condition, strategy, interaction, and consequence (e.g., when a patient is assessed as being at great hazard, then intensive monitoring is called for). Axial coding around tasks and task clusters had actually led the researcher originally to collect the data recorded in the fieldnote just coded.

Among the categories briefly discussed above was the monitoring of clinical safety. If the researcher were to have done further initial open coding, he might first of all have thought of other types of nonsafety monitoring through comparative analysis: for instance, the monitoring of increased skill done by a teacher of her students; or the experimental monitoring done by engineers for the limits of tensile strength of a new metal wire. Alternatively, the researcher could have made closer-in comparisons, thinking only of different kinds of clinical safety monitoring: for instance, monitoring the body reactions of a patient on a dialysis machine or undergoing various potentially hazardous medical procedures like an angiogram probe of the heart. Or the researcher could have made an even closer-in comparison of different subtypes of safety work: namely, assessing, and the rectifying of monitoring errors.

But if the researcher chooses to move directly into axial coding, he or she would focus on the specific kind of safety monitoring that seems associated with *this* set of data. Then either close-in or further-out comparisons could be used to further the immediate analysis. In such coding, the analyst must exert great discipline to stay concertedly on target, not allowing diversionary coding temptations to interfere with this specific and highly directed coding. Any other coding or ideas that come to mind should be noted briefly, but on a separate piece of paper for later consideration – not now!

To illustrate axial coding, an instance done around the category of monitoring will be presented next. The coding session and its products are reproduced as typed by the analyst for a team research project. To each coding item, a short explanation of what is transpiring in the analysis is appended in brackets. The relevant observations from the fieldnotes are also included, in quotation. There is a further commentary at the end of this section.

The coding items were done by the principal investigator of a project on medical technology's impact on hospitalized care. He was coding the first three fieldnotes written by a teammate, who had observed on a cardiac care unit, about eighteen months into the project, after several other kinds of units had been studied. The category of monitoring had by now been perceived by the researchers as repeatedly salient "in the data," hence important for its analysis.

Monitoring Items (off BS's cardiac care unit (CCU) fieldnotes, of Jan. 19, 23, 28)

Machine monitoring the procedural work Jan. 28, p.4: "She showed me the tiny plastic bulb attached to the catheter into the heart. The little bulb is there to move the tip of the catheter . . . . When they insert the catheter they take an x-ray so they can tell if it is properly situated" [naming a subcategory of monitoring, with supporting quoted data].

Monitoring the monitor, p.5: Charge nurse checking with her nurses, "No problems?" [same as above].

Monitoring by visual norm, p.5: Notebook with EKG patterns, so "can compare with if in doubt." That is, doesn't have to understand meaning of wave patterns, just know when they look awry; *lower-order knowledgeability* criteria for monitoring by EKG personnel. "The screens above the patient, which are constantly threading out wave patterns, all have patterns that look completely different. How does she know what is normal for the area being monitored, so that she can tell if the pattern is deviating? Margaret showed me a little notebook she carries in her pocket into which she has drawn a series of typical wave patterns she can 'compare with if in doubt' " [monitoring means; data quote; related to another category: lower-order criteria].

Sentimental work monitoring, Jan. 28, p.6: Checking up on a very sick patient even during dinner; deeply involved with him. (See also Jan. 26, p.4.) " 'God! You get so you can't go to dinner without worrying about whether your patient is being properly looked after. You keep runing back to check up!' Back at Gary's bedside a technician was working on him trying to get a blood sample, but without much success. Margaret appeared suddenly and leaped to the bedside of this nonsentient patient. She said that she would try to get the blood later. I'm having dinner now. Apparently she just came back to check up on Gary, the way Judith and the respiratory therapist had said it happens when one is deeply involved in caring for a patient" [naming a subcategory of monitoring; data quote; reference to other data].

Constant (virtually continuous) monitoring (BS's term, Jan. 26, p.7): On the CCU; in the ICN recovery room, too. Even when attention is away, when they think it's going OK, they have at least *peripheral attention*, and so can snap back instantly when aural or visual signs bring them alertly to attention. "Nothing, according to them, is more boring than scanning, but the scanners cannot be left unwatched at any time. Somebody always has to sit there – watching, watching . . . . Constant observation is what it is all about" [monitoring dimension: continuous–discontinuous; related to another category – peripheral attention; all resting on observed data].

Monitoring patients' behavior, Jan. 26, p.10: BS is referring here to the behavior of a *patient's undoing the staff's work*, wittingly or unwittingly. This fieldnote is replete with interaction of nurses seeking the patients' understanding and *cooperation when sentient*. BS notes that patient's behavior (clearly) can be an attempt to do what ordinarily one does for oneself (moving a leg, rubbing an eyelid), but which now is deemed injurious to do – either mustn't do oneself, but staff will do, or must not be done at all, especially taking the mask off, etc. "Another woman was being scolded by a passing respiratory therapist. She had taken her mask off. 'You should have oxygen ALL the TIME!' he said severely. The patient put back the mask. So, note: These staff people are busy monitoring not only the machines, but the behavior of the patients. Patients can do work, but can also undo the work of the staff. Eliciting their understanding and cooperation seems to be an ongoing task" [condition for effective monitoring; strategies for countering countercondition; condition for that occurring; observational data].

Monitoring trajectory reversibility, progress, off-course potential are *three things* which CCU is monitoring on those machines. On course – off course – reversed. And *a fourth – immediate action reversibility* (go save!) [four dimensions of monitoring category summarized after reading this far in the fieldnotes].

Aural monitoring complexities, Jan. 19, p.5 esp.: Say there are six machines and six different patients within earshot (including a nurse's own patient and own machine). The nurse has to be tuned in to all of those – either *peripherally or focally*. Each machine has its own noise pattern, and each patient his or her own pattern: She studies the resulting pattern, that is, the combined patterns created by unique machine and unique patient. That is something like 6 × 6 = 36? "I comment that I am surprised that the unit is so quiet and calm. Nora reacts with surprise. Although it is not an especially busy day, she 'hears' it differently than I do. To her, it is very noisy. It turns out that she is constantly doing aural monitoring. She knows the sound of every respirator and is immediately alert to any change. If there are six different machines, and five different patients on each, she learns to hear the various patterns: The machine has its own pattern and each patient has his or her own pattern, and she audits the resulting pattern which is the combined pattern of unique machines and unique patients. It seems almost incredible – an exaggerated boast! But I'm inclined, at least for now, to believe her. No wonder she perceives the unit as 'very noisy' " [conditions for difficulty in monitoring along aural dimension; related to structural condition of machine noise pattern × patient pattern; again related to peripheral–focal category].

Alarm adding to complexity, Jan. 19, p.8: The *perpetual false alarms* have to be separated out from the *true alarms*. "I asked her what's the worst thing about this unit. 'It's the noise,' she says. The noises really bother her: the sigh of the respirators and the everlasting alarms! For example, this week they had a patient whose condition exceeded the alarm level. There wasn't anything they could do about it, either by altering his condition or adjusting the alarm level, so they were stuck with that everlasting beeping. They HAD to respond, even when sure it was nothing they could deal with, because there was the off chance that the patient could have coughed and dislodged his tubes and therefore really be in need of help" [distinction between categories: true and false alarms; the latter as condition, re difficulty of monitoring].

Monitoring the patient's comfort needs, Jan. 26, p.3: But also, *patients monitor their own comfort needs* but, being in this instance unable to act themselves to satisfy them, they must *inform the nurse* of their monitoring . . . . She may or may not act to satisfy them, depending on various conditions: detrimental to trajectory, press of immediate work, etc. . . . When deemed detrimental to trajectory, then we see negotiation, attempts at persuasion, etc.; eventually, staff sometimes coerce to prevent (scolding, "put that mask back on") or correct. See bottom of p.4 for scolding, pleading, etc. "Elevated back into former position, patient began fussing again. 'Bear with us, John!' she pleaded. 'I know you don't like the tube, but we're not taking it out until tomorrow. Do you want to write something?' John indicated he did, and she pulled a pad of paper and a pencil out. In response to his feeble scribbling, she told him again that he can't have water until the tube comes out, that he doesn't need to urinate, even if it feels as if he does because 'there's a tube in your bladder that draws the urine out.' This sort of strange dialogue went on and on. She remained amazingly patient with him . . . . 'Are you cold? Do you need to cough?' The poor guy was obviously terribly uncomfortable. Among the things he had to negotiate was getting an itchy eyelid rubbed. It took several tries for her to get that right –

a real game of charades. His arms, spread out at his sides and supposed to be held still in that position, were a mess: terrible contusions. He had a catheter in his heart, a tube in his throat, a tube in his bladder, several IVs [a category – comfort needs, and both patient and nurse monitoring in relation to that, and the relation to interaction – informing – to the first two; plus consequences of his action, and conditions for that; plus strategies under conditions deemed detrimental to trajectory – which is the CORE CATEGORY of the research project].

In the code notes given above as well as those in the next pages, note the techniques that facilitate quick scanning and sorting later in the research project. These include heading the code item with a label naming the category or categories, and subcategories; in the coding, note whether they are new or old ones. Related categories may also be underlined in the text: often they occur to the analyst while typing the note. Sometimes the relevant lines of the interview or other document are referenced by page, or the data may be included in the text either in précis form or as a direct quote. Sometimes analysts draw on data so well known to themselves or their teammates that they do not directly reference to specific data. Note, too, that in the above coding notes, the coding sometimes brings out new categories without relating them to previously discovered ones; but sometimes a connection is made or suggested. Sometimes, too, the researcher's attention is drawn, in the lines or phrases that are being studied, to what they suggest about strategies or consequences or conditions in relation to each other and to the categories and subcategories. (The very last code item above illustrates that point very well.)

In this research project on medical technology and medical work, its two core categories turned out to be: types of work and trajectory. In the coding session reproduced above, one can see how monitoring begins to become related analytically to various other kinds of medical work: safety, comfort, machine, sentimental. Those linkages are made more numerous and increasingly complex, especially through further axial coding that focuses on one or another category. The same is true if monitoring in relation to the other core category, trajectory. How? By relating monitoring to subcategories of trajectory, such as: types of trajectory (problematic, routine), trajectory projection (visualization of illness course and varieties of tasks to be done in controlling it), trajectory scheme (immediate task clusters to be done), trajectory decisions and decision points, and trajectory phases.

Selective coding

All of this occurs before (and sometimes, considerably before) the researcher has decided which category is (or categories are) central to the research project. When they are decided upon, however, then the researcher moves into selective coding, when all other subordinate categories and subcategories become *systematically* linked with the core. In other words, although some of these links had already been established, now the search for them, and their coding, are done concertedly.

So, now imagine that some months have passed since the initial analysis of the data bearing on the cardiac recovery nurse's work, and that additional data had been collected and axially coded, also that two probable core categories (illness trajectory and work types) had emerged. How would the researcher do the selective coding, relating the new codes to the core categories further than had already been done through axial coding? Here are a very few of his thoughts and operations.

"In this fieldnote there are the following types of work: (1) *Equipment work* – here, tasks done *with* the equipment, not on the equipment itself, like its maintenance. (2) *Clinical safety work*, of course, but not other kinds of safety work, as in terms of staff or environmental safety. (How do these subtypes of safety work relate to each other?) (3) *Body work* with the patient; much of it with or in relation to equipment; the equipment work here is a condition for getting that portion of body work done. Other body work is done without equipment, so there are at least two subtypes, though I know that already. (4) But the monitoring is giving information. Call that *information work*. Equipment work is a condition for the subtypes of information work when information can't be read by eye or ear alone. But she is gathering lots of information by sense modalities. So, again, subtype information work, each depending on different sets of conditions. She has to do equipment work like punching the computer button to get readouts on cardiac function. Her *recording* of the readouts, and of other activities as well, is a subtype of information work. This, in turn, is related to her accountability to her superiors in a short chain of command, leading from head nurse to staff physicians to surgeon. (5) *Comfort work* – relaxing the patient with an injection, and switching off the light; this relates to sequencing the tasks of other kinds of work, for comfort tasks are slipped in somewhere. Possibly comfort tasks are less salient for her, given the grave importance of the other work here? (6) Each of those actions, but especially switching off the light, can also be construed as *composure*

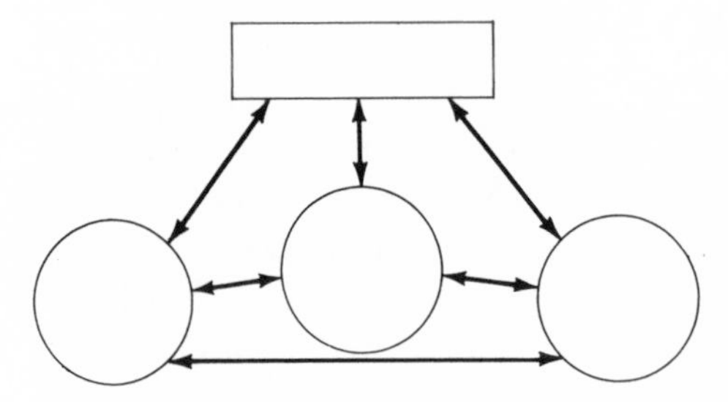

Figure 3. Selective coding and core category.

work, done to enhance this semisentient patient's sense of security. But composure work is a subtype of *sentimental work*, a complex type, as I already know. With the patient still in potential danger, and with all these other types of work taking priority, it is remarkable that she takes any time, however little, to do composure and comfort work! Is it because she is kindly or because of nursing philosophy? I would hypothesize that it is often because of the latter. However, she is remarkably unhurried in doing her tasks, which is a necessary condition for doing these lesser-priority tasks – assuming she defines them so."

Note that in this coding session the main effort was to fairly exhaustively itemize and relate these subcategories of the core category (work type). The session might have stopped there, to be resumed later with a consideration of comparisons with other data, some already coded, of course. For example, this body work could be compared in terms of the coding paradigm – for conditions, strategies, interactions, consequences – against the body work done in other situations: x-ray, telling the patients to move their body positions, or doing it for them; transporting the bodies to and from the radiology department – that is, transport work, a subtype of body work.

Alternatively, the next selective coding session might have focused more on relating previous axial codes around what now are seen as subtypes of the core category. For example, what variations in body work relate to what variations of equipment work? Comfort work? Composure work? And, in fact, in later phases of selective coding, Figure 3 best illustrates what goes on, for *clusters of analysis* are being related to the core category (or categories), as well as to each other. The circles in the diagram stand for clusters, and the rectangle, for the core category.

A relatively late, but not the latest, phase of selective coding will be illustrated below with yet another set of codes. Their presentation is designed not so much to show the process of selective coding (yet, something of how it is done will be evident) as its products. Here again, as in the earlier example of monitoring, the coding will be linking categories and subcategories, but more systematically now around a core category of pain management. The analysis is of data drawn from an earlier project on pain management (Fagerhaugh and Strauss 1977). The coding notes were written by the same investigator, Anselm Strauss, as those above. They were done over several days, about six months after the project's inception, and after a number of important related categories had been isolated and data relevant to them collected (through theoretical sampling) and analyzed. These included pain relief, pain minimization, pain expression, expression control, pain assessment, pain ideologies, negligence accusations, incompetence accusations, and balancing pain relief or minimization against other considerations. Some of those categories appear in the codes below. Note, as in the above materials on monitoring, the usual underlinings, occasional quoted material, and the frequent explicit relating of strategies, conditions, consequences, interactions to the category under discussion, and also the occasional research directives. However, here the relating of multiple categories – including to the core category – is considerably more complex now than in the case analyses presented earlier.

Patient tactic of handling expression versus minimizing pain (AS), June 30, 1975: On the obstetrics ward, we don't hear patients talking about their own techniques of minimizing pain, as elsewhere. What we get is tactics for handling pain expression. There is no *pain career,* so they can't develop tactics for minimizing. Compare with postoperative or back pains, etc. [Condition for tactic; condition is related to category, minimizing, concerning absence of the phenomenon; condition named: no pain career, and condition for that noted, too.]

Tertiary pain and tactics for minimizing (S), Feb. 20, p.9: Burn patient at home, all kinds of pain caused by burn itself, or secondary pain from skin graft, but from daily activities: Because of pain in moving her shoulder and elbows, she couldn't groom her hair and had difficulty putting on certain clothes. Soreness around neck meant she could only wear clothes with low necklines. Extreme tenderness of burned area of arm meant she could tolerate only short-sleeved dresses, but she was concerned about the ugly scars then upsetting others (i.e., *reactions to pain-related symptom*). A friend gave her clothes without sleeves and fixed some of her clothes so she could dress with minimum pain. *Minimizing pain agent?* [New subcategory, tertiary pain; related to another phenomenon, minimizing pain, previously noted, coded, and partly analyzed; tactic related to both.]

Staff balancing pain relief and maximization versus main job (M's CCU), March 2, p.5: *Inflicted pain*: Removal of intubation tube on cardiac patient for post-op coughing adds to the pain which is ongoing and will be increased by coughing. Some nurses find if they sedate or medicate just prior to tube removal, they will have problems of coughing, so they hold off and let the patient experience increased pain; also, *awareness if closed,* since the patient usually doesn't know this. [Major category, balancing, related to another, inflicted pain; and to nurses' tactic; also, the latter related to another category, closed awareness, as condition for doing it successfully; relations also with "main job" – that is, the core category of pain management.]

Patient's ideology and nurse helplessness – pain control: In one of C's fieldnotes, a nurse speaks of being or feeling so helpless because she can't help with labor pains – the patient won't let her because of her natural childbirth ideology. That's different than just being helpless in the face of pain you can't control, for this is actually controllable but the patient's ideology prevents the nurse from doing so, and so it is frustrating to her [consequence of phenomena, pain ideology; distinction drawn between this consequence and another type of consequence easily confused with it; also related to another category, control of pain].

Expression control by staff – main job (C's notes), April 2, p.2: A frank nurse says she moves in to lessen patients' expression; says other do, too, but we should check this out. She does this because "I think if she makes noise others will think I'm not caring for her – staff and patients, that is. Also, if I let her go on, someone will only come over and try to quiet her," meaning also that she will be viewed as incompetent. So, *expression control* is seen as part of her own main job. But another reason, sometimes, is that a patient gets so hysterical (she says) that they can't be calmed down; so they are stopped before that point – that is, before the *expression peak.* Also, expression in the last stage of childbirth will cause the laboring mother to push down before she should: *timing*, again, of the main *trajectory* [category, and conditions for related tactic concerning pain expression of patient and also pain control by the nurse].

Flooding pain and fateful options; balancing; pain relief: As in the case of Mrs. Abel, choosing between flooding pain and an operation that might lead to death; same with C's lady, who chooses amputation rather than endure her pain, although told it may not work. Fateful, irrevocable choices in the attempt to *relieve* or *minimize* pain, at least . . . . We have to ask about the processes that lead to the final decision. And, what are the consequences of making that decision? And Mrs. Abel? Other people's reactions are both part of the process sometimes and a consequence of choice. Parents may make the choice for a child, or kin may be involved in patients' choices (i.e., *option agents*) [category, balancing, related to pain control subcategories of relief and minimization; research questions about balancing; coining of term for another category, option agent].

Reverse control of expression (C), June 2, p.8: "There is a fourteen-year-old arthritic who is such a stoic that the staff has been trying to get her to express her pain. Finally, the psychiatrist talked to her and her family; found out her mother and grandmother are both in wheelchairs, one from polio and the other from rheumatoid arthritis. She had been taught to bear pain, and he has instructed staff not to interfere with family patterns." (C is quoting a doctor of physical medicine from that ward.) By *reverse control* I mean the staff is attempting to get the patient to express *more*, rather than the usual less expression. But then the psychiatrist discovers "why," and warns staff off their efforts [new subcategory, reverse control, amplifying previous incompletely dimensionalized category of pain expression; condition for subphenomenon, as well as for its alteration].

Regimen pain – conditions for increase (C), June 4, p.6: "We don't always have the patient's pain in mind. Sometimes we are sick of the patient, and we don't always have the patience of Job." That is: Focus on the *main job* of pain management, plus *negative mutual biographies* with that particular patient [inflicted pain category, related to two conditions, main job and previous bad experiences with patient].

Relief and pain messengers: There is the phenomenon of aides and nurses acting as pain messengers (researchers, too!) to get the staff there to relieve patients' pains. This is different from *legitimating pain,* so as to get relief by someone else. Under some conditions, there are no messengers handy – such conditions being pretty obvious (e.g., patient is alone) – or others won't carry the message because they don't believe the complaint, or are too busy. Also, the phenomenon of the light flashing or the buzzer buzzing, which is the patient's way of *acting as his or her own messenger.* Note that the legitimation problem crosscuts the problem of getting a message to relieve, but is not identical. Also, a nurse, sans legitimate order for medication, in turn has to act as a secondary messenger only to the physician: She can't give the medication unless that has already been okayed [new category, pain messenger, in relation to another, pain relief; distinguished pain messenger from legitimating agents – two types of agents; conditions where no messenger agents; conditions for patient being own agent; condition for relieving agent rather than messenger agent, as related to category, legitimation].

Inflicted pain and negligence, negligence accusations; awareness context: Burn patient, who warns staff she is allergic to a given drug; then, after some days, it causes painful blistering, since they persist in using it. So they have, through negligence, actually inflicted pain (whatever the condition for that "negligence"). More important for us is that she knows the cause of this inflicted pain, and so can make her accusations stick with the staff. Sometimes, however, there can be *closed awareness,* so the staff knows its negligence, no doubt, but the patient does not – and so cannot accuse them. Or, under still other conditions, the patient can be suspicious, but they deny negligence and the patient cannot prove it (*suspicion awareness*) [category, negligence accusations, and condition for it; condition for awareness or not, or maybe; successful consequence of the action; condition for staff's answering tactic].

Pain assessment conditions – linked reputations (C), Feb. 7, p.4: See how patients are typecast from the moment of arrival just by being one of this particular doctor's patients; especially if they seem true to type. Then you see the staff using a *discounting assessment*: "Oh, they had some pain, but it was psychosomatic, they were using it. The kind who don't want to get well, who hold off having sexual relations with their husbands, or use it not to go somewhere they don't want to go." (The head nurse): "The nurses hated them. They would groan, 'not another psychiatric patient' or count how many there are on the ward." The nurses had a psychiatric scale, patients getting points depending on how many qualifications for the type they filled: aggressive, demanding, dyed hair, etc.; consequence: The head nurse claimed the reason for high staff turnover was the preponderance of this type of patient on the ortho ward (but check whether she said this, really) [condition; new subcategory, discounting assessment, and conditions and consequences of latter].

Discounting assessment, conditions (C), Feb. 7, pp.5–6: Placebos – patient on physical therapy unit (not exercising when has an option to go home on weekends); orthopedic patients (loving to talk about their pain); ortho patient responding with; "That's great!" to ultrasonic treatment (which physical therapist thinks is a placebo); and the psychologist (rehab case, Jan. 29, p.5), whose testing showed that a patient was "monumentally dependent, exhibiting an exaggerated inability to do things"; so his assessment was that she saw this as "a chance to use her disability." All of this is, of course, related to the patient's failure to *legitimate*, and nonrecognition of the necessity to legitimate [several conditions re category].

Pain ideology; main job; illness and social trajectories (S), Feb. 9, p.2: Dialysis transplant staff have a pain-alleviation ideology that is very much interactionally oriented (reducing patients' anxieties, using comfort measures, etc.). The psychosocial sensitivity of staff is quite outstanding. Another way of putting this is to say that part of the staff's main job, as they conceive of it (compare with the burn unit!) is the alleviation of pain – it has high priority. Among other things, because anxiety increases probability of transplant rejects and other complications. Another way of putting it is that the *social trajectory* concerns of patients are also to some extent in the purview of the staff, and its job: because those concerns can affect the *illness trajectory* adversely . . . . Concerning failure on illness trajectory: Two patients who had vague complaints of discomfort were managed by back rubs, talking with them, etc. Later, they were discovered to have very advanced infections which fulminated into septicemia; so, wrong *relief measures* and *misreading the pain indicators* (p.8) [relating multiple concepts plus relating two subcategories].

Controlling pain expression and interactional disturbance; awareness closed: Colitis patients sometimes have to control public expression of their pain, or they upset people around them. They have to keep the awareness context closed (just as do heart patients who have angina, or people will rush over to help them or get upset). . . . In fact, colitis people (who may simultaneously have to clean themselves up) who route themselves to public restrooms have to be careful of expressions of pain, which will bring people rushing over, which will then interfere with the main task of handling wastes: They will only complicate your problems, get in the way. NOTE: keep an eye on *interactional disturbance*, for this is likely to be relevant category also [relating four categories; a research directive also].

These codes illustrate selective coding because they all relate to the core category of pain management. It can be seen in the analysis that the codes (categories and subcategories, too) densely relate to each other, many of those relationships being brought out above in conditional, consequential, interactional, and strategical terms: and all of that related to the core category of pain management – whether it be relief, minimization, or prevention. So one can sense in these codes a considerable degree of cumulative integration taking place, even though the project is only six months along. (Integration is further discussed and exemplified in two later chapters.)

These codes also exemplify the list of coding functions noted on the first page of this chapter; though the raising of generative questions has not been especially prominent, except in the coding of the cardiac recovery nurse's work. Remember, however, that after some coding the analyst will write theoretical memos, both to summarize some of those codes and to include research questions raised by the codes. It is worth adding that although codes may be handwritten on the margins of the document being analyzed (and probably most qualitative researchers do that), they tend to be far less detailed and less easily sorted than the typed alternative. Also, nowadays, with the increasing use of computer retrieval, the typed version is much more efficient on the usual masses of data collected by qualitative researchers (Conrad and Reinharz 1985).

Utilizing codes in writing for publication

How do codes get incorporated into the final drafts of manuscripts written for publication? To show this, here are a few paragraphs about monitoring work, taken from many more pages on monitoring, in *The Social Organization of Medical Work* (Strauss et al. 1985). Recollect that the core categories in this study were "trajectory" and "work types." So, in the publication, monitoring should be related not only to other categories but to these core categories. Of course, there are various ways to do this in discursive presentations, but here is one way:

> Monitoring is a term much heard in today's hospital, but given the various contexts in which personnel use it a thoughtful listener can be confused about exactly to what it refers. Attempting to avoid the analytic tangle inherent in their use of the term, we begin this section by noting the various types of monitoring involved in highly skilled work done with high-risk premature infants. . . .
>
> Now if we ask what kinds of monitoring are going on, the answer typically must take into account the following – aside from any monitoring of the mechanical functioning of the equipment itself . . . . she is monitoring both the machine's information and the child's temperature, recording all that information, and acting in accordance with her interpretations of information . . . . the nurse is monitoring the equipment, also the equipment in relation to physiological functioning, plus the information given by the meter . . . .
>
> A fourth type of monitoring is the paying of close and almost continuous attention to signs yielded by the infant's body and behavior: movement, skin color, temperature, respiratory rate, and the like – this reading being done by

eye, ear, and touch . . . . A fifth type of monitoring might be termed "second order" – exemplified by the physician or head nurse who listens to or reads the nurses' reports of *their* monitoring: That is, the latter monitoring is closer in to the machine and body functioning; while second-order monitoring is more distant, being laid on top of the other . . . . Just to keep the empirical record straight, not all monitoring may be of equal importance, for that depends on trajectory phase and the infant's immediate condition – hence decisions are being made about how frequently to monitor what, with what degree of alertness, and so on.

All of this monitoring, including by or with machines, is designed to keep the staff abreast of one or more of several things: Let us call them *dimensions*. First, there is the monitoring of trajectory stabilization or change, whether negatively or positively, and how much change has occurred. An important aspect of that is "present condition," meaning precise location on the trajectory. Second, if the negative changes are drastic then clinical safety is at stake, and that is being monitored, especially for high-risk trajectories or during dangerous phases. Third, there may be monitoring along at least two other dimensions, neither strictly medical although each may greatly affect the medical course: one pertains to the patient's comfort (for instance, does the machine cause undue discomfort); another pertains to the patient's "psychology" as affected by the machine and its operations. In fact, each of those dimensions may take precedence over strictly medical monitoring during some moments or even entire days of the patient's hospitalization.

Different trajectories call for different totalities (or arcs) of work, including monitoring work, the implicated tasks varying according to phase of the trajectory. Therefore, depending on trajectory and phase, different machinery will be utilized, whether for therapeutic or monitoring purposes. What makes the staff's work both variable and potentially further complicated are the many properties of the machine–body monitoring itself. A listing will immediately suggest why this is so. These properties include:

frequency of monitoring
duration of monitoring
intensity of monitoring
number of items (including body and body systems) being monitored
number of dimensions being monitored
clarity or ambiguity of signs being monitored
degree of discrimination required in sign interpretation
number and kinds of sense modalities involved in monitoring
sequential or simultaneous monitoring of the signs.

Typical monitoring in ICUs, cardiac units, and dialysis units is then contrasted with additional theoretical points as well as with vivid illustration, all in relation to the previous points and to different types of work and trajectories.

One other actor in the monitoring drama, the patient, should not be overlooked. Machine-wise patients, familiar with equipment from their repeated hospitalizations, need not be taught monitoring chores and usually require no urging to do them. They know the machines and they also know the vagaries of their own bodies better than anyone: And that combination can make them valuable partners in the monitoring work. However, that combination makes them impatient or critical of staff members' monitoring work when perceived as incompetent or negligent (we saw instances of this in the preceding chapter). By contrast, patients new to their diseases or to particular equipment may require persuasion to engage in some measure of monitoring by staff, who wish thereby to either share the work or increase the clinical safety. Nurses will quite literally size up patients in accordance with their probable trustworthiness to do, and learn to do, this monitoring. Those who are deemed too sick, unintelligent, or unmotivated to monitor themselves are likely to be placed in rooms closer to the nursing station. Teaching the patients how to monitor themselves is usually done on the wing, rather than through formal instruction. Of course some kinds of monitoring, like reading cardiac waves which dance across the screen, require too much medical sophistication for most patients to monitor even if they had the requisite energy and motivation.

To return to the personnel's work: An immense amount of transmission of information yielded by their monitoring is characteristic of any section in the hospital where monitoring goes on. The transmission takes the form of verbal or written reports, or both . . . . Transmission of information laterally and upward is, then, a major industry engaged in by nurses, technicians, residents, attending physicians; and, for the machines themselves, additionally: bioengineers, safety personnel, and various other calibrators, maintainers, and regulators of equipment . . . .

All this transmitting of monitoring information is, ideally, in the service of allowing the physicians to make informed interpretations bearing on patients' trajectories: location, movement, and relationship with past medical interventions. Future courses of medical action – options perceived and chosen – depend primarily on these interpretations, pyramided atop the information gathered by technicians, nurses, residents, and the personnel of specialized labs. At the bedside-operational level, transmission of monitoring information, as from nurse to head nurse or to a resident or attending physician, may result in decisions bearing immediately on a patient's safety, comfort, or anxiety. It is analytically useful to make a distinction between these two levels of information transmittal . . . .

For those who do the operational monitoring, there are consequences too: perhaps principally boredom, excitement, and stress . . . . By contrast, monitoring is challenging and rewarding under a variety of conditions: when the worker is first learning how to monitor, or is learning about a new machine; when the trajectory phase is at high risk and so the monitoring is vital; when the monitoring indicates that a worsening trajectory is reversing itself and the monitoring is indicating good news or has contributed to it; when the monitoring itself challenges craft or professional abilities (including when those are associated with ideologies that emphasize the importance of monitoring, as with comfort or psychological dimensions; or as with the physician's joy in his "sixth sense," composed in part of craft and part of ideologically based satisfaction); also when monitoring tasks are varied because the trajectories worked on are varied, hence the monitoring agent is somewhat in the situation of an orchestral musician confronted by contemporary music – difficult but interesting – rather than playing the same old music.

But whether boring, exciting, or stressful, monitoring in the service of trajectory work is a very large and important aspect of all medical production work. Increasingly, visibly and dramatically, nowadays, it involves the monitoring of body-related machines.

Now that you have read or scanned the above material, it should be useful, later, to compare the original code items for monitoring with the final written version. The earlier codes do not usually get incorporated as such – sometimes the later ones do – but the ideas and the relationships specified may find their way into the published write-up.

Coding for structural and interactional relationships

This section illustrates briefly the initial steps in linking larger structural conditions with the interactions among actors, and between them and their institutional settings, who and which appear quite directly or are reflected in the interviews and fieldnotes. Often in contemporary qualitative research the emphasis on interactions (and on immediately contextual aspects in relation to interactions) is so strong that it overwhelms or prevents attention to the larger structural conditions. Yet all of that, as noted earlier, needs to enter the analysis. Minimizing or leaving out structural conditions, whether more immediately contextual or "further away" (or, as some social scientists say, the *macroscopic* or *structural*) short-circuits the explanation. Doing the reverse, overemphasizing structural conditions, does not do justice to the rich interactional data that put life and a sense of immediacy (or as some say, *reality*) into the analysis.

How to bring both into conjunction involves thinking *both* structurally and interactionally. One can examine and collect data about the structural conditions. One can examine and collect data about events, actors, interactions, and processes. Eventually, however, the grounded theory researcher must engage in coding that results in the detailed codes connecting *specific* conditions with *specific* interactions, strategies, and consequences. When examining the data bearing on the structural conditions, a researcher must ask: "But what difference do these structural things make for interaction and interactants?" When examining the more interactional type of data, the researcher must ask: "But what helps to account for these phenomena, including not only the more immediate structural conditions but the larger, macro ones?" It is understandable that making such linkages takes much time and thought – using directed data collection through theoretical sampling – and that the associated skills also take time to develop.

The examples of codes reproduced above reflect the making of connections between the interactional realm and various close-in *contextual* conditions, such as the properties of a hospital, a hospital ward, or a medical machine. The examples do not make explicit, however, the comparable relationships with *larger structural* conditions. But to continue with the instance of coding for the impact of medical technology on medical work: Eventually the researchers on that project explicitly and in detailed ways linked the personnel's work with and interaction around medical machines with various structural conditions flowing from the properties of the health industry, the equipment industry, the populations using the hospitals, the health occupations, professional careers, the explosion of medical knowledge and technology, the government's role in health care, and the contemporary social movements that are affecting that care.

Examples of coding for larger structural conditions and interactional consequences (mediated almost always through more immediately contextual structural conditions) included coding for such matters as those following: Government regulations bearing on safety are interpreted by hospital safety departments to staff people working on the wards, but much of their interpretation must be proffered diplomatically and in an advisory capacity because the safety departments have little power over how the wards are run or how medical and nursing care are given, or indeed over how machines are utilized. For equipment that utilizes nuclear materials, the advisory role tends, however, to become also a more closely monitoring one. Or, again: Because medical equipment must ultimately be tested at clinical sites, there has grown up a relatively close relationship between the sales representatives of equipment companies and the users of new equipment or new models. The users frequently report back about "bugs" in the equipment and may even make suggestions for improving certain of its features. This linkage between users and manufacturers is increased by the fact that much equipment is invented by medical researchers working at the frontier of their particular practicing specialties and basing their innovations on the latest specialty knowledge. Many of the equipment companies are relatively small, producing for specialized types of medical care. Coding of interview and field observational data brings out these kinds of relationships between structure and interaction, because both tend to appear in the data (either they just do, or the researcher collects specific data bearing on suspected connections – if possible, utilizing theoretical sampling to direct the data collection).

Analysts also code for any impact from the interactional level on the contextual and larger structural levels: For example, safety department representatives funnel advice back to governmental safety regulators, indeed sometimes they are nationally known experts themselves; and we have already remarked on hospital staffs, physicians, nurses, and the bioengineers, all being requested or taking it on themselves to affect details in the manufacture of specific medical machines.

Rather than give specific codes as examples, we have elected to discuss the general procedure of coding for what seems to many researchers rather disparate levels of analysis. Grounded theorists do not think of structure as something "up there" and as more or less fully determining of interactions. Nor do they assume that given structural conditions, whether economic or political or sociological (the latter would include class, gender, occupation, capitalism, etc.) must necessarily be relevant to the interactional/processual phenomena under study. Rather, the researcher must search for *relevant* structural conditions, which means they must be *linked as specifically* as possible with the interactional/ processual. The structural conditions can be at any level – whether more immediately contextual (like the institution in which people are working or living) or more obviously macroscopic (the class system, type or state of the economy, and government legislation).

The *rule of thumb* for the researcher is to be alert for what in the collected data bears on the more microscopic as well as the more structural. For both levels, the researcher should also be developing categories, following the usual coding paradigm. Analysis should relate those categories (as always). And, as always, the emerging analysis should guide the further data collection, through theoretical sampling, as it bears on the hypothesized relationships among the major categories being developed throughout the course of the research.

Otherwise the researcher ends up with a choice of the following options: first, either a structural study or an interactional study; and second, a bit of one and a lot of the other. If the second choice is made, the connections between both levels will tend to be nonspecific. Metaphorically speaking, either the macro forms a backdrop for the *real* drama, or the backdrop becomes the drama and a few puppets go through rather unreal, undramatic sets of gestures. Sociological monographs are replete with examples of these choices and their variations.

Proper coding can surmount the dilemma represented by these choices and still allow the analyst to put more weight – because of personal interest, substantive knowledge, research skills, or contingencies affecting the research project – either on the macro- *or* the microanalysis. In any event, proper coding within either level will make for more effective theory about phenomena at that level. Thus, one can study negotiation among nations without looking at the minute details of the negotiative interaction among them, rather than making a study of one or two specific negotiations, in standard case study style. The focus can, instead, be upon nations interacting through their respective political or economic institutions, their political maneuvering, their negotiative representatives, etc. On the other hand, if researchers choose or are forced to study interaction and/or process, they should still be systematically searching for and analyzing structural conditions that are more immediately contextual, even if they eschew detailed pursuit of the more macroscopic ones.

Rules of thumb

One last note: The examples of codings and the commentaries on them in this chapter can suggest, by a slight bit of imagination, several rules of thumb concerning coding procedures. These include:

1. Do not merely précis the phrases of a document, but discover genuine categories and name them, at least provisionally.
2. Relate those categories as specifically and variably as possible to their conditions, consequences, strategies, interactions: That is, follow the coding paradigm.
3. Relate categories to subcategories, all to each other: that is, make a systematically dense analysis.
4. Do all that on the basis of specific data, and frequently reference them by page, quote, or précis right into the code note itself.
5. Underline, for ease of scanning and sorting later.
6. Once the core category or categories are suspected or decided upon, then be certain to relate all categories and subcategories to that core, as well as to each other: That is, open coding moves through axial into selective coding. In this way, integration of the individual bits of analysis increasingly can take place.
7. Later, the totally or relatively unrelated minor categories, with their associated hypotheses, can be discarded as more or less irrelevant (albeit often interesting, as such) to the integrated analytic product; either that, or the researcher must attempt to specifically relate them to the major core of his or her analysis.

4 Seminar on open coding

This chapter is not a discussion of research procedures as such, but is an extended illustration of open coding, with commentaries on particular instances of it. Again, readers who are eager to move directly to procedures should defer reading this chapter until later, although it is placed here because to most readers it should be useful to at least scan it, especially the analytic commentary in its closing pages.

The chapter consists of one case: a research seminar session which was recorded on tape. The format of the presentation is this: first, a short introduction to the case; second, the analytic discussion itself; third, an analytic summary, with a detailed commentary on each phase of that unfolding discussion.

In the long extended case, the seminar participants are seen working together on the very real data of a researcher–student. By contrast, the pain-theory case in Chapter 2 illustrated a very active teacher, at a very early session of the seminar, "getting across" various elements of grounded theory methodology, using not a presenting student's data but only the combined experiential data of himself and the class participants. Here, while experiential data come visibly into play as an element of the analysis, the chief data are not collective data. Besides, there is the additional, if invisible, drama of a presenting student who is deeply concerned about the outcome – the product – of the class discussion.

Of course, the materials in this chapter are designed not only to illustrate the teaching–learning of analysis within the seminar setting, but to clarify further how qualitative analysis, especially open coding, is carried out in the grounded theory style of analysis. (For a further look at teaching it, see Chapter 14.)

Case

This seminar session – the participants are graduate students in sociology – had a specific purpose: to explore for the presenter (A.C.) aspects of interview data that she had not focused on, thus expanding the possible scope of her future analyses. She would not necessarily be committed to the lines of inquiry explored in the seminar, but would follow through only on those that turned out to fit her data best, and with seemingly greatest exploratory power.

The student had interviewed several women diagnosed as possibly or actually having dysplasia (a form of cancer of the cervix). As is usual, the presenting student is able to supply useful or necessary background information at certain junctures in the seminar discussion; while other students are able to draw upon their own relevant experiences and sometimes on experiences comparable to those of the interviewee. The instructor (A.S.) opted for close examination of the interview data, but not necessarily for a focus on each line examined sequentially: So, he will be seen here directing the seminar's attention to selected lines and paragraphs in the opening pages of the presented interview, selecting them so as to bring out potentially interesting features of the data.

In this particular session, the instructor was especially active, "talkative," doing much of the actual analysis, though sometimes taking his cues from students' remarks and insights. He chose this style both because he wished to cover maximum ground for the presenting student and because at this point in the class development (about seven months of training) the participants were judged able to follow his (or their) techniques of dimensionalizing, utilizing comparative analyses, and to slip easily into the line-by-line mode of analysis. So, this seminar discussion, as it developed, illustrated for the class the rapidity with which diverse lines of inquiry, generative questions, and initial categories could be developed even from the first pages of a single interview.

For the reader of this text, other points are especially worth noting:

The use of experiential data by the participants;
The use of the interviewee's terms to form in vivo categories;
The coining of provisional terms to form other categories;
The posing of directing questions by the instructor, for various purposes: to keep the analytic discussion from wandering, keeping it on target; to force attention on the potential meaning of certain lines or terms in the interview; to push the students' thinking further in specific analytic discussions; etc. The instructor chose not to explore any category, or any dimensions, in much depth, in view of the presenter's request for expanded scope of her analysis;
The instructor assumed that: (1) more areas of exploration were desirable, rather than fewer in depth; (2) the student herself could add greatly to the analytic depth and she would do so if analysis of further data warranted it for any given category or dimension.

Other points are brought out in the detailed commentary which follows this text.

As the student herself later stated, the analytic discussion in this session "skimmed along and captured the tops of important waves." Later, too, she found herself following through on many of the main lines of inquiry and categories developed through the discussion. In the discussion below, the various students are indicated by M. for male and F. for female.

Phase 1

1. A.C.: This is my Dysplasia Study – the first time that the seminar is going to work on it.
2. A.S.: I think we need about five minutes to scan the interview. So why don't we scan it.
3. Apropos of this data, let me first give you a rule of thumb. If you know an area, have some experience, as I have said before, you don't tear it out of your head. You can use it. Now, with things like illness, we've all had a fair amount of experience, alas, either ourselves or somebody else's. So we can talk about the properties of the illness, and about the properties of signs of the illness – that is, the symptoms. Or we can talk about the properties of the regimen. Without even reading one interview. In other words, we have plenty of information in our heads about symptoms, diseases, and regimens. We don't know about this particular illness. But by dint of reading the first page of this interview, we can begin to make some guesses about the properties of each of those kinds of things. And so, let us just start out.
4. Now, your response is somewhat biased because of already having read the interview, but you can do the task. For example, diabetes, with its regimen – we all know about the regimen. It involves daily work, it's complicated, it involves more than one person, you know, and so on. And the symptoms: We know a little bit about them. The symptoms go all the way from catastrophic shock to havine urine show too much sugar, and so on. And the nature of the disease – but let that go, you get the general idea. Diabetes can be, for example, terminal. In the sense that you can pop off quickly – in crisis. But, you know that some diseases are terminal, some are not. You mention you've got a disease and somebody says "What's that?" You mention that you've got cancer and everybody knows what that is and what that means. That's as public as childbirth: Everybody knows about it and there are stories about it.
5. So, if we take just the face sheet and the first page and work on them, you can see some interesting things. For example: I'll give you a few points to get you moving – that it's cancer, therefore it's a touch of the terminal – it's got all the public imagery of that. Then, there's a question about whether it's treatable and how long you can keep terminality away, with all the remissions and so on. Or, is it a one-time thing, and they can get rid of it? Okay? And then the diagnostic stuff seems to be really brought out. You'll notice the first page doesn't tell you very much about treatment, except for the one paragraph which we get to in the middle. The diagnostic line makes clear that the diagnosis is uncertain. It can come as a surprise to the patient. There could be mistakes in it, undoubtedly. It doesn't say that, but we could

guess there might be. And there's a question as to whether this kind of diagnosis is new or old to the patient. Has the patient had experience with diagnostic procedures like this, or is it a new kind of experience? Most of us have been through x-ray tests for example, but maybe very few of you have been through this kind of business. Pap smears are probably pretty common. For most women nowadays, this is so. Now if you consider next the symptoms – I don't know – are there any symptoms in this illness?

6\. A.C.: Not at the dysplasia stage. No.

7\. A.S.: See, so it's really fascinating. And in terms of the regimen, notice that it's solely in the hands of the medical professions. The woman doesn't do anything. It's one of the most passive things you can imagine except to get to the doctor. In other words, the question about how long does it go on – my guess would have been they would do the treatment right away – bingo! It's over. The interview shows something different. But, you know, regimens can go on for a lifetime or two months, and so on. The amount of money it costs is another dimension of the regimen. And so on. So, you can begin to kind of lay these kinds of issues out ahead of time. This already begins to raise questions, if you want to, even before the first interview; or they'll come out in the first interview. About, you know, what are the attractions, what are the experiences of people in this? So, for example, you could know ahead of time: They have a lot of Pap smears and then suddenly one shows cancer – you can just imagine what happens. If there were symptoms that preceded it, it might not be such a surprise: But there are no symptoms!

8\. The exact analogue of this is that somebody goes in with a broken leg, and they do a blood-pressure test, and they ask: "Did you know you have high blood pressure? Hypertension?" Some people don't even know what that is. They have to have it explained to them. They see no symptoms. That is a complete surprise. To some, that can be pretty upsetting. Others don't know what it is all about – but it is hard to conceive of any women who would not know what cervical cancer meant. Though it could be; there might be some variation in response.

9\. Anyway, in this kind of illness you can begin to guess beforehand what some of the reactions would be – some of the meanings, and all the rest. But again, you do comparative analysis to highlight this. Now, I'm ahead of you because I've been doing this for many years. On the other hand, as I've said, you've all had experience with various kinds of illnesses.

Phase 2

1\. All right. Adele, what do you think the best way of handling this would be for you? What do you want to get out of the session today?

2. A.C.: I'm not sure whether it would be best to do it line by line or be more impressionistic. I mean, use both approaches. I don't know.
3. A.S.: What do you want to get out of it?
4. A.C.: I see certain things in the data already, and from hearing the next thirty pages of the first interview on the tape. But I would like to see what other people see, rather than me say what I've seen.
5. A.S.: To see if it comes out the same way?
6. A.C.: Yes.

Phase 3

1. A.S.: All right. Well, since she doesn't want it line by line, let's take it a third of a paragraph by a third of a paragraph. Let's start, say, halfway through the first page. What do the first four or five lines tell us?
2. F.: Well, that this had a beginning. It wasn't something there forever or . . . .
3. A.S.: What had a beginning: the illness or the discovery?
4. F.: It. The problem. Her cervix as a problem had a certain time to begin, and that was August, 1980. Although there is some confusion there, for some reason, about this very important date. And it was not a thing over which she had any control.
5. A.S.: What else?
6. M.: There was a break in a routine aspect of life. You go in and you do this thing and you expect nothing to happen. And you go in and do it and expect nothing to happen and then there's a break.
7. A.S.: And do you want to give it a word?
8. M.: No.
9. M.: Nothing comes to mind.
10. M.: Crisis?
11. M.: A break?
12. A.S.: It's a routine diagnosis. Right? In other words, this tells you right away that it's something you keep going in for; you know, at regular intervals. It's a routine, relatively scheduled diagnosis. There are very few diseases that are like that. Well, that isn't true – you go in for regular checks of your teeth. What else?
13. F.: TB tests, every year.
14. M.: Yes, if you go to school here.
15. M.: Blood pressure every time you go in to see a doctor.
16. M.: People get x-rayed every now and then – chest x-ray.
17. A.S.: But there's an interesting feature to this. I, myself, go for a check every month or two on my heart; now three months, maybe later, five months. That's a routine, diagnostic, scheduled check. But there's something behind it. Whereas *here* there's nothing behind it – behind it in the sense that I have an ongoing disease and I'm having a routine check about it. But here, there's no

disease. But it's still a routine check. So there is . . . . What's the matter?

18\. M.: Well, when you go in for your teeth, it's with the understanding that there might be a problem. It's really a preventive process. So, it's the same kind of thing?

19\. A.S.: Yes. The dental thing is like that. But when I go in for my routine check for my heart . . . .

20\. M.: Oh. I thought you made the statement that you were . . . .

21\. A.S.: No. In other words, with an ongoing disease, even if the disease has vanished, so to speak, but is repeatable, it's returnable; there's a difference between that and a dental check. All right, let's take the dental check. What you have is a routine, scheduled diagnostic. But no disease experience behind it. So we've already made distinctions between routine diagnostic scheduled and the nonroutine. And obviously there are other nonroutines. You've got a symptom and somebody checks it out. That's quite different. So this is a symptomless . . . you ask yourself, how is it you have this scheduled, routine, diagnostic check when there are no symptoms and no disease? What's it all about?

22\. F.: It's similar to having your teeth checked, because it's preventive, but it's qualitatively different, because you know that the chances are, the percentage, or whatever is probably going to be less. Most people have cavities when they go in and have their teeth checked.

23\. A.S.: Just what do you mean, "chances"?

24\. M.: It's a bigger risk. If something goes wrong with this exam, it's potentially more damaging than if you need a tooth repaired.

25\. A.S.: We already know this is a high risk. We know that's the nature of cancer. But *chances*: What does that mean?

26\. F.: It means that you can't do anything about it?

27\. A.S.: Do anything about what?

28\. F.: Whether you're going to have the cancer or not. I mean, your teeth, you can do more than having them checked up. You can brush them and, you know, not eat sweets and all that.

29\. A.S.: That simply means that you are active in the preventive treatment.

30\. F.: Right.

31\. A.S.: So there is a difference between the onset of the business. Some, the person can prevent some; and some not. I thought you used chances in another way. Why chances, for this scheduled, routine diagnostic?

32\. M.: It's against the risk that there might be something wrong. But the risk that there might be something wrong, in this case, is a lot less than the risk that there might be something wrong in a dental exam.

33\. A.S.: What's fascinating about this is that the developments of dental cavities are riskier, in terms of statistical probability. On the other hand, once the problem really confronts you, you'd rather have cavities than symptoms of cancer. Anyway, you're playing a statistical game here. So there's some sort of – whatever word you coin for it, this is a diagnostic procedure that has to do with probability.

34. F.: I think that's people's experience of it. That most women go for Pap smears all the time and absolutely don't expect when they go that anything's going to be wrong.

35. F.: Something you do every once in a while.

36. A.S.: We'll come to that in a minute. But let's take the other part first. First of all, there's the statistical game, probability. The other is that if it's caught or not caught, there are degrees of fate. That will come out in the interviews, because some women will tell you that they skipped for ten months when they shouldn't have, and then they got hit. "Why did I play with fate?" If you could predict that. All right, now we come to what Gayle has pointed out. Say it again.

37. G.: Well, because of the regularity, because of the scheduling, it just becomes a sort of stage in your life, something you just do, maybe like going to the dentist (but the experience is so different, I don't want to compare it) – you come, after a while, to actually believe that nothing will be wrong. And that's the assumption under which you go; you don't schedule any extra time for it, you don't think I'm going to need extra time after I go today because I might need to recover from what it's like to do this. You just do it. It takes minutes.

38. A.S.: Lots of people go to the doctor for an annual checkup, really not expecting anything. So, you have a distinction now between checkups – with more or less degree of expectation of something going wrong.

39. F.: I think it's interesting – and it comes out in these very first few sentences – Adele asks: "You went for a regular physical?" No, she didn't. Women don't do that all the time. But they always go and get a Pap smear. I mean, that's medical habit. That's been institutionalized in this society. You get a Pap smear.

40. A.S.: But can't you imagine a variation? Imagine some woman having read about this cervical stuff, you know, a couple of weeks before she's scheduled. Then she might go on – there are some people who would then be apprehensive. "You mean it's twenty-five percent?" You know, whatever, something like that; or let's say your best girlfriend has just come back from a "bad" diagnosis and you're coming up yourself. You might, then, even speed up your schedule. So, it's mostly routine, like an annual checkup or a dental thing, but you can suspect there might be variations that a routine becomes nonroutine, psychologically, let's say.

42. A.S.: So that's – let's call that the diagnostic durational span – something like that. How much time in between? Two days ago somebody said to me that his spouse is going for a check – his spouse had a mastectomy a while back – his wife had another

lump. And I said: "When will you know?" And he said, "In about a week." I said: "What! You should know within twenty to thirty minutes. Get on it." He said: "You don't know Kaiser." I said: "I don't give a damn, if you curse like hell." Well, it turned out to be benign. It turned out that they did the test (and it's ninety-seven to ninety-eight percent accurate usually), and it's OK. So that durational business is a – especially when there's a fatal disease, is fantastic. So that's not only an element in the interviews, but it's part of the analytic picture. Now, what else?

43. F.: Another part of this experience that I would think of before I even read this (about what the circumstances are under which this could lead to the dysplasia) are the issues about how you're treated when you go for this kind of examination. Most women have lists of experiences in their lives about being treated in terrible ways when they go for Pap smears and any other kinds of gynecological examinations. And it turns out later that she has had such experiences. Before I even read them I thought, you know, Who's going to shit on her? Who's going to do some unreal, unbelievable thing during the course of her going in even for these things? no matter what happens. . . .

44. A.S.: You're running ahead of the story, you're reading ahead of the data. I had that down on my list, but decided to hold it: that one aspect of diagnosis is, it is done by medical people. And you can ask about the medical ways of doing this particular diagnosis. But that obviously varies a very great deal in different kinds of diagnostic procedures. Even if they're all medical. OK? But it will also make a difference if it's a small town where you know people and "U.C.," where you don't. You could foresee some of this ahead of time. Hold that point, because it does appear later.

Phase 4

1. A.S.: Now, let's do the next four or five lines: "Actually, I normally go every six months because I have herpes," suggests that she knows there's some statistical probability that of an increase because of the herpes. So maybe that people with herpes – in terms of the variation of the population – people with herpes should be more expectant of their high probabilities of having cancer, than people without it.

2. F.: I think the regular routine, normal and abnormal – you know, it's almost in every line there of the first seven or eight lines: "I normally go," "I usually do," "I . . . ," and so on. "This is the first time I'm abnormal." So there you have a little – I don't know what abnormal is and I don't know what normal is, and what the terms mean to her. If it's normal, then you just don't worry about it; you don't do anything; and you don't really know what normal is, except it's not a problem – nobody says it's a problem.

Then, all of a sudden, you've got to define *abnormal* – I mean, you didn't have to do anything about *normal*; but when you find abnormal, and all the way to, I mean, she's putting this . . .

3. A.S.: That's very good that you've picked normal/abnormal; but, if you just don't take the line – "It was the first time I had an abnormal, all the way down to three" – think of the difference comparatively of when you go to a doctor and he does an EKG on you, let's say. He will give *you* the information about what it means, what it might mean – very specifically. Or you have an annual checkup and he discovers something. And he says, "Look, you've got anemia. You don't feel it, but you have it. Well, a touch of it. Better do this now." And he gives you pills. Nice, clear diagnostic. Here, it's not only ambiguous, but the patient is thrown back on herself, in a certain sense, to interpret it. So you have the patient's interpretation of diagnosis.
4. F.: But it's masked by all these elaborate medical categories. And each place has a different category system of terms. But even here, when she says "all the way to a three," she already had to learn the system that they have imposed on this to make that statement meaningful to her. They told her she has a *three* – what does that mean? It doesn't mean anything. She had to learn the major difference between a *two* and a *three* or a *two-and-a-half*.
5. F.: Although you don't really know what they actually told her.
6. F.: They're really ambiguous, but they're faking the diagnosis, by pretending with this elaborate, precise, quantitative-sounding system.
7. F.: But I had a question about that because she seems to know a lot about what these twos and threes and all that mean, and I was wondering if she knew that before . . . .
8. A.C.: Yes, that's what I said at the top of the interview. She was a para-, worked as a paraprofessional in a family-planning clinic, so she had some health background.
9. F.: So we don't really know what they told her. And if she's a paraprofessional, they may actually have told her what it meant, and then she uses it, too.
10. A.C.: I think she likely knew what it meant before, because she was already aware that herpes is associated with it.
11. A.S.: That gets into variations. Some people know it and some of them don't. That's built in. You don't have to worry about the absolute truth of this particular lady.

Phase 5

1. A.S.: I think "all the way" is an interesting phrase. I mean when you talk about "all the way" you talk about completion of the sex act, you talk about "he made it to the top," all those kinds of things. So here she's using her diagnosis and the fact that it came back

abnormal to indicate it has reached a certain definitive kind of stage. That's a dire way of talking: "All the way" – as far as a diagnosis of dysplasia is concerned.

2. And then, what do you do when you find a bad diagnosis like that? You immediately look for a second opinion – which is just what she did. First she "freaked out," and then evidently she began to think about it, and went right back down to somebody else to get an opinion, and then another opinion. So she's really checking out what resources are available: to have gone all the way to a three.

3. F.: I think there is also surprise there that it wasn't more gradual. It was the first time and it was already more than a two all the way to a three. It's like she goes every six months; she should have had a one, then she should have had a one-and-a-half.

4. A.S.: What does the phrase "all the way" mean? Converted, coded? "It was more than a two and all the way to a three." If it came from zero. Right?

5. M.: An extreme development.

6. A.S.: Right. Extreme development along a continuum – and what is the continuum?

7. F.: Noncancer and cancer?

8. A.S.: You can't say noncancer and cancer. It's a continuum of from zero to catastrophic. So, it's a disease continuum. If I have a bad back, I can say, "Well, it's not so bad, it's mild." Or I could say, "It's killing me." That's a symptom continuum. So I have words for both. OK? Here, they're not talking about symptoms, they are talking the disease itself. Well, they're symptoms – but they're invisible to the patient.

9. The other thing that is startling about this is – well, I'll give you the counterexample. When I had my episode in the hospital last October, the doctor, after I got out, did not tell me how bad he thought it was. I totally surprised him by doing far, far, far, far better than he ever expected. But he didn't tell me. It wasn't until months later when I asked him to tell me what he really had thought back in October. But here, they're telling: "You've got a three." So what do you want to do about that one? If you have a tooth cavity, the dentist tells you, doesn't he? He says, "It's bad enough. I think we'd better keep an eye on it. Maybe next time, we'll do something about it."

10. M.: Why are we considering the results of a test on a disease and not the symptom? Isn't it just a symptom made manifest?

11. A.S.: I say it's a disease continuum.

12. M.: Carcinoma is on a continuum: carcinoma in situ and long-term carcinoma.

13. M.: The results of the test, though, could show that a disease will occur or won't occur.

14. A.C.: In this disease? No. All the tests can tell you is whether the cells that are there now fit into a normal range or varying ranges of abnormal: "Today."

15. M.: But even if they vary into the abnormal range, it does not necessarily mean that the person will have the disease. Or is that itself a disease?
16. A.C.: That, well, it does mean that dysplasia is a disease. Because it has a, yes . . . . But it can remit. I mean, that's where, that's what the weird part . . . .
17. A.S.: But we haven't talked about remission. Hold the remission . . . .
18. M.: Then I have a problem, too, because if you say that, then, why isn't back pain a disease, rather than a symptom?
19. A.S.: If I have diabetes – OK? – and I begin feeling really badly; and my tests show that I'm in bad shape; I may even go into the hospital for a few days, if they put me back. On a continuum of symptoms, the symptoms are obviously much more severe. I know it. And if you want to say, well, all right, the disease is more severe; but the doctors wouldn't really talk that way; they'd say, "It's acute now." Yet, they don't really mean it's any worse than it was. You can here blur this symptomatic thing and the disease, but it's clear that with dysplasia the symptoms are not visible to the patient. But, the doctors are talking about a disease continuum, that is what it is.
20. M.: I think that one part of the continuum is very nebulous – what they are talking about. It could mean nothing. And on the other hand, it means cancer straight out.
21. A.S.: If it's up in the upper regions diagnostically, that's cancer. So, what you want to talk about when you talk about the disease continuum is that, in fact, the upper regions of it are nonambiguous; but the lower regions are ambiguous. That gives us a distinction. All right: How are we doing?
22. F.: Of the disease diagnosis?
23. A.S.: Of the disease diagnosis.
24. F.: But not of the symptom diagnosis.
25. A.S.: Yes. Then we come back to the business of telling or not telling. This is a direct telling. Isn't it? [Pause.] A very direct telling, but it's a direct telling of exactly what the physicians know, or suspect. Sometimes they'll say you've got cancer; but they won't tell you how bad it is, which is very usual. But here they are telling her – directly. So, there is something very, very different about this announcement. So, let us call this a *diagnostic announcement*. OK? And there is no withholding of information. (This is the whole "awareness context" business that Barney Glaser and I talk about in the *Dying* books.)

Phase 6

1. F.: Did you ask, at that time when she said, "all the way to three" whether what had gone on between her and whoever did the Pap

or whoever reported to her? What information she had available at the time? I mean . . . .

2. A.C.: Well, she tells you about that later in the section you have. That she found this first diagnostic, and that it was in her home, by a friend who worked in the place where they do their diagnoses, and sat down and talked with her about it.

3. F.: This first one was in "U.C."? I know the first one was not done there, but this one that we're talking about – "all the way towards three" – Was that the one that was done in Ukiah?

4. A.C.: Yes.

5. F.: Well, that was a warm, friendly clinic, so she should have had somebody there; at that time, did she talk to somebody about becoming a three, and . . .

6. A.C.: Yes. She says that later on.

7. F.: . . . and that was the assurance, that, "It may not be anything," and so on?

8. A.C.: Yes.

9. "And the person who told me my Pap smear was weird was a friend of mine and she told me when she came to my party and was very sensitive and nice about it, and told me not to worry, it could be anything."

10. A.S.: Before we go to the last third of this page, the phrase – "freaked out" – by the way, she's given you . . .

11. M.: An in vivo code.

12. A.S.: She's given you an in vivo code. This is a category stemming from a class of responses – "freaking out." Which clearly has to do with: What are the conditions for "freaking out" here? Well, the imagery of cancer. That is, terminality. And surprise. You don't freak out when it's fed to you in small doses, or you begin to suspect it, yourself, before the diagnosis.

Phase 7

1. Now, the last five or six lines. "So they referred me to U.C." Now, what is that all about?

2. M.: There are some places out there that just tell you the information and some places that do something more about it. You've got to go around, you can't be taken care of in one place.

3. F.: It kind of looks like there is a career that you go through – I mean, she did have something definitely wrong. But that isn't the end of it, there. Then you go somewhere else and this keeps building up until you finally . . .

4. A.S.: Let's do this comparatively. Let's say that you have a very strange disease. For example, I know someone who had lead poisoning and nobody diagnosed it, although she went to dozens of doctors. Finally, by accident, she happened to talk to an English doctor who was an expert on this, who was an expert on sailing, as was

she. He said, "My God," and he gave her tests, and she had lead poisoning. So, it took her three years to find out that she had lead poisoning and she went through a lot of doctors, and had a lot of tests. So, *diagnostic career* is your word and it's absolutely right. But sometimes you don't have a diagnostic career. As when, if you drop in at the doctor's and he says, "My God, you have cancer," on a routine test. Or, you have one symptom, and they put you on an EKG and you've got heart problems. There's no career at all. The referral business from the G.P. or the internist to the specialist is a very short diagnostic career. Bang, bang – it's finished. But here you have a diagnostic career, which seems to have more direction. And it is more complex and has more steps in it.

5. F.: Isn't the reason that you have that kind of career because the test itself is inherently ambiguous?
6. A.S.: Under what conditions will you have a longer diagnostic career? Well, severe disease; or, if you are in the wrong place and people aren't very experienced in detecting it; or, it's ambiguous. Now, if this were not a very serious disease, they wouldn't send you to six different doctors or four different clinics.
7. So, the conditions for this diagnostic career are clear – some of them. The consequences begin to show up later in the interview; the consequences of having a long diagnostic career combined with a dread disease . . . . What else is in these few sentences?

Phase 8

1. F.: But after you go to all these different doctors and you're supposedly trying to get a more precise statement from them, you get a statement like, "It looks a little weird," which is a colloquial expression – it's not medical language; it's not a specific statement; it's not precise in the slightest. Aside from this whole thing of having no compassion, it's . . . .
2. F.: I was wondering about that . . . .
3. F.: . . . "Gee, it looks a little weird."
4. F.: . . . talking about lumps and weird: Is that *her* terminology, or is that what they actually said to her? I can't imagine . . .
5. F.: She used "weird" earlier, so I think it's her term . . .
6. F.: . . . her way of saying, interpreting what he or she said.
7. F.: Really? I thought that was what . . .
8. A.C.: Well, it may be. I will have to ask her.
9. M.: On the last page, where that same person, I guess, is clipping away, trying to get tissue to examine, probably it means that they want to do a further test to find out, so it won't be weird. We don't know that for sure. It's just . .
10. F.: And then I thought that in the last section she was getting a little – I mean, she says, "The doctor was just wanting to cut away."

That's kind of a – I think that goes with "weird" and "lump" and that sort of thing, that kind of terminology.

11. M.: Well, "D&C" – it's like it's a rationale for further testing, further procedures.

12. A.S.: The main point is not "weird." "Weird simply means that they've discovered a diagnostic sign which alarms everybody. So she could be high up on the disease continuum. So what you should be focused on is not the weirdness – that's simply another way of saying she's high up.

13. F.: That's like that the doctor told her that she didn't know what it was.

14. A.S.: But that simply means that the ambiguity goes on.

15. F.: Right. But the doctor could not have told her that. I mean, the doctor could . . . .

16. A.S.: Did not have to tell her. The doctors are still telling her "We don't know" – but it might be dangerous. It's still going on, just like it did in the beginning. So there's a repetition of that. So, the diagnostic career consists of a series of diagnostic steps. It isn't that she's just going from clinic to clinic and doctor to doctor, it's that each person is doing something additional. Right? And so, what I am saying – convert it analytically.

17. M.: Progression? Different stages in treatment?

18. A.S.: No, we're not talking about treatment; we're still talking about . . .

19. M.: Diagnosis. But there are stages there.

20. A.S.: Right. So the diagnostic business is in stages – or, if you want, cumulative, or whatever language you want to use. In other words, they're adding tests, so the tests are done sequentially. First, they do one, and then they do another one. They may repeat – but they do other kinds. Now when they did this other kind – that is, when they examined the uterus – they didn't, the first time. Now they examine the uterus; now they discover, what? Another diagnostic sign. This is common sense: It is hard to see it analytically. The fact is, the additional diagnostic test is supposed to show additional diagnostic signs which might have been missed by not using such a fine test, or such a supplementary test. Right? That's what it's all about. When they give you a CAT scanner after an x-ray, they're looking for new signs that were missed by the old test – or couldn't be shown up on the old one. But there are diagnoses that aren't like that. Although this is pretty common in the medical business. But, say, if she had the Pap smear and it showed up as potentially dangerous, then she should have been ready for somebody to say, "Well, the uterus is cancerous," or "the uterus is" something. So she shouldn't have been that totally surprised. But she was. You know, variations in responses are possible.

21. A.C.: Those two phenomena of a bad Pap and a lump can be found very independently. They're not necessarily linked at all.

22. A.S.: Sure. But what I'm saying is that on the face of it, somebody shouldn't be so surprised: having a three and now to find something wrong with the uterus.
23. F.: Except that forty percent of the women – at least those over forty years old – have a lump in their uterus anyway, a fibroid.
24. A.S.: Do they know that?
25. F.: Well, if she's a paraprofessional and she worked at Planned Parenthood, she should.
26. F.: But she's not forty years old, either.

Phase 9

1. F.: You know, what's interesting here to me is, alongside of the career of the diagnosing of the disease, is Goffman's "cooling the mark out." What does that mean to that patient's concept of herself as a patient? I mean, she is increasingly becoming committed to patienthood as a possibility for her – a cancer patient.
2. A.S.: You can't tell that from the first page. That will show up later, maybe. You can't tell that from the first page, can you? I think you're right. It's going to show up. What you're doing is perfectly right – you're making a memo to yourself – "this is a possibility." But, you can't count on it yet, not from this data. It's perfectly permissible, as you know, to write a memo to yourself, to remind yourself to look for it on the next page of the interview.

Phase 10

1. All right, let's move on. The business about the "compassion" and so on is so obvious that I don't think you have to tell Adele about that . . . . Let's take half of the next page. The question here is: What is this paragraph mostly about?
2. M.: It's a comparison. It's a comparison between her city experience and her country experience.
3. F.: In terms of interaction.
4. A.S.: In terms of what?
5. M.: In terms of the interaction with the people who were treating her.
6. A.S.: That's OK. But it's a little crude. What is the interaction really all about? The interaction is humane in one place and impersonal in another. But what is the interaction all about?
7. F.: If she's not sick, it's a comparison. She could be fine. The whole rest of the thing, her saying that she was . . . .
8. A.S.: We already know that it could be ambiguous and all that. But what is the interaction about?
9. M.: The diagnosis. No?

10. A.S.: Well, let me put it this way. What has her friend said to her? Initially. Let's take it step by step. "And the person who told me" – which is about five lines down – "is a friend of mine." "And the person who told me my Pap smear was weird . . . ." And then she gives you the property of the person who told her: ". . . was a friend of mine." And then she tells you how she told it. So she's telling us: She told me. The person who told me had this property. And the interaction had this property. And who told me my Pap smear was weird? What does she say? What is that interaction?

11. F.: Well, those are different strategies of doing the same kind of thing.

12. A.S.: Yes, but what is the "same kind of thing?"

13. F.: Breaking the news.

14. A.S.: Breaking the news. It's an announcement. This is a diagnostic announcement. "Adele, I'm glad to tell you that your symptoms mean absolutely nothing. It'll go away in a week. Just leave it alone. Don't do anything." That's an announcement.

15. F.: Also telling her what to do about it.

16. A.S.: We'll hold that for a minute. Just plain announcement of what the diagnosis is. So that has various properties, if you stop and think about it. We don't have to spell them out – Adele can do that. Think comparatively. It can be gentle, it can be severe, it could be surprising, etc., etc. *You* work it out . . . . Then the woman gives you properties of the announcer: stranger, friend, acquaintance. But the real emphasis, obviously, is the way it was told. Adele can surely handle that.

17. Now, at each one of these places along the diagnostic career, somebody's going to make an announcement. And the announcers get to be more and more strangers, as they get further out. And they tell in different kinds of ways, with different kinds of consequences. Notice, the consequences here are not the consequences of where it is on the continuum, but the consequences of the way it's told. And you can see this comparatively. If somebody just tells you, "Look, I'm afraid you're going to die in six months"; or somebody puts his arm around you and says, "I just hate to tell you something. You'd better brace yourself. You've got about six months to live" – it might not make any difference – six months is six months. But it may make a big difference. Right? And this is what she's getting. She's talking about the quality of imparting the information to her.

Phase II

1. A.S.: There are some other things like she sends this woman along the diagnostic path, career.

2. F.: And you could call Ukiah and U.C. and General Hospital in San Francisco *medical stations*.
3. A.C.: I should like to make one thing clear. It wasn't that she was coming to the city only for diagnosis. She was going to law school in the city and her husband was in Ukiah. So she would commute to Ukiah. I mean, she was living in two places, so she wasn't . . . . In other words, if she had been living in Ukiah, she may well have ended up at U.C. But it might not have been so rapid.
4. F.: Except that when she freaked out, she immediately went to another medical station, another place, for a second opinion. And I think she would have done that whether she was living in two places or not. Wouldn't she?
5. A.C.: I think she might have repeated the Pap very quickly, but not necessarily in a different place. But I think your point is well taken anyway, you know, the medical stations.

Phase 12

1. A.S.: What about the last two lines? "Get another one right away, within two or three weeks." What's that? Convert that analytically.
2. F.: Although I have been very gentle about this and given this information in the nicest way, nevertheless it is urgent that you attend to it and do something about it. It's still a serious . . .
3. A.S.: So the doctor is giving a diagnostic directive: Get another one. She's moving along the diagnostic career path. She's giving a directive. A directive, in this case, is not where to go, but how quickly to do it. A directive could be various kinds of things: Go here; go anywhere, but do it right away, etc., etc.
4. M.: Also, the way the person interpreted it: "She asked me to follow up on it." She didn't say, "You've got to go tomorrow and get another one." It was less of an imperative. More – leaving more control in the hands of the person on whom that diagnosis had been done.
5. A.S.: Scheduling, then; the duration of time between now and the next diagnosis; as a terminal day: two or three weeks. She's not pushing her to do it right away. It's not like a doctor saying, "OK, would you please go down and get an x-ray right now? It's in the same building."
6. M.: It may be a difference in the way it's told to her, because they're friends.
7. F.: Right. Like I had a friend with cancer of the prostate and another friend who was a radiologist. And the friend with the cancer of the prostate – who was young – wanted to know what his chances were. And what was involved in the radiology. So, I called the radiologist and an oncologist and talked to them. Well,

the radiologist was a good friend and he said to me: "What difference does it make? If it's positive and he's got cancer of the prostate and it has metastasized, what are you going to gain by giving him all this information right now, as his friend? Let him have his weekend free. Don't give him all the bad, bad possibilities. Let him find this out from the doctor. Because, one way or another, it's going to happen, but why do you have to be the one?" It seems like there's a similarity in letting down gently or have it taken care of, but not kill him off right away.

8. A.S.: Yes, but what are you talking about? In the first place, who gives the announcement? The doctor? The nurse? But there's also the question of the timing of it. The doctor knows, for example, you have cancer – right away. But he'll wait a week to tell you; and he'll feed it to you gradually before the real announcement. He wants to put it in the general way, first. But notice what I did: I took your phrasing and I split it in two.

Phase 13

1. I want to stop here a minute and give you a comparative example, and give you an analysis along similar lines here. I go to New York with my wife. (I'm going to narrate the story just to give you the general idea.) My sister-in-law is having angina. She goes to the doctor, who says to her: "There are changes in your EKG. You might have to have a bypass. Maybe. Someday. No urgency." He sets up the next diagnostic test, it's a stress EKG test, for three weeks later. Here, we're seeing her only one week later. And I say, "That's ridiculous. Speed it up. It makes no sense. I don't care how crowded the diagnostic lab is." She drags her feet. Two days later she has a lot of angina. She calls her doctor. The doctor gives another EKG. It's bad. He sends her to a cardiologist, who orders a stress test the next day. The day after that, they do an angiogram. The angiogram says you have to have a bypass or live as a cardiac cripple at home. She has the bypass. All right: You hear the same language about a different disease, different kinds of pacing, and so on. Now do a quick analysis of that, in terms of the concepts that we have developed. Let Adele try it. Some of it. Just do it crudely.

2. A.C.: Well, we have the diagnostic career and the patient career running alongside one another. In terms of the diagnostic career there is a moderate revelation. You know, that yes, there's something wrong, but don't worry about it, and you'll have this other test coming up along the line. And then in terms of the patient career, there is an intervention or a change in her career with your arrival, provoking a reanalysis on the patient's part of her situation. In the return to the doctor and more decisive bad result . . .

3. A.S.: Speedy, rapid diagnostic work. Sure. Repacing of it.
4. F.: . . . narrowing down your diagnosis?
5. A.S.: But you're also getting additional information from increasingly sophisticated techniques.
6. F.: And you're also educating your patients.
7. A.S.: But also a key turning point is that the symptoms are also increasing. They're coming more quickly, every four or five days.
8. M.: In fact, the only language that doctors listen to is "symptoms."

Phase 14

1. A.S.: What about the "fitting" of the information to the patient? What does that look like?
2. F.: It seems to vary with the situation. I mean, if they think it's urgent, they'll lay it on you – heavily and quickly. But if they can avoid it, they won't. I mean that's how it strikes me from that story.
3. A.S.: The announcement about the real surgical bypass was very direct, in fact, graphic. They brought her diagrams, immediately after the angiogram, and started talking about what the options were. They were very clear. So, you know, talking about the properties of that particular announcement: were made under what circumstances – slow, muted, direct, etc., etc.? Anyhow: You can see how different this is. But one thing, it's up to the doctors to treat it. There isn't much you can do except go through with the operation.
4. F.: And then the whole thing also is a career in loss of self-control of the patient. I mean, at first when you go in you have a decision to make. And you can't. Increasingly you are overwhelmed by technology and tests and diagrams.
5. A.S.: Well, you have decisions right then, up to the moment they knock you out and put you into the operating room. Up to that point, you can opt to die, you can opt for medication instead, or you can opt to put it off for three weeks. There are lots of decisions you can still make. Ordinarily you don't. But until the point at which you make the decision and turn it over to the doctor, a lot of it is in your hands, unless you're weak or "out of it."
6. F.: I think there are junctures where, yes, you can make a decision to do *x*, *y*, or *z*, and even counter the doctor's recommendation, but there's something about the frame of that decision, and that the framing of it gets narrower or constrains the patient's experiential vision of alternatives.
7. A.S.: Well, let's give it a word, like *medical funnel* or something like that. I shouldn't give you these terms, you should make them up yourself. But you see how they're coined. I mean, essentially, the narrowing control – or whatever term you want to use. It doesn't

make a difference. Later on, if you don't like the word, you can change it. You've got to coin the words in order to cope and pin down exactly what you mean . . . . But you can see what the properties of the funnel are. How fast? Is it reversible? Is there really a funnel like that? How quickly? Etc., etc. How much of a funnel? Etc., etc.

Phase 15

1. A.S.: Do you see what I did? Having got as far as we did, I simply opened up another comparative example. To see whether the categories made any sense. But also, it now begins to tell you some of the differences. So that we can be even more aware of the niceties and some of the unique features of this particular diagnostic process. Maybe they are not totally unique, but certainly patterned in ways that are different than a lot of other diseases. And if A. wants to do it, she can just try it out a little bit on other kinds of diagnoses.

Phase 16

1. A.S.: I want to move quickly, now: that business of this lady telling her about the lump in the uterus. This is simply another example of announcement. Of another diagnostic sign. It might be cancer. You get the results of the Pap smear and they won't tell you what to do with it. It's simply a rather nasty way of saying what the next step to the diagnostic series will be. And again, the duration. The duration would be the diagnostic wait.
2. F.: How about the comparison of the lump and the pimple? I mean, she's doing something there, that kind of contrast.
3. A.S.: Where is that?
4. F.: About, "U.C. is extremely professional. I mean she might have been telling me I have a little pimple, and then she says I have a little lump in my uterus. A little pimple on your nose is the small end of things and the lump in the uterus is the big end of these things. And yet she's treating me like this great big thing is just really a little thing." I don't know how you would generalize this, but it seems to me that that contrast is something that you would want to think about.
5. A.S.: Let's use Hughes's distinction between some people's crises that are only routines to others. Maybe she's really not announcing this as a sign of real cancer. Maybe she's just saying, you've got a lump so we have to look at it. Possible. In a real interaction, that's what someone could be in a position to do: Find out.

Otherwise, *you* don't really know what the interaction is about. You only know the way the patient interpreted it.

6. M.: She sees it as a lack of concern on the part of her doctor.

Phase 17

1. A.S.: I want to move quickly because time is elapsing, and I think A. could get much of this by herself. The first half of the interview can be covered by the kinds of concepts we have now.

2. It isn't until the middle of the interview that you get anything new. And that has to do with prices. The properties of regimens. This particular one is not exhaustingly expensive, but it's pretty expensive. So you can play with the properties of that one, and the patient's responses. Variations of it – if it's $1,000, it's all taken care of by insurance.

Phase 18

1. A.S.: OK. Go to the bottom of the paragraph.

2. F.: The last paragraph? It seems to me there is a lack of fit between the patient career and the clinic, career of the clinic, pacing context in terms of the people rotating the doctors through.

3. A.S.: There are some funny things about the diagnostic process, the diagnostic career. Sometimes you see the same people, sometimes you don't see the same people, depending on circumstances.

4. F.: And for me that all seems to be, all part of pulling the patient apart – whether it's deliberate or not – you are then focusing on that clinical problem and what is presented to them.

5. A.S.: Well, if you do a study of the medical system at U.C., you'll probably discover that there wasn't anything deliberate about it; it's just that there is no system, in that different people do the different tests, do them in different parts of the hospital. So you're dealing with the structure of the test making.

6. F.: Yes, I think so. But there's also that the patient isn't really and truly as important. Like, I know a little leukemia patient, and he's had leukemia since he was two and he is now ten. And they want to do tests on him. The concern is not so much the patient as what they can learn, what they can gather together from his experience. And I wonder if these people rotating through, and so forth, if the patient role is different.

7. M.: You might have the same person all the time, you know, rather than different people doing the tests.

8. A.S.: Now look: What's going on here is that she's taking additional kinds of tests, right? So, structurally, in terms of the hospital,

sometimes the same people are going to do these different tests. For example: The same people who do an x-ray also do a scan. And sometimes, it's done in the same department, and sometimes it's not. Depending on the hospital. If they are going to do an echocardiogram, that's going to be in a different part of the hospital, different department. So, it looks like she's hitting different people because of the institutional structure. Gayle's point is well taken, in the general sense, that if you want to ask what the staff attitude and handling of this test is: That they're really concerned about this patient with leukemia, that's one thing; but just doing tests for the sake of the research effort – then the attraction is likely to be something different.

9. F.: But I think what she's saying, too, is that there is no communication between these people who are doing the different tests. So they're all, you know, atoms out in space with nothing keeping them together. Whereas . . . .

10. A.S.: That gets you into analysis of the structure of the hospital and the structure of test taking. What happens, if you want a picture of that, is that here's the doctor in the middle and there are bits of information funneling back to him or her. So you get a division of labor which is pointed to him or her, rather than their talking to each other. OK? There is no communication, unless there was an x-ray or a CAT scanner in the same department. But, under routine testing, they wouldn't bother talking to each other. So what you're talking about is a division of labor in the diagnostic process.

11. F.: But they could talk to each other if they made an effort. And I think that might make her feel better.

12. F.: The hospital assumes that communication occurs through the patient's records. That's the supposed vehicle.

13. A.S.: Look: If here you want to get into the structure of the hospital, the way it is done, it looks like this: Here's your patient. She's on the department's escalator. People coming through, in the course of the day, she's number 691. That's the way they're set up, to do x-rays. Now, what's the likelihood of their telling anybody about this unless it's exciting to them?

14. The analytic point is that the information is going this way and that way. And there are days in between. What's the likelihood of communication when their talk is this way, and not this way? It's only the person at the receiving end who gets the information.

15. M.: Is that the same person whom she can call her doctor: Is there a continuity of the doctor as well as the patient – or is it, you know, she goes to the clinic and she may see the person she saw before, or may not. And they may have the tests from previous times, and they may not.

16. A.S.: Now we're back to the announcement. If you have a doctor: Let's say, you go to a doctor because you have something wrong

with your heart. Then he sends you out for these tests. The angiogram guy doesn't talk to the stress guy. The stress guy doesn't talk to the other guy. These test results all will be funneled back to the cardiologist, who tells her. So, this has to do with the announcement. And how it is done. And the degree of severity, and the content, and so on, of the announcement.

17. F.: There might be a change between what goes on in the diagnostic process and the dividing up of the people and no communication, and what happens when she finally gets a definitive diagnosis and treatment. Frequently, you will find that, OK, you have a connection with all these people afterwards. You do get a network going. You do know the people in x-ray and the people who do your CAT scans and the people up in the unit, and so on. Even if they don't communicate with themselves, or with one another, you have a network going. I don't know whether that shows up.

18. A.C.: Not in the Dysplasia Clinic, if it is still organized the way it was then. Those people do a two- or three-month residency "training," or whatever they're doing.

19. F.: OK, that's the diagnostic stage. But when she gets her diagnosis and something happens, and she's treated – is it different then?

20. A.C.: Oh. There are supposedly two head doctors, one of those left during the process . . . . Later in the interview – I mean, it seems to be an extreme end of that continuum of continuity, where there's very little communication.

Analytic commentary

In the pages that follow, the class discussion reproduced above will be divided into sequential phases of development. Commentary is addressed to what is going on – in terms of analysis – during each phase. Points will be numbered 1, 2, etc., while the associated paragraphs or lines of text will be numbered within parentheses; (1), (2–3), (4–10), etc.

Phase 1

1. (1) Injunction to scan interview for five minutes.
2. (2) A rule of thumb: When you have experiential data at your command, in your head, don't ignore them. Use them. Here use them, via comparisons of different diseases, to sensitize yourself to features of this particular disease.
   (3) An example.
3. (4) First lines plus instructor's knowledge of cancer allow him to suggest dimensions of the disease (visibility, treatability, etc.).
   (5) Line in the interview about diagnosis raises questions about another dimension (certainty) of diagnosis.

(6) Student gives relevant information.
(7) Discussion of regimen duration, and statement that one can raise questions about a dimension like this even before reading the first interview (because of experiential data). Anyhow, the questions will be raised by reading the very first interview (example: re cancer).
(8) Brief comparison with blood pressure: invisible symptoms and consequent surprise after diagnostic announcement.

Phase 2

1. (1) Question: What do you want to get from the discussion?
   (4–6) Student's response.

Phase 3

1. (1) Decision on lines of interview and directive to seminar.
2. (2–11) Focus on break in routine aspects of life expectation.
3. (3) Coining a term for category: routine scheduled diagnosis. This raises questions about other diagnoses for comparison.
   (13–16) Examples.
   (17) Cardiac comparison raises questions about routine, scheduled diagnostic check with or without disease.
4. (21) Routine versus nonroutine diagnostic check. Question: Why nonroutine check when there are no symptoms?
   (22–32) Answer: dimensions of risk; comparison with dental situation.
   (33–36) Risk probabilities.
5. (37) Student raises issue: routine checks and nonexpectation.
   (38) A.S.: a continuum.
   (40) Questions: the issues of variations?
6. (41) A.C.: Gives data.
   (42) Diagnostic durational span. Questions: amount of duration? comparisons with other cancer; what else?
   (43) Experiential data: bad interactional treatment during diagnosis session.
   (44) Wait! Later in the interview we shall see.

Phase 4

1. (1) Directive: Return to interview inspection.
2. (2) Normal–abnormal ambiguity.
   (3) Comparison: EKG, to bring out specificity of M.D. interpretation and diagnostic announcement, versus ambiguity and patient interpretation here.

(4) Difficulty of interpretation because of ambiguity of "numbers" announcement.
(6–10) Fact: What actually told her, and what would she know about "numbers" in general?
(11) Don't worry about the truth for her: The issue is variation.

Phase 5

1\. (1) Discussion of "interesting phase" . . . stage of illness.
(2) What then to do? shopping for other diagnoses.
(4) Directive: Code the "all the way" phrase.
2\. (5–16) Disease continuum (versus symptom continuum); discussion.
(17) Instructor steers class away from a potential digression.
(18–24) More discussion about disease continuum.
3\. (25) Issue of the physician's diagnostic announcement.

Phase 6

1\. (1–9) Questions and answers about information; consequences of "freaked out".
(10–12) In vivo code, and conditions.

Phase 7

1\. (1) Question: What are these lines about?
(2–7) Discussion about diagnostic career . . . variation . . . some conditions for.

Phase 8

1\. (1) An ambiguous announcement.
(2–11) Discussion of "weird"; what it might mean.
(12) Instructor relates it to disease continuum, trying to get discussion off a fruitless path.
(16–22) Diagnostic steps re diagnostic career.

Phase 9

1\. (1) Student's suggestion.
(2) Wait! Maybe that will show up later in the interview. But write a reminder memo.

Phase 10

1. (1) Question: What is this paragraph about?
   (2–3) Comparison concerning interaction
   (4–9) What is "interaction" all about?
   (10) Spelling that out via the interview lines; and again, a focusing question.
2. (13) An answer.
   (14–16) Diagnostic announcement . . . properties of announcement style; properties of the announcement.
   (17) Consequences of properties of announcer's style.

Phase 11

1. (1–5) Information about diagnostic step.

Phase 12

1. (1) What does that line mean? Convert it analytically.
   (3) Diagnostic directives and diagnostic career path.
   (4–5) Scheduling of the next step of diagnostic career.

Phase 13

1. (1) Comparison case; now analyze it.
   (2) Student: analysis re diagnosis and patient career relationship.

Phase 14

1. (1) Question about the announcement.
   (2) Answer: variation.
   (3) Properties of the announcement.
2. (4) The issue of control by patient . . . and decisions by patient.
   (5) Medical funnel; properties; an aside on provisional naming of categories.

Phase 15

1. Summary

Phase 16

1.	(1)	Issue of another diagnostic sign and announcement.
2.	(2–6)	Discussion of lump–pimple contrast re the interview data.

Phase 17

1.	(1)	Let us move along quickly because of today's time constraints.
	(2–3)	Data again . . . issue of costs, variations of that subject.

Phase 18

1.	(1)	Back to the interview.
	(2–8)	Staff's focus on clinical problem, not on patient; discontinuities in the diagnostic process.
2.	(9–13)	Student raises issue of staff noncommunication. Instructor gives experiential data.
	(14)	An analytic point.
	(15–22)	Discussion of discontinuity, again.

5 Memos and memo writing

In this chapter, a number of theoretical memos written by researchers during their various studies are reproduced. Before reading and studying them, it is requisite to at least scan the earlier discussion in Chapter 1 about memos and their indispensable functions in discovering, developing, and formulating a grounded theory. In the previous chapter on the student seminars as well in later chapters, one can frequently sense the hovering presence of memos which arise out of codes and ideas generated in seminar, consultation, and team sessions. In fact, one explicit rule of thumb is that such sessions must soon be followed by a jotting down or typing out of the summary or the thoughts stimulated, just as individual researchers need to interrupt their data collecting and coding to write memos. Furthermore, recollect that waiting for the muse to appear is not the model here. Although there are periods of intense memo writing, grounded theorists are trained to write memos regularly – often from the first days of a research project – and in close conjunction with the data collecting and coding. (See discussion of the triad, Chapter 1. See also Glaser 1978, pp. 83–92.)

The initial memos tend to look a little like those written by novices at this general style of memo writing: at first, a high proportion of them may be operational (what data to collect, where to go to do this), or reminder notes (don't forget to . . . , or don't forget this point), or scattered "bright ideas," or fumbling around with a flood of undifferentiated products of coding, or just thinking aloud on paper for purposes of stimulation in order to see where that thinking will lead, and so on. Later memos will incorporate the results of the (early, frequent and later, occasional) microcoding; focus on emerging major categories and their relationships with each other and the minor categories; struggle with whether to choose one or more core categories; integratively summarize previous memos and coding; suggest pinpointing bits of data to fill out last points in the analysis; and so on.[1]

These, then, are some of the varieties of memos, varying considerably by phase of research project and given additional variety by the personal styles of the researcher's thought, as well as by his or her experience with the phenomena under study and with the research itself; also, by whether the researcher is working alone or with a partner or teammates. (See also the presentation of the summarizing memo, Chapter 6.)

All of these points are easily observable in the illustrations given below. Each memo or set of memos will be introduced with a commentary which locates it in a context that will make it readily understandable – not necessarily in substantive detail, but in purpose and overall style. Note also in all memos how the data are drawn upon, are interwoven with, and inform the analytic content of each memo.

There is one further point about the memoing process. Even when a researcher is working alone on a project, he or she is engaged in continual internal dialogue – for that is, after all, what thinking is. When two or more researchers are working together, however, the dialogue is overt. In any event, the memos are an essential part of those dialogues, a running record of insights, hunches, hypotheses, discussions about the implications of codes, additional thoughts, whatnot. Cumulatively, the memos add up to and feed into the final integrative statements and the writing for publications. This kind of highly cooperative, even closely collaborative dialogue has also been emphasized by the American Pragmatists (especially Dewey and Peirce), whose thinking pervades the grounded theory approach to qualitative analysis. Of course, this working together, discussing continually together, does not at all preclude disagreement, sharp debate, even full-fledged argument. It does put a premium, however, on the ultimate faith in the working agreements to result in "payoffs" for all the partners. (This is true even when the partners are all in one researcher's head, as he or she works alone.)

[1] Researchers trained in other analytic traditions probably (though there is no reliable literature on the point) do what Becker and Geer (1960) suggested some years ago: fairly quickly after a study's initiation, formulate a few initial hypotheses, write them down in memo form; then they are verified, qualified, or discarded in the next phases of data collecting. Meanwhile, new ones evolve and are similarly worked on. Some of these memos may look like ours, although presumably some will not, especially in conceptual density and in drawing explicitly on intense microcoding.

Memos, memo types, and commentaries

Here is a series of memos written by researchers who will appear in the team-meeting session of Chapter 6. Each memo was addressed to all of the other staff members. The memos constitute different types, composed for different purposes, written early in a four-year project about the impact of medical technology on work in hospitals. They represent useful items in the total memo file, each helping in the final systemization of the analysis.

Memo type

This is an example of an initial, orienting memo.

Intent

1. *Produced during the first week of the project by its director, on the basis of many months of exploratory interviewing and field observations; his main intent was to give his staff a sense of the overall scope of the project, while pinpointing various areas to be looked into.*
2. *To raise questions and issues for the staff to think about and collect data on.*
3. *And, not incidentally, to summarize for himself what he knew or could foresee as potentially important for the evolving study.*

Comment

Eventually all of the outlined areas were looked into, in depth, and proved very relevant to the final analysis and writing up of the research.

This kind of introductory memo is written only during the first phase or phases of a project, and can be then thought about by the research team (two or more members). Of course, if a researcher is working quite alone, an initial, orienting memo will still be useful. Some of its contents may get overlooked or ignored in the excitement or evolution of the project, but other items usually will prove to be invaluable.

9/16/77 – A.S.

The most general memo

Something of the range of *areas to be looked into,* other than what happens on the ward floors. (Other memos re the wards themselves will be written.) And a few comments and guesses strewn in.

Scientific medicine, its ideology, and its technological thrust. Ideology of machine use.

Chronic illness and the halfway technologies to handle it.

The range and variety of machinery, and its utilization, along with other technologies (drug, surgical, procedural, etc.).

Is machinery largely used with prenatal, elderly, chronic illness? Check out the geography of machine location in the hospitals; then, by number and cost.

Structure of the machine industry and its market. Lots of questions here. What companies, how many, what kind? What trends? Marketing, sales?

Innovation processes: Who, how, when, to whom, etc.? Specialized versus generalized (mostly specialized)? Competitive, monopolistic, etc.? Maintaining position; breaking the market, etc.?

Varieties of personages involved with machinery (besides company people). Hospital administrators; M.D.s, nurses, technicians, engineers, patients, families. Also, bioethics people, sometimes newspaper journalists, etc.

M.D.s: as inventors, purchasers, users.

Nurses: as machine tenders and managers. Learning and teaching issues. What is the role of school, if any? Women and machinery issues? Job mobility issues?

Technicians: re learning, teaching, using. Especially relations with nurses whose wards they work on, or who come to their units, or whose patients travel between? Licensing, certification, professionalization?

Hospital administrator perspectives and actions. What are they juggling? Cost, departmental pressures, etc.? How decisions and allocations are made? Relationships with the machine companies? Issues of restructuring hospital spaces, costs, obsolescences of machinery, "keeping up" status re other hospitals. Same for administrators of large machine output wards (x-ray). Centrality vs. decentrality issue: rhetoric and decisions.

Interdepartmental relationships? Borrowing machinery. Fighting for scarce resources. Patient traffic between, etc.

Funding issues (see hospital administrators): Who, how, juggling, negotiating and other processes?

Government considerations: codes, limitations on who can have what machines, safety, cost. Also, insurance companies' relations to this?

Cost–benefit calculus.

Machines in relation to other machines. To procedures.

Patient on the machine. But, also patients as part-time workers on the machines. (See memos on this.)

Bioethical issues: These include – dehumanization, prolonging life. Saving the damaged (including gene pool, retarded, injured). Questions of equity (dialysis choices). Cost–benefit: saving the elderly vs. cost to the young. And lots of others.

Among the relevant general issues are the following:

Expert vs. the layman.

All our reliance on technology (progress) vs. human consequences.

Questioning of the technological escalator – Where is it leading us, etc. (and medical science ideology as a subvariety of this).

Among the sociological issues:

Body handling: machinery, procedurally, drugs, spatially, temporarily, etc., etc. And patient's responses to that handling re identity: viz., dehumanization, humiliation, etc., etc.

Task analysis: This involves not only machine tasks, but procedural ones, managerial, policy–political, division of labor, etc. Issues here are not only notational and relational (for us) but the important processes in relation to those tasks.

Memo type

A preliminary memo, done a month later. The researcher is beginning to lay out, here, bits of analysis around possibly important categories.

Intent

To focus one's own and teammates' attention on these items, thus to stimulate further analysis and data collection along these lines.

Commentary

The memo functioned well, so that eventually much of this and its later follow-through were incorporated into both a monograph on clinical safety and a chapter (on safety work) in another monograph concerned with the social organization of medical work (Strauss, et al. 1985; Fagerhaugh, et al., 1987; see also Chapter 6, "Summary Memo").

This type of memo is likely to be written repeatedly during the early phases of a research project, also each time that the researcher embarks on examining new facets of the project's terrain. Sometimes, as here, the memo can be quite extensive and conceptually detailed.

10/13/77 – A.S.

Danger (a preliminary memo)

The danger, usually thought of in terms of the patient himself, can come from *five sources*:

1. the machine, including parts, like drugs, used within it;
2. connection (hookup) between machine–body;
3. "patient" as body systems;
4. patient as person (moving, willing, refusing, etc.);
5. other therapies combined with or supplementary to the machine.

Signs, of *forthcoming* or *immediate* danger
Signs have various properties:

- visible–invisible;
- expectable–nonexpectable;
- . . .
- etc.?

Signs are related to "state" or condition of machine-connection, patient, person.

That is, reader reads signs in terms also of the state; also *stage* of either treatment itself totally or today's treatment (first hour, second hour, etc.).

Conditions for "correct" reading include, at least, skill, experience, spatial proximity to the sign, physical conditions like light that make the sign visible, etc. Negative conditions are the reverse of that, plus conditions that "distract" or "take attention away," like work elsewhere, too much work, tired worker, etc. (Reverse those again for positive conditions.)

But, correct reading is not so much the point for us (see below). DANGEROUS TO WHOM? (It is necessary to distinguish these carefully):

the machine itself;
its connection;
the patient's condition (can be single or multiple dangers, of course);
the patient as person;
others (worker, other patients);
the environment.

Also, sentimental order and work order (at any scale). Also, dangerous to what part of body, machine, etc.?

Differences of definition

Among: staff, patient–staff, family–others, etc.
Dimensions of difference (agreement) include, at least:

expectation of danger (0 to 100)
awareness of danger (0 to 100)

Locating chart, to show each interactant vis-à-vis other on each dimension.
Recognition of Agreement or Discrepancy: That is, what each is aware of about others' definitions of awareness of danger, expectation of danger. (That's *a most important point for us*, because of conflicting or cooperative action.)

Prevention of danger (i.e., how to lower the risk)

1. WHO is to do the work? (How many people; together; sequentially, etc.?)
2. HOW is it to be done? (i.e., *what* is to be done and *how*?)
3. WHAT is needed to get it done (resources)? People–money–space–skill–materials–time, etc., etc.?

S.'s footnotes and memos on dialysis bring out all these issues very clearly, especially the patient as worker, his work, and the needed resources.

Conditions which mitigate against prevention (i.e., raise the risk of danger in general)

Wrong whos, hows, and whats; few requisite resources.
CONDITIONS PRO: include opposite of that, plus "motivation" to have right people, resources, means, etc.
ORGANIZATIONAL CONDITIONS, of course, are central. And we should *CHECK THEM OUT SYSTEMATICALLY*; including industry-to-ward linkages.
The breakdown, emergencies, bungling, etc. will especially bring out:

1. precise nature of the *necessary tasks*, and
2. *requisite organization.*

And not incidentally:

3. what actors are *taking for granted*, which can't always be taken for granted.

Patient himself as source of danger

I've not emphasized this above; just a word about it. The patient, unless insentient, is supposed to do things: posture himself, lie still, move around, keep tabs on (monitor) his reactions on machine, etc. If he does these, he keeps risk down. If he doesn't, he raises risk, sometimes terrifically. Conditions Pro include . . . (skill, experience, "motivation", etc.). Conditions Negative . . . .

Degree of danger

Forgot to note that it can be from 0–100 (total destruction of whatever is in danger: patient, kidney, machine, etc.).

Degree of malfunctioning

This can be total, or partial, as in machine or connection – which relates to degree of danger, of course.

Balancing danger vs . . . .

That is, risk is always a possibility, but you balance degree of it vs. considerations: *cost, time, energy, risk* to other patients, *sentimental order*, risk to *machine*, etc., etc.

That is, you are balancing, in some part, risks to various objects: machine vs. patient's functioning; functioning vs. person; functioning vs. staff work, etc., etc.

Alarm systems

The purpose is either to reduce risk, to reduce work; therefore changes the nature of the work (which we will CHECK OUT).

(I will write a separate memo on alarm and fail-safe.)

Memo type

The next memo, then, is focused on the very visible and striking ringing of alarm systems on machines.

Intent

To raise a series of questions about "alarms" and to think aloud about both the phenomenon and the questions, themselves.

Comment

This was a very useful memo which fed into further data gathering and their analysis.

10/13/77 – A.S.

Alarm systems, fail-safe (and danger – see danger memo also)

Purpose of alarm system is to reduce risk probability; to reduce work; or both. That changes nature of staff work (which we should CHECK OUT IN DETAIL).

As we shall see, under certain conditions, alarm system may even *increase* probability of risk, and certainly probability of increased work. That is, alarm system may work in reverse! But, ideally not.

Alarm objects

Again, the alarm may be *monitoring various objects*: the *machine* itself, *the connection, the patient*. (Rarely, if ever, the person *qua* person? The staff is supposed to do that!) Perhaps, *the environment*. (And certainly, not the work or sentimental orders, etc.)

I don't know, yet, but assume the monitoring alarm can be set to go off at various *degrees of danger* or hazard. Is this *automatic* or can it be decided by someone? *By whom*? How far in advance? How often? At what cost, etc.? (Those may not be salient questions – we shall see.)

(Certainly they vary when it's a person rather than an actual alarm that is monitoring.) But, I visualize that *alarm adjusting* will be affected by anticipated degree of danger, margin-of-error calculations, etc.

Which one?
The person who responds to the alarm has to decide which one of the systems being monitored (if they are multiple) has gone off. Is it the machine, the patient, the connection?

Which part of it?
May also have to decide which part of the machine, which body systems, etc., unless there is a *fine degree discrimination of alarm itself*: That is, multiple discrimination built into the system. "Something is wrong with the machine" is not the same as "It's in the electricity" or "a bearing has burned out."

Priorities
Given multiple alarms, worker has to make decisions as to which one has to be *corrected first.*

Or, judgments as to degree of danger, so it has to be done right now, or an hour later, etc.

The patient alarmed
That is, the patient *qua* person can be alarmed. This may take priority, actually, since then his fright may shoot up endocrines or blood pressures or mobility: so, he has to be taken care of first. Or, he may not. Or, he may not be read correctly.

Multiple alarms: increased safety, increased danger
Some machines have just one alarm, or did when state of the art was simpler. Many machines have multiple alarms. One fascinating condition today is that there are machines now that are multiple machines: So there are *multiple-multiple* alarms.

This may make for *increased safety*. If they work. If people can read them accurately. If right priority choices are made, etc. (That depends on skill, experience, physical availability, etc., etc.)

Increased danger is another possibility, just because "things are now so complicated."

This ties in with the replaceability–maintenance issue (see below).

Replaceability: maintenance as condition for danger, safety
For alarms to work correctly, they must be maintained. (They may also not work right as with false alarms – because responding not to true signal but false one, like patient mobility. That's another issue.)

So, one condition for proper alarm is proper maintenance – an *organizational* matter. But also a staff-preventative one, since they have responsibility for either forecasting breakdown or recognizing it short of breakdown – not alarm itself, necessarily, but machine or patient connection. (That is organizational, too, of course.)

Maintenance ties in with replacement, since you have to have both a *maintenance organization* AND a *replacement organization.* The best skilled mechanics, best motivated, lots of time, without replacement parts or resources, in time will fail. (Underdeveloped countries lack both, but even with good skilled mechanics will fail because of replacement problem.)

On the *ward,* the replacement issues involve other things. They have to link up with maintenance–replacement organization, or else!! Or they have to be able to do the maintenance–replacement themselves (that is, be that organization, at least in part).

They have to make decisions about what may be wrong, and replace fast (innovative connection, etc.). Or decide whether replacement needed, or adjustment, or was it the patient's bodily movement, etc., etc.

Replacement can be fast–slow, available–not available, etc.

Replacement, in emergency or even temporarily, can mean not replacing a part, but replacing the whole machine. Which means replacement–organization, again: Can you borrow from within ward, or interward?

We need much more thought about replacement–maintenance in relation to alarms, danger, etc.

Failsafe

These alarms can't be foolsafe, failsafe, because signs may be misread, not seen, wrong priorities chosen, maintenance–replacement organization may be defective.

And it is S.'s insight that the more complicated the systems of alarm become – the more functions they are monitoring – the more hazardous the situation. That is, the less foolproof, more failproof they may become. But I have tried to spell out some of the conditions for maximizing–minimizing fool- and failsafe!

Memo type

A brief memo, "sparked" by a previous memo.

Intent

1. *To elaborate aspects of one category (machine storage).*
2. *To raise specific questions about conditions, consequences, process (dispute regulation), etc.*

Comment

A useful memo, later elaborated much further by further fieldwork, coding, and analytic memos. Eventually, the analysis found its way into a chapter on machine work in the monograph on medical work.

This memo-sparking type of memo can be written during any phase of a research project. Why? Because readers of any memo can be stimulated by its whenever they happen to read or re-read it, and then can respond with a memo-sparking commentary. Indeed, it is wise periodically to review preceding memos for exactly that reason.

3/19/78 – B.S.

Memo sparks (as 2/28/78 on machine storage)

As we all know, it doesn't end with finding a place to store equipment – difficult as that may be in itself. Then you've got to be able to retrieve the darned stuff when you need it. If it's too hard to get at, may even forget it, and improvise. (I'm thinking of household storage – gadgets, etc.; special equipment, but I think it's reasonable to extend the idea into hospital.)

Retrieval: what kinds of problems?

How is it decided what gets stored nearby, what probably won't be used much? And, here, what – even if it isn't used much – is absolutely necessary when it IS needed, as compared with less urgently needed equipment, say: the "would-be-nice-to-have," expeditious stuff as compared with essential stuff. So, is there some sort of storage protocol? And who has charge of it? Some sort of general-storage file clerk?

In the data from hospital (12/77) when patient vomited in nurse-call device and short-circuited it, one nurse knew there was a replacement in storage unit nearby, the other nurse did not. This could be a problem: If you don't know equipment is available, might as well not have it!

So, with nurses floating in and out, isn't possible for everyone to know how the household is arranged: where what is, even IF there is a what. (My kids come home and put dishes away for me, and I can't find ANYTHING.)

Also, when several units have access to same storage areas, seems likely there are going to be housekeeping disputes. Are there? How are they regulated? Who knows what?

Memo type

This memo, first in a series about comfort work, was written two years later by another team member, a sociologist, who is also a nurse. She had finally realized that so-called comfort care had been profoundly changed by contemporary medical technology. This memo represents the opening phase in the team's attack on the phenomenon.

Intent

To put down first thoughts on issues like: What is comfort work? How is it different in the hospital than at home? How has it been affected by medical technology? What is its relationship to other types of work?

Comment

This memo and succeeding ones became the basis for the directed observations and further analyses which fed into a monograph on medical work (Strauss, et al. 1985). That is to say, this memo illustrated thinking about selective coding, in this instance done in relation to the core category of types of work.

This type of memo can be written at the outset of an attack on a phenomenon not yet focused on, though it is much more likely to be written during earlier phases of a project. As the illustration reflects, however, it can be composed much later, when much of the analysis done on related phenomena will inform it.

2/20/80 – S.F.

Comfort work

Comfort work includes a wide range of medical and nursing work, but mainly involves nursing because much of nursing includes tasks relating to relieving discomforts. Comfort work may be very specific to very ambiguous and murky. Take the definition of comfort.

Definition of comfort(comfortable)–discomfort(uncomfortable). Definition of comfort–discomfort includes:

VERB: To impart hope to; give help to person in sorrow or pain; implies comfort, console, solace. Comfort, the homely intimate term implies imparting cheer, hope, strength, as well as, in some degree, the lessening of pain; console emphasizes the alleviation of grief or sense of loss; solace suggests a lifting of spirits that means relief of loneliness, dullness, etc., as well as pain.

NOUN, suggests: easy, restful, reposeful; implies enjoying or providing conditions that make for contentment or security; or cozy – suggests comfortableness derived from warmth, shelter, ease, and friendliness.

Discomfort – to distress the comfort of; make uneasy; mental or physical distress.

Pain–discomfort work

Comfort work might be seen as the less acute end of the pain continuum. Like pain, discomforts are highly subjective, so there are problems of assessing discomforts for the staff, and problems of legitimation for the patient. There are wide variations in discomfort toleration, meaning, expression, etc. from patient to patient. This idiosyncratic nature of discomforts is due, in part, to the fact that discomforts are tied up with biography as well as illness trajectory; but more later.

Like pain, there are discomfort tasks for the staff and patient as outlined in the pain book (p. 244). The tasks include: (1) assessing (diagnosing); (2) preventing; (3) minimizing; (4) inflicting; (5) enduring; (6) relieving; and (7) expressing. And these tasks are balanced for their consequences on (1) illness trajectory; (2) life and death; (3) carrying on; (4) interaction; (5) ward work; (6) sentimental order; and (7) personal identity.

Pain work is extremely difficult, but in many ways discomfort work may be more complex. The difficulties stem from many factors:

1. Discomfort or dis-ease includes a wide range of physical and psychological states, sensations, and moods. For example:

   Physical *discomfort sensations* may include itching, tingling, soreness, pressure and fullness, burning, coldness, hotness, stiffness, dirtiness.

   Discomfort physical symptoms may include dizziness, headache, flatulence, constipation, thirst, ringing of the ear, weakness, upset stomach, etc.

   Discomfort psychological states and moods may include "feeling blue," "out of it," and a whole set of feelings of insecurity and even anger by interactions which makes the person feel ignored, slighted, embarrassed, etc.

   In order words, discomfort and dis-ease states, sensations and moods can come from *many sources*:

   from the illness itself;

   treatments and procedures, drugs, etc., in the service of the illness trajectory;

social interactions;
environment (temperature, tidiness, etc.);
organization of hospital.

2. The most striking feature of these discomforts is their *mundaneness*. They are physiological and psychological states associated with everyday life and bodily activities. They are related to everyday sociability, eating, drinking, body posture and ambulation, defecation, urinating, and so on. Everybody has had many of these states and lots of individualized ways of managing these discomforts.
   a. The very mundaneness poses difficulties in management because it is visible and yet not visible; it is murky and very subtle. Yet, these everyday sensations, physical states, and moods are at the heart of oneself – are highly idiosyncratic and personalized; thus, when discomfort work is neglected, patients feel they are treated as non-persons and feel dehumanized. Indeed, patients' angry criticisms of hospitalization are an accumulation of unmet and unrelieved discomforts.
   b. The very subtle nature of comfort work, and its commonsensical quality, makes it difficult to distinguish this as work. Comfort work is being applied simultaneously and sequentially in any given area of work, sometimes bordering on sentimental work, sentimental gestures, biographical work. *I need help in thinking this through.*

Body work and comfort work

A large part of comfort work is body work. Categories of body work include:

tasks directly related to trajectory course which includes diagnostic and treatment procedures. They include body positioning, body movement (gross and fine, e.g., transporting or moving body part); and doing things to the body such as injecting needles and drugs, putting down tubes into various body orifices. Properties of this body work are it's variable, painful, embarrassing, requires lots of skill by staff, dangerous, requires lots of patient cooperation, etc.

tasks related to bodily sustenance such as feeding, drinking, ambulating, hair care, mouth care, defecating, etc. They include a whole host of tasks which are *everyday body housekeeping tasks*. Depending upon the illness and the trajectory phase, the properties of this work are variable. How patient reacts to neglect varies with the biography.

tasks related to psychological well being which are essentially sentimental work, but may also include biographical work.

Technology and comfort work

In the past two decades, comfort work has drastically changed. The changes are due (1) the complexity of hospital organization due to the overall technological changes; (2) the technologizing of comfort care.

As hospitals get bigger and more complex, comfort *care services* become more complex. Comfort care services include laundry, dietary department, central supply, cleaning services, etc. A tremendous amount of coordination is necessary to get enough linen, clean the rooms, etc., etc. In other words, there is a whole line of departments and complex task structures required to get goods in order to do the comfort work. Results in lots of delays. Also, many people are involved and the work tends to get routinized, such as time to pass out drinking water, time to pass out dinner trays. Hence, individualized care gets to be very difficult. *Scheduling of comfort work* gets all gummed up.

Since the hospital is committed to diagnosis and treatment, and there are now many more diagnostic tasks and treatments, so that patients' time is spent more and more in these tasks, comfort care gets lower priority.

Nurses' time is spent more and more on diagnostic and treatment tasks, so that comfort work, which may be seen as servile work, gets handed down to aides and orderlies. So, *dirty work* is part of issue in comfort work.

Because there are more treatments and procedures done on the patient, there is an increase in *inflicted discomfort.* Tubes stuck in every conceivable orifice; patients having to be still to avoid dislodging a tube; irritations of tubes, etc. In fact, ICU patients are a mass of discomforts, but monitoring and saving of life are the high priorities.

Technologizing of comfort work

Comfort work, such as body positioning, back rubs, sponge baths to lower fevers and decrease discomfort are all being technologized. Beds are electric, so patients can lower or raise the bed. If there is a potential for bed sores because of inability to move, there are air-circulating mattresses, or gadgets such as sheepskin, cooling mattresses to lower fever, etc. There is a whole array of gadgetry of various kinds. In the old days, nurses used to invent all kinds of stuff for comforts; now this is all commercialized: "ouchless" tapes, "comfortable" restraining belts to tie patients to wheelchairs, etc. Flipping through the nursing journal, one finds ad after ad on gadgets to make comfort work easier and more efficient. Also, there are special sections devoted to new technology for comfort care and "creative nursing care" which all have to do with gadgeting comfort care.

Ideologies related to dependence–independence, that is, a drive to make the patient independent, pushed patients to "do for themselves" more and more. Nurses get furious if patients persist in wanting to be "waited on." So, there are arguments about how much, when, how, of comfort work.

Comfort work can also have therapeutic implications and there is a whole body of knowledge and skills – and art – to properly position a patient, etc., in order that physiological function not be compromised. But this work seems so commonsensical or artless that patients don't see this as a technical matter. The nurse is just "being nice."

Increased number of options in comfort work

Drugs play a large part in comfort work. There are all kinds of drugs for relief of itching, flatulence, constipation, and so on. The array of drugs is immense. Take constipation for example: in the old days, there used to be enemas and a few laxatives. Now there are packaged enema sets, stool softeners, suppositories, laxatives with different chemical reactions. Nurses have to know a lot about what kinds of enemas not to give in certain kinds of illness conditions, forcing fluid intake, etc.

TOUCH THERAPY is a movement against lack of comfort care, but in order to have legitimacy must have a theoretical base and a technology. A large part of holistic health and self-help groups is related to body comfort.

The next memo was written by Juliet Corbin, one of two researchers studying interaction between chronically ill persons and their spouses. It was written for her co-researcher and herself.

Memo type

Announcement of a new category

Intent

1. *Early in project or when new sample populations (in this instance, paraplegics and quadraplegics) are studied, new categories are discovered: So this memo is tendered to announce and discuss that category.*
2. *And to distinguish it from another category (attendant work versus the more general "wife" work).*
3. *And – again as is typical with new categories – to raise a series of initial questions about these associated categories.*

Comment

This memo precipitated further discussion and memoing about "attendant work" and its relation to other categories, as well as to a detailing of sets of consequences flowing from this type of work.

J.C. 7/2/82

Wife work vs. attendant work (husband vs. patient)

Jumping out from this interview is a concept of work that never hit me before. It has to do with wife work and that there are certain types of work that belong to a wife. These are different from those of a simple attendant. I haven't worked it through my head yet, but it seems from this interview that when the body work becomes the focus, with work an attendant could do, that identities become blurred. The wife is not sure where her identity as wife comes in and where it leaves off; where the attendant begins and leaves off. She also becomes confused as to where the identity of the husband comes in and leaves off and where his identity as patient comes in and leaves off. Conversely, the ill mate has the same problem keeping these four areas straight.

There are separate tasks involved in each and they can be done by separate people or by the same. How does one keep them separate? How does one integrate them? Can one? What are the separate tasks of each; how and when do they overlap? Can one successfully do both? If so, how and why? What are the consequences of each possible combination for each partner? It looks, from this case, that when the wife tries to do both, then the work of wife and attendant becomes blurred, confused, for both her and husband. What would normally be a division of labor becomes all mixed up. Not only is the body resource work the focus, but the mutual sustaining work is missing. The attendant gets paid, gets time off; but the wife doesn't even get a compliment.

She is taken for granted. Not working together, widening of the marital gap. Not a mutual give and take; reciprocity is missing. One can also see the movement in this case, the gradual blurring of identities, the crisis, and the couple trying to sort these identities out and to keep the marriage going.

Next are three memos written by the same researcher in quick succession, each pertaining to the "busywork" in which wives of chronically ill mates are sometimes forced to engage. After several months of interviewing, the researcher was struck by these activities, and began to organize her thoughts around a relevant, but minor category (minor in relation to the core categories already conceptualized). The memos are addressed to the co-researcher.

This kind of memo sequence of course rests on a certain amount of prior analysis, otherwise it would be much more like a set of cogitations written during an initial phase of the project.

5/82a

Memo type

An initial "discovery" rumination

Intent

1. *To call attention to a possibly relevant phenomenon (busywork) in relation to specific data from a recent interview.*
2. *To suggest the contribution it can make to "overload" (an important category, previously developed in the research).*

5/82b

Memo type

Additional thoughts on the new category – a memo note:

1. *While thinking about "strategies for getting the work done," busywork comes into focus again, and a memo note is written to relate to this category to three others.*
2. *And to distinguish this phenomenon from "other work."*

5/82c

Memo type

Memo distinguishing between two categories

Intent

1. *While adding to "strategies for getting the work done," the researcher thinks through some differences between two easily confusable types of work, and jots down her thoughts in order to distinguish between those types.*
2. *And to relate one type (busy work) to core and major categories (trajectory work, overload, overwork, error work). (See, for some details of this, Corbin and Strauss 1985.)*

Comment on all three memos

These memos constitute a series of brief analyses, which together begin to elaborate the complexities of the category, and to provide grounds for discussion between the two researchers.

5/82a – J.C.

Busywork

In talking with G., Wednesday, I was struck by the amount of *busywork* required to do the work. I mean all the running around, repeat calls, questions that need to be answered, negotiations, threats, etc., that go into getting a type of work, like financial work, done. For instance, G. found out from her tax man that the amount of money they will receive from Social Security is influenced by previous earnings. What she needed to find out was how the Soc. Sec. estimated this; combined incomes, single income, for how many years, etc. Who would know this? She had to ask at nurses' station, who then referred her to social worker, who referred her to someone else. Perhaps when she calls someone else, they will refer her to still someone else, until finally she gets an answer. Then she has to get this information back to the tax man so he can figure out income tax. So much time and energy are spent dwelling on these little details, or busywork.

I am not sure what this all means, but think that with all the other problems or work that must be done, it can contribute to *overload*. At the same time, she is trying to get the house remodeled to meet his needs, learn what she needs to know to take him home, run up here to see him from Watsonville a couple of times a week (visiting is sentimental work); her yard needs work now that spring is here. "How am I going to find time to do all of this?" Especially since she must travel East this month to buy furniture and other antiques for her antique business upon which they are now dependent for their income. Things compounding. She seems less relaxed, much more pressed for time than any time I had seen her previously.

5/82b – J.C.: *Strategies for getting the work done*

1. *Rearranging.* There are lots of things that have to be done related to management of the illness, all the trajectory work plus the home work. Integrating the two means that activities, routines, etc. will have to be rearranged. What gets done, when, why, by whom, how, with what consequences, are all relevant. Rearranging seems to relate primarily to *scheduling of work* and *setting priorities about work.*

2. *Making an arrangement.* When ill mate can't do the work because of limitations of body, energy, etc., and spouse can't because physically unable, lacks knowledge, is too busy, etc., then someone must be hired, or the services of friends are enlisted. This relates to the division of labor and the *distribution of workload.*

3. *Busywork.* Busywork has to do with *work flow.* The day-to-day activities that are involved in keeping the work going. If these little things are not done, then the work could not progress. *Work stoppage. Sentimental work* may involve driving, coming from there to here so that the ill mate can be visited and the sentimental work done. *Regimen work* involved a lot of busywork. Without the proper food to prepare a low-sodium diet, the regimen work gets interrupted. The busywork involves running around trying to find these foods. There can be a lot of intensity about busywork and the person doing it can become weighed down and eventually overloaded from the demands of it. *Busywork* is not the larger work, it differs from the big contingencies. It *involves all those extra things, those little tasks that are necessary for work to flow.* Demanding and time consuming.

5/82c – J.C. *Supplement to memo on strategies for getting the work done* (after a team discussion of the last memo)

Making arrangements and rearranging. All kinds of arrangements are made to get the work done. When the arrangements break down because of interactional problems, shifts in regimen, or illness phase of trajectory, life style, other contingencies coming in, then new arranging or *rearranging* has to be done.

Detail work and busywork. Detail work refers to all the little odds and ends one has to attend to in getting the work done properly, grinding it out day by day, that makes the difference between degree of success or failure of the work. Paying attention to the details prevents or minimizes incidence of errors or having to pay more taxes out than need be, etc. While some people may label detail work as junk, it is necessary and expected, a *routine aspect of trajectory work.* It can be very time consuming and tiresome.

Busywork, on the other hand, refers to added work resulting from the errors, forgetfulness, oversights of self and others: an added burden, "a pain in the ass." *It is unexpected.* For example, when an error is found on an insurance or hospital form, it is necessary to make phone calls, offer explanation, follow up to see that the error is corrected, etc. Not a normal, necessary part of the work.

The next memo was written by Elihu Gerson to S. Leigh Star as part of a study of the work of scientists. It represents a working through of some implications of the political nature of robustness *– a concept derived from the writings of philosopher of science William Wimsatt – which has, by now, been worked on successfully by the researchers in terms of their own materials. (Something is robust when you get the same answer back using different methods – for instance, experiment versus field observation.)*

Memo type

Extending the implications of a borrowed concept in terms of your own research finds

Intent

To raise further data collection and analytic issues about this important phenomenon.

Comment

Note how a literature-derived category can be utilized if *it fits one's data; also how this memo draws on others in the same and related research projects.*

This kind of memo illustrates well how related literature – indeed, a concept – can be drawn into and further direct one's research, once the central and some of the minor categories are firmly in place – providing the newcomer is genuinely integrated, and not just merely added to a conceptual pile or completely dominating all of one's own discovered categories.

1204/M167 1 June 1982 E. M. Gerson

Robustness of theories and the persistence of conceptual artifacts

It is a routine point in the philosophy of science that the strength of an idea comes from its simultaneous role in *many* theoretical contexts, not just one. Usually, this is phrased upside down and backwards (i.e., that an idea gains strength as it is "corroborated" in many different theoretical contexts; and that it becomes a "core" idea as it is picked up, used, and further corroborated in numerous different lines of research). As often as not, this notion is used to *explain away* the equally well-known fact that ideas are often *disconfirmed* by experimental results, and scientists still persist in retaining and using them despite the disconfirmation. This phenomenon is viewed as an anomaly, and the "support in multiple contexts" notion is used to explain it. . . .

If we look at theory construction and maintenance as a matter of work and work organization, things look somewhat different. The first point of course is that robustness is political (1212/M30, 30 April 1982; 1204/M163, 28 May 1982), and that an "idea" is a commitment to organize work in a certain way. Let's draw some of the implications of this.

Suppose we have some theoretical conception or relationship, which has been adopted in several different lines of work (presumably, these lines are closely related). For example: the speed of light is constant, or acquired characters are not inherited. There is evidence supporting the notion in each (or most) lines of work, and several lines of work use the notion as a taken-for-granted package, without being very much concerned with its justification (e.g., plant breeding and inheritance of acquired characters). Suppose, finally, that the notion is disconfirmed in one line of work, and claims are put forward that the idea is no good/needs revision/etc. What happens?

As we know from many examples (and the philosophy argument), the claim is often ignored/ridiculed/etc. Cf. Star's work on anomalies in neurophysiology used to impeach the localizationist position or Steele's current claims about IAC. In fact, this process is common enough to have become a significant problem in our work – how do we explain the *persistence of conceptual artifacts* in the face of disconfirmatory evidence, etc.? Why is something known to be no good (I'm extending the point now) held anyhow? Star has been talking about "inertia" in this sense.

So we want to look at *patterns of ignoring threats to robustness of theoretical ideas across multiple lines of work*. In practice, something like the following is happening:

1. Anomaly appears in a line of work; a well-established notion is attacked as inadequate/no good/etc.
2. Within that single line of work, a debate starts; the idea attacked may have a lot of support or a little; the data impeaching it may be good, not so good, etc. The attack on the theoretical notion makes some degree of headway. Suppose it convinces at least some people.
3. Within *neighboring* lines of work (sister specialties), three different things can happen:
   a. The attack gains support, and the weakness of the idea starts to spread. We aren't very concerned with this one here.
   b. Lines which use the idea in a packaged way are not likely to be very supportive; first, because they aren't familiar with the issues; second, because they have a vested stake in keeping the idea intact, because they have built *their* commitment structure on it, at least in part.
   c. Lines which are also concerned directly with the idea are likely to defend it if it works *in their context*. That is, if the idea works in *my* house, I'm not too likely to get excited if it appears to be flaky in *your* house, especially if I'm going to have to rip out my plumbing if it turns

out to be no good. Before that, I'm going to make very sure that the flakiness is real, not apparent.

So, a generalization: *neighboring lines of work act as points of resistance* to idea-impeaching claims when they've built the idea into their own work. In particular, the easiest thing to do with such a claim is to *ignore* it or ridicule it (as per *c*, above). As long as I can claim validity for the idea by *pointing to its robust character* (supported in all the *other* neighboring lines of work) I don't have to take the impeachment claims too seriously – after all, the notion is robust. Something like this seems to have been happening in the units-of-selection debate (Wimsatt), in which several different approaches converged on the same artifact, and fell into the trap of using the same bad heuristic assumptions in their work.

Rules of thumb for memoing

Here are some rules of thumb for memoing developed over the years in our research, as suggested by Barney Glaser (1978, pp. 81–91).

1. *Keep memos and data separate.* Thus, memos should not be written into recordings of fieldnotes, since when the fieldnotes become somewhat abstract the memo may appear like the conceptual perspective of an informant. Later, when reading the notes, the analyst may not be able to tell the difference. By the same token data should not be put in memos, with the exception of clearly demarcated, useful illustrations, referenced to the fieldnotes from where the illustration was taken. All memos should be referenced to the fieldnotes from where they emerged, so the analyst can check grounding and draw illustrations when needed.

   While the same incident can indicate two different concepts, it is advisable to use it as an illustration for only one, and find another indicator as an illustration for the other. With adequate references and illustrations the analyst can write straight from the memos, with occasional forays back to the data, with a sense of complete grounding.
2. *Always interrupt coding or data recording for writing a memo, when an idea occurs,* so the idea is not lost. If you cannot stop, jot down a short memo on what to write a memo on later. If another idea comes along when memoing on an idea, weave it into the memo, but also write a brief memo to do a memo on it later. Set aside a block of time for coding and memoing when you will not be disturbed. As always, it is best to bend to the dictates of one's own personal pacing recipe.
3. *The analyst can bring a memo, literally force it, by starting to write on a code.* Such writing is very likely to open up the output stage of creativity. Sometimes analysts need to press themselves this way to start a memo flow. Conversely, analysts should not be afraid to stop writing a memo, if it is not flowing. The code will occur again in the data if it is relevant. So whether to press or not is problematic, but the general rule is to press as little as possible, as memoing emerges easily enough.

4. *Do not be afraid to modify memos as your research develops.* It can usually lead to a better memo. Remember the data is more precious than the theory. The latter must fit the former. Memos allow this freedom.
5. *Keep a list of the emergent codes handy.* In the later stages of coding, when memoing is at a peak, refer to the list for possible relationships you have either missed or not thought of.
6. If too many memos on different codes seem the same, *compare* codes or their dimensions for differences that are being missed between the two codes. If they are still the same, collapse the two into one code.
7. *Problematic digressions should be followed through on a conceptual, elaborative basis for the purpose of theoretical sampling or to indicate an area for future research.* These digressions should be grounded and referenced as much as possible, as well as points indicated that are ungrounded, coming from hunch, inspiration, or insight. The memo should be quite clear on data vs. conjecture, because when returning to it later the analyst might forget and think it was grounded. It happens.
8. *Run the memos open as long as resources allow,* to develop the rich diversity that they can afford for doing various pieces out of them.
9. *When writing memos, talk conceptually about the substantive codes as they are theoretically coded*; do not talk about people. This maintains the conceptual level of analysis as relationships among concepts, and it gives the analyst practice for the final writing. People occur in the references as indicators, but the analysis is about conceptually generated patterns which people engage in, not about the people, per se.
10. *If you have two burning ideas, write the ideas up one at a time.* This will keep them clear, straight, and not lose either. To write them up together is confusing and hinders clear relations between the two.
11. Indicate in memos "*Saturation,*" when you think you have saturated the category.
12. *Always be flexible with memoing techniques.* Analyst's techniques should serve them, not hinder or enslave them. Each analyst's memoing has a personal recipe involved, and this is always emerging and forcing change of techniques. Follow those changes which are worthwhile.

Perhaps you can think of other rules for memoing. There are more, but the above give us enough to work with, while personal rules will emerge to supplement and change them.

Summary

The memo types reproduced in this chapter have been given names to bring out their central respective features. In order of their presentation they were:

initial, orienting memos
preliminary memos

memo sparks
memos that open attacks on new phenomena
memos on new categories
initial discovery memos
memos distinguishing between two or more categories
memos extending the implications of a borrowed concept

Of course, this does not exhaust the entire range of memo types, but it suggests something of *how* and *when* varieties of memos are written, as well as how they function in research projects. Other types include additional thoughts memos, taking off from previous memos. One may even code anew after rereading a previous memo and being stimulated to fill in gaps or to extend points made in that memo. Following that new coding, another memo is written. Another type of memo is the integrative memo, which will be discussed in Chapters 8 and 9. Another important type is the organizing, summary memo, presented at team meetings in order to prompt discussion, the meetings themselves constituting a form of theoretical memo. Such an organizing, summary memo and a portion of the discussion that followed its presentation are given in the next chapter.

One final point: Selective coding – coding in relation to core categories – was seen in the memo on comfort work, and will be seen again in the next chapter, in a summarizing memo written about clinical safety, as well as in the memo sequence reproduced in Chapter 9. In each instance, the core categories for this particular study (trajectory and types of work) are more in the nature of reporting on or sparking off of the results of open coding, because the core categories had not yet clearly emerged for the researchers.

6 Team meetings and graphic representations as memos

Team meetings as memos

There is a special kind of memo writing which can occur when two or more researchers are discussing either data or just ideas that pertain to joint research. In effect their exchange can result in coding (new categories discovered, relationships among categories discussed). Or, a number of generative questions are raised, hypotheses are suggested, comparisons are made and perhaps explored. This kind of discussion can even occur between a solo researcher and an understanding colleague, but usually it has more focus and thrust if it occurs repeatedly between or among research teammates. Thereafter, one of the participants often will write a memo based on notes or memories of the session. Sometimes the tape recorder yields a transcript, a kind of theoretical memo, albeit it tends to be a less concentrated form of memo than if the reporter had deleted excess phrasing, asides, and other irrelevancies.

The memo reproduced below has two parts, both taken from a transcript of a team meeting concerning the impact of medical technology on hospital work. Part 1 consists of approximately the first hour of the meeting, during which the principal investigator presented a *summary memo*. This dealt with main themes and categories touched on or developed in previous memos pertaining to safety, risk, and error – memos written during many months of data collection. This kind of summary presentation is a useful device for forcing an interim sorting of memos and achieving analytic order from that sorting. The typed summary could have been handed out before the actual meeting (sometimes they are) but in this instance the task of summarizing was finished only shortly before the meeting, so the summary was presented orally.

Part 2 consists only of the initial two topics discussed by the team after the summary had been given. There is enough material in these relatively few minutes of discussion to illustrate the kind of interaction that goes on in experienced teams, and to allow for commentary on that discussion. Both the summarizing memo and the entire discussion that followed made contributions to the eventual writing of a monograph on clinical safety in hospitals (Fagerhaugh et al., forthcoming. See also Appendix, and Strauss et al., 1985).

The summary memo

Several points are noteworthy about this type of presentational memo:

1. It is based on a scanning or quick re-reading of team materials (in this instance only sessions occurring between two researchers around safety and related topics). The speaker named the exact supporting items, giving their respective dates.
2. The analyst was putting all of this material into some kind of order – analytic order as well as order for his teammates.
3. He chose this task because, as the principal investigator, he sensed that now was the time to think more deeply about this important area. The team had agreed at the last meeting that he should do this summary.
4. He began the review with the largest (macro-)structural conditions for machine impact on patient care, because he wished the team to keep its collective eye on those conditions, since they might easily be overlooked, given the vivid events being witnessed or talked about daily on the wards. The presenter thereby was also keeping the team focused on connections between macro and micro levels of analysis.
5. He was drawing both on concepts developed during the project and on others that had emerged during his prior research, concepts such as arenas and trajectories. (Trajectory is a course of illness, along with all the work of people involved in trying to control that course.)
6. He had begun during his re-reading of materials to recognize the need for linking the phenomena of error and safety, and so was introducing that issue here.
7. He next moved to the issue of risk, and attendant issues like assessment and action.
8. He connected up safety issues with routine and problematic trajectories, the division of labor, and types of monitoring.

The presentation: safety, danger, and risk

(Anselm): Here are the memos that I've pulled. Interchanges with the sessions with Shiz – I looked at no coding and no fieldnotes. The three sets of sessions with Shiz, November 8, 9, and 27; there's a memo on monitoring February 1, 1979: there's a memo on risk, June 27: a memo on monitoring, February 1,

1979; (there's a long typed team meeting that we did on monitoring about three months ago that I didn't look at): another memo on risk written by Carolyn on the ICN, December 9: a memo on alarms which I didn't look at but which is relevant, November 13, 1977; an interesting fragment written by me off Aaron Smith's fieldnote on pediatric physician–patient–parent interaction: The memo is 8/24/78.

I. I'll do my best to lay out the areas. You break in at any point you want to. The place to begin, it seems to me, is away up in the stratosphere of macroconditions, but eventually we get down to the ward. So *I start with the idea of safety work, which is a particular kind of work.* The whole point of safety work is to minimize the consequences of potential or actual danger. And then there are questions: Consequences for what? And I take these straight out of the memos. To begin with, for the patient's trajectory. Second, for other individuals, mostly those people working around me. Third, for the whole ward. And fourth, very interestingly, the organization itself, in this case, the hospital. But in other cases, it could be the whole social world. So danger can exist at those four levels.

Of course, what we're primarily interested in are the organizational conditions which make for those consequences and what happens when you get them (those consequences) and try to prevent them, and all of that.

Now comes the interesting surprise – if you think on a macro level about the area of safety helping to minimize danger, you find, in fact, that there are very wide, broad safety areas. What I'm about to say comes directly out of the kind of thing that Carolyn did in her thesis: That you can do an arena analysis starting with the broadest possible scope and then coming all the way down to the ward. And in memos of conversations with Shiz, running all through, you get this feeling of fights over regulation, fights over what is safe and what is not safe, all kinds of social world representatives in all those fights. And, of course, they were fighting about a multitude of issues. It's not just one arena – there are many arenas. It can be over machinery, over certain kinds of nursing care, over the way hospitals are built, all kinds of things. And different worlds are involved in different arenas. With different kinds of strength, and persuasiveness, and power, and resources, and everything else. In any case, these are debates over safety, and of course, the nuclear stuff going on now is a beautiful example of it. So, they're fighting over issues, definitions, priorities, degrees of risk, you know – all of that.

Now, as a result of that debate, you get legislation having to do with safety issues, regulation, and so on. Then, of course, there have to be decisions as to who should do the enforcing and how, and when, with what resources. Mind you, at every step they're debating, negotiating, fighting it out. The next step is that even when they're enforcing, people are opposing it, breaking the rules, trying to change it.

Finally, in terms of the hospital, you get these rules and regulations bearing in on the hospital and so you get a hospital setting up its own rules and regulations in accordance with that legislation/regulation from the outside. Plus any adaptation they make to it. I'm going to have to say more about that later.

Shiz: It's interesting that in each of these, in legislation and enforcement, they bring in all different types, different social worlds.

Anselm: Different social worlds are likely to be involved at every level.

Shiz: Some of them are in the foreground, some of them in the background . . .

Anselm: That's correct. The additional point that I wanted to make is that somewhere in one of those sessions we use the term, *organizational sedimentation*. Which means that these outside rules and regulations and general decisions are built literally into the organization. And become part of it, and sometimes aren't even noticed.

Then, the next part, you'll find keys to the first Shiz–Ans session I gave you, p. 13. We notice that there are five structural conditions that affect safety work in hospitals. These are the largest structural conditions you can find, offhand. There may be more. First, the outside arenas are in turmoil. Secondly, the safety departments of all these areas tend to be weak, or nonexistent. Third, the technological change is very rapid. Fourth, the hospitals are decentralized places, as we know. Fifth, and especially as a consequence of number four, the whole articulation game is very problematic. So, I submit, when you get all five of those you have a situation where you wonder how anything ever gets done, without things collapsing in terms of safety. OK, now if you begin to look at what goes on inside the department or ward, it seems to me that you could do the same kind of arena analysis. Because, first of all, there are people there who represent different social worlds. And, either implicitly or explicitly, they will act toward danger in terms of the perspectives of those worlds. And they will also engage in a great deal of debate, discussion, criticism. What they're talking about are the organizational conditions which they think are for or against maximum safety. In addition to which, of course, you've got this organizational sedimentation I talked about. It's *literally* built in, in the forms of rules and regulations about spaces between beds, where the oxygen shall be, etc.

Those things which are really regulated are given a regular tour of surveillance by governmental people.

Then, I'll point out as a parenthesis here that in one of the sessions with Shiz I'm pushing her very hard about the nurses, who seem to be very practical in the matter of danger. And I keep saying that; I *know* there's ideology there, but Shiz is saying, they're very practical. By the end of the memo, it's very clear that nurses are very ideological, too. Some are ideology bearers of the other worlds, and some of it is that they represent the nursing point of view. But those struggles go on in the nursing profession itself; they don't necessarily look like safety issues, but often, certainly, bear on them.

Now the next step – I'm just going to touch on this and come back later – is that what we're really interested in is clinical safety on the ward, that is, the safety of patients. When you're talking about clinical safety you get into the whole area of trajectory, and that's a special issue which I'll talk about separately.

II. As I thought about some of the distinctions having to do with trajectory, I thought I'd fit error *into it,* because in fact when I started to do my homework I picked up the stuff on mistakes and error and then did a double take because it's clear that error contributes to danger. The condition for it or result of it will magnify it, or all the rest. But it's different, because this is endangering safety. So, I said the following:

If you think about trajectories and where the dangerous sources are, you think of the following. First of all, you get contingencies of the disease itself. Secondly, the regimen may give contingencies, and in problematic diseases often does. Third, there are external contingencies of all kinds: like earthquakes, or the roof falling in, or a patient falling out of bed. Fourth, there are biographical contingencies. While there are more, I ended up by saying there is a fifth one, which was the contingencies of error (error is a completely separate subject, so I'll keep that aside for a moment). Moving ahead just a little bit, think about the trajectory. I have the following notes: That either with the disease course itself, or the contingencies of regimen, if you go back to the notes on trajectory you'll see where we're talking about foreseen or unforeseen consequences. And if they're foreseen as possible or probable, then the staff stands ready with methods. Now some consequences have to do with danger, so you really have all kinds of methods ready for people to jump into the battle to lower the degree of danger. On the other hand, it's just logically possible there is no method of handling some of those consequences. But *I think* you can also see that you could do the same sort of calculus with unforeseen consequences; suddenly they just pop up and you may have methods for handling them, and you may not.

III. Now we get on to the business of risk. And my ideas here are not quite so organized. I'll give you what I'm thinking about, which goes something like this. Danger has lots of different kinds of dimensions. The dimensions are written into our memos. People can, of course, see the same dimensions, or not. They may disagree on which dimensions are important. They may agree or disagree even on the degree of the dimension. And where risk seems to come in is that people are assessing the degree of danger on some sort of continuum, with explicit dimensions built in.

Now comes a five-point note on the assessment process.

1. The business of assessing danger – that is, its degree
2. The figuring out of the actions and resources necessary to minimize the danger
3. People are balancing the perceived risk versus other kinds of issues: cost, effort, etc.
4. Unless there's just one person doing it, that person has to convince somebody else. Often, of course, there's a debate on that one.
5. Even after they've acted on it, they're going to reassess their action often, because there will be unforeseen consequences for other people; their activities, and even for the organization.

So, without saying very much more about this, it's quite clear that assessment and balancing and juggling and all of that is central to the whole risk business. And so you get languages like this: "acceptable risk," "calculated risk," "balancing" – common language like that.

IV. Now comes the question of trajectory, problematic trajectories. As we did the other day, we begin to see certain kinds of things. There are all kinds of routines built into these wards for handling those types of trajectories which are common. There are standard operating procedures. And there are, also built into those wards, organizational sediment. I talked about where you still have a division of labor of people doing those things, like calibrating the machines and checking out the temperature and all kinds of things like that. So you get a lot of people coming in, doing the work on the ward, who are not concerned particularly with *a* trajectory, but they're concerned with keeping the environment and the equipment in ready order. And all that is routine stuff. In addition to which there is a lot of routine activity on the part of the nurses. Nurses are central because they are primarily the people dealing with the day-to-day, minute-to-minute trajectory action. And I don't have a great deal written down about this, because I figured we would talk more about it, but if you think of the issue of monitoring from the standpoint of everything else we've talked about, then you get the nurses being the primary monitors of danger, along that trajectory. I say, "primary," because the physician can enter in and see something, or the physician can review things and sort of say, "Oh, my God, we're in trouble." Meaning, that the patient is in trouble. But on a minute-to-minute, hour-to-hour basis, the nurses are doing all that monitoring, and make judgments about how much danger there is, 0–100 percent emergency. So, she's on the line. But as we say in the memos, in and around her activity you see all these other people marching in and out, doing these standard other things or doing things that have to be done for the particular trajectory. Like the x-ray technician, and the RT, and the engineer, and the doctor, and the resident, and so on.

A lot of the nurses' activity is prevention of danger. That is, as decisions are made, options are chosen and clusters of tasks are laid out, and at every one of those points the nurse is not only doing those things, but she's monitoring how it's being done, by herself or somebody else – she's their monitor also. They can make sure the connections don't get pulled out or something else happens that's quite dangerous.

When there's an emergency, anybody can be in on the act. And often is.

On certain wards, like the ICN, the nurses are essentially doing continuous monitoring. That monitoring includes monitoring for danger, either potential or right on the spot, or what's done already, either by error or something else. And you had better pick it out. So that idea of *continuous monitoring* on these kinds of wards suggests very much what Shiz brings out in her notes: There's a fantastic emphasis on that kind of ward for keeping your cool. These people are explicitly and avowedly and almost continuously involved in composure work. To put it another way, if you have composure work, you don't panic and face the danger; that is, they don't make errors. If composure breaks down, the error potential is very great, with consequences for both the patient and, mostly, for themselves. So you find them working very hard at composure, and there are three reasons for it.

1. One is keep the sentimental order up.
2. The other is so that they can do their own work, literally.
3. The one we're interested in, it impacts directly on the amount of error.

I think one of things coming out is the central role of monitoring for anticipated danger, but also for unexpected danger. We mentioned first- and second-level monitoring here, because we say the physician is sometimes monitoring in a kind of secondary way, because he's reading charts and machine-product information, without necessarily seeing the patient, or the machines.

One other point: Certain phases of the trajectory and certain miniphases are seen as potentially much more hazardous, so the monitoring and other kinds of work are much more intense. But I want to point out again that it's not just the disease they're monitoring, and noticing; it's the actual work of implementing what she must do in those miniphases that can also be extremely hazardous. And so it isn't just that this person is in danger of dying; it's also that we have to do these things *now*, in order to make sure he doesn't die, at *this* point.

Last, but not least, here again you get the potential for a great deal of argumentation. Because so many of the signs are ambiguous, so many of the actions are not entirely predictable in terms of consequences. And there are just hosts of conditions there that make for that ambiguity, that uncertainty. And you can figure out most of them when you put your head to it.

V. This is the issue of error. Interestingly enough, I think, when we first started out in very early memos we were just talking about mistakes and errors and not so much just in terms of danger. And we didn't, until we got to the safety department's rules and regulations and we got to the larger issues of safety.

If you think about why errors are committed in hospital work, you start out at the widest box in the Chinese box system: nesting boxes, that is. How come you get a lot of error when you work with patients? So you can start out with the widest conditions, like the nature of the industry and how people are trained, and kinds of hospitals that we've inherited, and we work it all the way through. So that it's quite clear that the macrostructural conditions for error are very, very great. The kinds of things you would notice in a commonsense kind of way, that people could be unskilled, or inexperienced, or that the trajectories could be uncertain, and all the rest. Simply translate, and simply say: Why are they unskilled? What kinds of trajectories do we have today that they should be so unpredictable?

Anyhow, everything I've said about danger and error fits here because you can make errors at every decision point, option point, at every cluster task that involves in judging whether error has been made. Now, the point of this is that some of those errors increase danger failure. Clearly, most errors have nothing much to do with danger. So you get errors all over the hospital, and nobody suffers very much. But they can, and they do. So if you think of your own hospital careers, you'll see that lots of errors were made, but they didn't do anything except to your mood, the way you felt about things. But some errors are really bad. So you distinguish between these errors and the others. Then, without going into all of this, there are very good notes on the fights over errors, the accusations, the definitions of it. And all the ways of preventing it, how you rectify it when it happens, how you make it invisible to other people when it happens, cover-up, including the team-meeting tape on mistakes at work.

One last point is about taking into account the larger context. If you say there's an error that they made, they made an error in judgment as to how bad this was, and then they compounded it by doing this, and so on – we can't talk about that just in those terms. We clearly have to talk about more macrostructural kinds of conditions for the errors. I will now have some tea.

On the discussion

In the discussion that followed, some analysis was done. In Phase 1 (Hospital and Danger), the team began to recognize the enormous focus of hospital staffs on patient safety, and the centrality of that safety work in this particular – and peculiar – institution. This is not only because illness may be dangerous but because the great numbers of medical interventions put patients at potentially greater risk. In Phase 2 (Danger and Types of Work), the team began to relate safety work to other types that it had already discovered as important in the total complex of medical work: These included comfort, sentimental, coordination, machine, and information work. This was also the beginning of the next steps in making linkages between them very specific through coding – the core category being trajectory work. Phase 3 (Teaching) consisted of a discussion about teaching and training hospital staffs, as well as a dilemma or two attending these activities.

A few general points now about this kind of verbal and theoretical memoing, as illustrated by the team discussion:

1. These memos are *collaborative*. Their effectiveness, or sometimes lack of it, rests on the participants' interactions. During given sessions, some participants may be more active, more alert, make greater contributions. But, withal, these are cooperative sessions, in which all are working to produce good ideas, concepts, hypotheses, codes, and so forth.
2. In this *collaborative* work, there tends to be some division of labor for each session, depending both on properties of the participants and on the purpose and direction of the session. Participants may be more or less verbal, confident, and analytically or substantively experienced. These kinds of characteristics will understandably affect not only how much they talk but what they talk about and how they talk about it. Again, a session may center around a participant presenting data and then having others ask questions about this data, being stimulated by them, analyzing them, making comparisons with their own data, and so on. Or a session, or two in sequence, may center on clarifying the nature of one core category or subsidiary ones. Sessions may have several phases that differ in unanticipated functions, directions, moods. The principal investigator is likely to be taking responsibility for setting some directions, raising generative questions, and summarizing the discussion from time to time.
3. The principal investigator is likely also to take responsibility for keeping the discussion on course, steering his or her way, if wise, between allowing great latitude for creative discussion and yet keeping the participants' eyes firmly on the main topic or topics of the session. If necessary, he or she does this by using various interactional tactics, sometimes directly reminding them of the chief discussional thread but usually by more gentle hints, and even by cutting some of the talking short if it seems to be getting nowhere. The leader may also take main responsibility for raising the generative

questions that direct or redirect the discussion or promote decisions to follow up with future fieldwork. Of course, he or she is also likely to act as the chairperson of the session, if more than two people are involved, as well as to keep in mind where the session fits into the flow of the research project as a whole.

4. The problem of participants having different viewpoints should, at least ideally, present no great problem. Radically different viewpoints, of course, must be dealt with openly and as sensibly as possible, in terms of the overall purpose of the research. Of course, where the team members have had very similar training, differences among them are not likely to be as much of an issue as when they are differently trained. (Teams have broken up or have requested members to resign, when differences could not profitably be utilized or amicably settled.) More of an issue, sometimes, is that because of different experiences when collecting data, as well as doing different portions of the coding and memoing, team members can talk past each other during the team sessions. Everybody must be alert to that possibility, but the principal investigator has the responsibility for sensing discrepancies and getting the proper discussion going that will resolve them.
5. How to generate compatible codes from different sets of data is, again, no great issue, providing that the team, by dint of working together for a while, has learned to code together, to draw upon their own data occasionally for comparative purposes, and to be willing to take the common or another's coding items and to code their own data with reference to them.

The consequences and some of the common interactional features of the discussions, before and after, regarding its specific content or form during a given session, include the following. (1) The discussants stimulate each other, the stimulation deriving from the "actual" and experiential data elicited from each of them. (2) Issues also derive from the ideas that are enunciated and developed during the flow of the discussion. Some topics are barely explored while others stimulate more discussion and genuine, if minimal, analysis. (3) In consequence of such team discussion, a number of actions may follow. The original summary is immediately added to, in scope or depth. Team members may choose or be assigned to follow through with data collection or data analysis along paths opened up by the discussion. The discussion itself, when transcribed, becomes part of the memo file, and is reviewed later in the research when sorting memos on the categories touched on, or somewhat delved into during the team meeting (in this instance, safety, danger, risk, error, patient work, biography, trajectory phasing, etc.). Team discussions and primitive analyses may be used also during the final write-up of papers or chapters that deal with any of those categories, since the discussions are likely to have contained a combination of rudimentary analysis, useful raw data, and felicitous phrasing. If the writing does not occur too late in the project, this use of a team discussion plus the memo sorting may suggest the need for further data collection or additional analysis to fill holes in the analysis and so in the writing itself.

There are three further important consequences of such team discussions. The researchers are forced to confront common team issues, when ordinarily they might be pursuing somewhat different lines of inquiry because each is collecting data at different research sites (hospitals or types of wards, in this instance) or because each has at the forefront of attention somewhat different phenomena. A second consequence of regular discussions is that a shared analytic framework is insured, even if the team members fall behind in their reading of each other's memos – as indeed they are likely to do during the busiest phases of their research activities, or during the most hectic moments of their own lives.

A more hidden consequence is that each researcher is likely to be somewhat differently affected by every group session – either by the entire topic under discussion or by particular segments of the discussion. Indeed, a scrutiny of the transcript given below quickly suggests that team members often are not attentive to remarks made immediately before they themselves talk because they have been pursuing their own thoughts, stimulated by the preceding conversation. So, as in many other kinds of discussions, participants sometimes focus on common items of interest, albeit coming at them from different perspectives, and sometimes they "push" their own ideas and views, in at least temporary disregard of others. In terms of the forward thrust of the entire research project this means that the team discussions not only ensure commonality of perspective, but also the possibility of individual growth and a measure of autonomy in the further pursuit of ideas: pursuit – it is important to emphasize – within the common framework of analysis. In this particular research project, everyone contributed to the data collection, to the evolution of ideas and actual analyses of safety/danger; but most of the later analytic development was done by two researchers, one of whom took the responsibility for writing a monograph on clinical safety, which reflected her own extension of all further analyses done after this particular team meeting. In short, such discussions – although sometimes they may not rise to high analytic heights – are nevertheless an essential ingredient in the final analyses, which eventuate in integrated, dense, and of course grounded theory.

The discussion, itself

Phase 1: Hospital and danger

Anselm: The central work of the hospital, leaving aside the pursuit of personal careers and the running of departments and all of that is, after all, the trajectory

stuff. So, all its work should integrate at some point or another, no matter how far people are away personally, for patient care.
Carolyn: All these words, "error," "safety," "risk," can suggest a kind of negative, critical stance that isn't necessarily bad once seen this way.
Shiz: It's like the whole thing is organized around danger, every danger you can conceive of. They do that with interns and with nurses, too. It's always uppermost.
Barbara: But that seems perfectly correct; philosophically, that's what the whole hospital's organized for in the first place, anyway.
Carolyn: Well, I wonder. Suppose that the nuclear industry is organized in the same way? That they scare them, too? I mean, wouldn't you imagine that? I can't believe that those engineers didn't have dangers in their minds, as well.
Barbara: I think that the basic argumentation that the nuclear energy plants have is not to minimize danger – it is to produce energy. And, therefore, I think that minimizing danger is a secondary function, which is important, but not primary. In the hospital, one way or another, it is there to give care to those in danger, so that the focus, in a sense, is on danger just in being.
Anselm: Somebody said in this morning's paper that the commission was primarily interested in the health of the industry, not with the health of people – which is what you are saying.
Barbara: Yes, that's what I'm saying.
Anselm: See, you're raising a very interesting question. Is hospital work different from any other work in that sense? Everything else you can think of has danger involved, and you can try to minimize it, but it's always in balance against some other things.
Shiz: One is, you're creating a product. The other is, you are working on a product. And there's a whole history of profit making.
Anselm: I suppose if you have an enterprise, like raising a child to be a good citizen, that enterprise is fraught with danger, too. Is that central? The way that hospital work is?
Shiz: No, you're always concerned in the hospital with being in trouble in lots of ways.
Barbara: "Concern" is an interesting word there. There is always the peripheral possibility that you have to be concerned with.
Shiz: When you have a toddler, you're always concerned with danger.
Barbara: But that's not the primary thing. What you're trying to do is get that toddler through that stage, and I suppose you can say the same thing about its trajectory. The simple process of bringing up a child is not dangerous.
Shiz: No, it's not the central thing by which you are constantly absorbed; but it's a constant absorption in hospitals.
Barbara: Yes, because a trajectory, by definition, is dangerous.
Shiz: Patients come in at a dangerous state.
Anselm: Well, they don't always come in in a dangerous state. And you don't always expect them to get to a dangerous state.
Shiz: But the things that are being done to them can be dangerous.
Barbara: And the fact is, they wouldn't come in, if they weren't in a dangerous state of some kind, at some level.
Shiz: Somehow their trajectory is anticipated to be going off kilter, or something is wrong; the trajectory is somehow out of kilter, and that's when they come in.

Anselm: So, what you're saying is that the course of illness is either actually/potentially dangerous, and the regimens can also be actually dangerous, or potentially.
Shiz: Yes, so physical exams are once-a-year safety work.
Anselm: So, you're also adding there's a lot of prevention in it.
Shiz: Which is the other connection I was making, that safety work is really prevention work; danger-prevention work.
Anselm: Not all of it.
Barbara: Isn't it?
Anselm: No, well, emergencies, and post hoc, after it happens. You try to minimize it. Then there's cover-up work, so that other people don't see it. Right? And a lot of the safety work is argumentation. So, it's not just prevention. You see that, because you're working with nurses.

There are obviously various kinds of organizations whose work centers around physical and physiological danger: fire department could be one, police another, hospitals, and so on. And there are a lot of organizations that encounter danger, but it may be secondary or peripheral. I think that's what you're moving towards. I want to point out that it's not just physical or physiological danger that we're concerned about. There are aspects to this that have to do with danger to the hospital and danger to the world of medicine. But let's keep that to the side, even though it's part of it. This, in common with other safety organizations, really has safety work at the very core of what it's doing. It's not peripheral or secondary or episodic.

Phase 2: Danger and types of work

Shiz: I was looking at the types of work that are listed here, and every one of them has danger.
Barbara: In sentimental work: overinvolvement, loss of composure.
Shiz: Yes, I can see it. It occurred to me in looking at it that there's some element of danger in each of them, because a lot of them require clusters of tasks.
Anselm: Try it on for size: Let's see if you're right.
Shiz: Machine positioning . . .
Anselm: Option work, therapy, body position, error, obvious; life support; even teaching; diagnostic, very much so; even comfort: You can do it wrong. There are some that seem not to be so involved with danger, but they are like cleanup – clean up the nuclear waste. Comfort seems less so, but it could be. Composure is very much so, but in that indirect sense; because the breakdown means error. So, in other words, some of these things seem very obvious. The breakdown means error. Some of them are clear, but more subtle.

Ideological certainly has to do with it, because that consideration goes into the defining of what is dangerous or not dangerous.
Barbara: And also the risk element – juggling the risk in Jehovah's Witnesses' resistance to blood transfusions, and so on.
Anselm: So, Shiz, your point seems to be very well taken.

Phase 3: Teaching

Carolyn: I'm interested in the point that it's so central to the teaching. It's in and out of focus while you're working.

Shiz: That's why, when you look at a manual, there's always a "Dos and Don'ts." Usually it's the don'ts which have to do with either risk or danger.
Carolyn: I'm saying that then, in the practicality of it, the most risky kind of work can become routine.
Barbara: But that's a danger in itself.
Carolyn: Yes.
Barbara: Your concern, Shiz, keeps going back to the concern, the motif – "What nurses want to be is safe, clinical nurses above all" – and your impatience with a lot of the modern educational requirements, because either they divert from, or are not, well, relevant to safety.
Shiz: Yes, because that's what I hear students yelling all the time: "Make me a safe nurse."
Barbara: So, what's going on in the training and the education that's leading away from that, in your opinion?
Shiz: Oh, a lot of stuff on theoretical frameworks, and things of that sort. They want the nitty-gritty, you know. I remember a book, way back, about tenderness and technique, you know, saying that nurses were focused too closely on technique and not on psychosocial factors. But given the technology, the nurses are saying, "Teach me more technique." So they can be safe persons when they work on the body; and the expectation, too, of the employing agency.
Anselm: You can line that up with that of the reactions of nurses who've been out of the market, raising babies and doing other things for ten to fifteen years, and then they go back into nursing. And they're so frightened of mistakes.
Shiz: Oh, it's incredible. Everything from cleaning the skin to giving a hypodermic – you don't use alcohol anymore, you use something else. That kind of stuff. And it's danger, again.
Barbara: And the theories, you feel, that are absorbing the students' time really are not related to the work directly.
Shiz: Well, I'm sure there must be a lot of danger talk. I haven't looked at it or sat in a class, especially in medical areas.
Anselm: Well, see, they're balancing risks. The ideological nurses, let us say, who want to raise the level of nursing will say, "Well, yes, the first year of nursing those hospital-trained nurses are better than our women. But after that, our people are really more intelligent, more competent, give more imaginative care; they're going to save more lives, make fewer mistakes that are really dangerous." So you see, they balance.

Phase 4: Summary

Anselm: Conversation so far is very good because it's added one point more to the essential puzzle, and that is, we now know why the danger work is so central. It seems obvious on the face of it, but it isn't, until you think about it, because all of these types of work have their dangerous aspects. Or, these kinds of considerations can affect the degree of danger and the kinds of danger work that are done or not done. Over every one of those types of work there are danger rhetorics and danger debates. Very simple things like body positioning and how high the railing should be, and how close to the machine, should they go to the machine or should the machine come to them – all of that.

Theoretical memos and visual devices

Operational visual devices

Various kinds of visual devices conceived while doing analyses can be incorporated into the follow-up theoretical memos. Among these devices are diagrams, matrixes, tables, and graphs. Some are standardized, part of the repertoire taught in graduate school or absorbed while reading technical literature, and so qualitative analysts may use them too, although a great many seem to think without them. Whether one uses them or not undoubtedly reflects personal thought styles and predilections for various types of imagery. Other visual devices are invented while struggling with how to give specific data a greater conceptual order. They are designed to handle, or at least to get more understanding of, a particular problem.[1]

The operational visualizations that researchers use and compose at various steps of their analyses are often likely to be of this latter kind. Among their working functions are the following. Even as spontaneous scribblings, they can suggest ways to get off the ground during various stages of the research. They can give visualizations of what's going on with the phenomena under scrutiny. They can yield rough working models in visual form. And they can jog faded memories about, "Where was I?" after several days away from one's desk. Also, these operational visualizations can sum up the gist of a given work session, so that later one can more easily start from there. Others help locate visually and conceptually the different sites, or institutions, which must be visited or at which one must interview. Still others suggest new concepts and holes in conceptualization, just because the researcher is able to stare at and be stimulated by a diagram, a matrix, a table of items. These all help our thinking about comparisons and theoretical samples.

Illustrations of this type of graphic means are familiar to researchers who do them, especially those who do it flexibly, allowing their analysis of data to call out in themselves diagrams, etc., which "fit" the particular bit of datum under current scrutiny. For those who do not diagram and have never paid much attention to those of others, here are two illustrations, both used in analytic memos by members of my research teams, hence accompanied by commentaries. The first (Figure 4) is a

[1] See Miles and Huberman (1983), especially Chapters 4 and 5 – the bulk of their book – for discussions and illustrations of what they term *displays*. This material strikes me as potentially very useful.

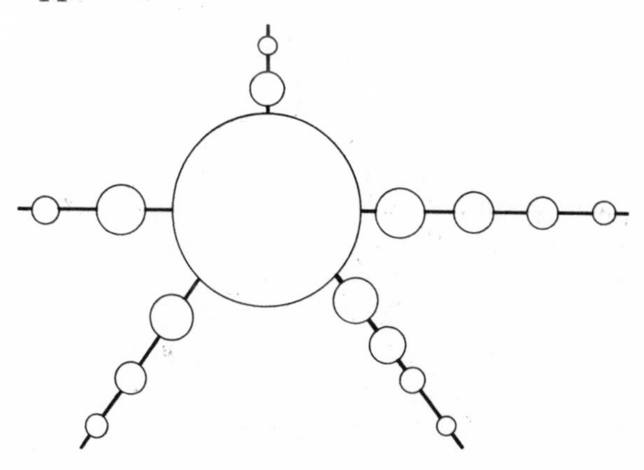

Figure 4. Orbits, hospital–home axis; technology sparse, and clustering consequences.

very simple diagram, sketched at the beginning of a team meeting concerned with the geographical siting of medical equipment and other technology. This particular meeting was focused on recent interviews done in a rural area. The diagram is typical of spontaneous blackboard graphics, which help researchers to visualize quickly, and suggest "next thoughts."

1. In developing countries, the medical services are all clustered in the central star city, and maybe two or three subcities. Anybody who needs relatively high technology either must be living there or travel there; otherwise, lives without, or dies.
2. In industrialized countries, the model gets much more dense out to the rural areas. But, as the interviews show, people still must travel to next-up centers, at least for much technology and services. Or, they must move there permanently. Physical circuit riding by bioengineers is one way they try communally to beat the problem, but that's only done weekly, with no backup services at the hospital, etc.

The visual means need not, of course, consist of an actual diagram: It can, for example, be a conventional table. Such tables can be quite simple like the traditionally helpful fourfold table used in the first memo fragment (Figure 5), or they can be more complex, as in Figure 6. Many research problems are quite complicated, and working out graphic means for helping to understand them requires innovative imagery and careful consideration.

SF/AS 5/5/78

Models for work analysis

We have been playing around with various 2 × 2, 3 × 3, etc. tables to analyze the work. General work in different places can be seen in Figure 5. Where

	Homogeneous patients	Heterogeneous patients
Easy work		
Difficult work		

Figure 5. Homogeneous/heterogeneous patients × easy/difficult work.

Phase of illness	Number of machines Few-Many		Frequency Few-Intermittent-Often			Duration Short-Forever	
Early							
Middle							
Late							

Figure 6. Illness course: machine–time dimension.

work is generally easy and patients homogeneous, work organization is simple; gets a little harder if with various 2 × 2 , 3 × 3, etc. tables to analyze the work. General work in different places can be seen in Figure 5. Where work is generally easy and patients homogeneous, work organization is simple; gets a little harder if patients are heterogeneous.

Where patients are homogeneous and work difficult, the work is still relatively manageable because work can be routinized and standardized, and also, built-in ways to anticipate things sometimes go awry. Dialysis units and coronary care units are examples of this (see, e.g., Figure 6). The most problematic places are where patients are heterogeneous and work difficult (i.e., neonatal intensive care).

However, operational visualizations not only function to further immediate analysis and often are incorporated into theoretical memos;

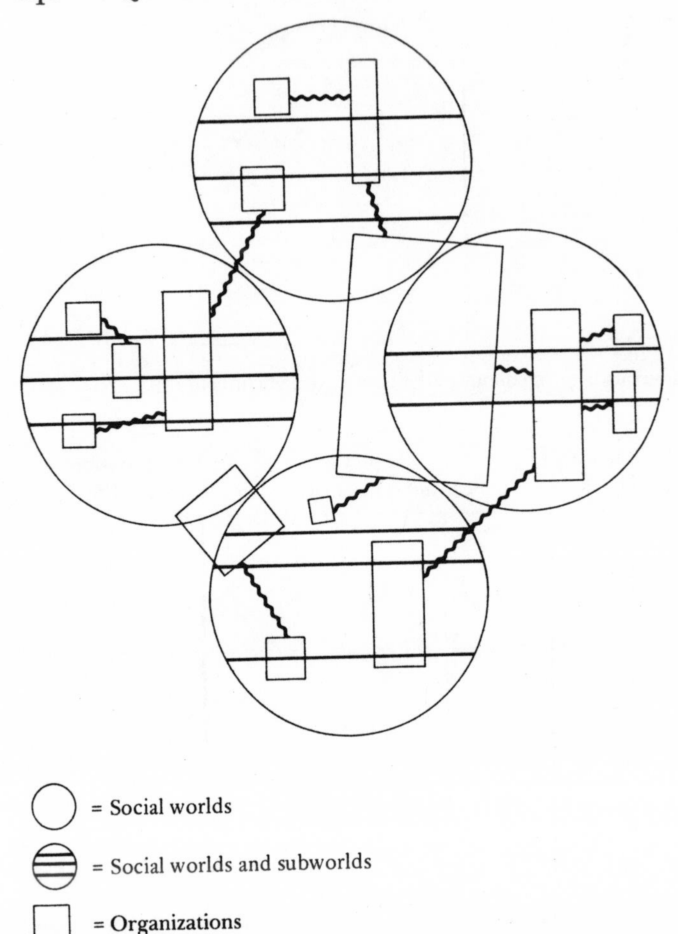

Figure 7. Social worlds, subworlds, organizations, and negotiations.

they may also be used in publications from the research project. Here are two examples. The first (Figure 7) was used in an article dealing with interorganizational negotiations (Strauss 1982). It utilizes the concepts of social worlds and their internal divisions, termed *subworlds*. Social worlds refers to "a set of common or joint activities or concerns, bound together by a network of communications" (Kling and Gerson 1978, p. 26). Social worlds vary considerably in size, types, numbers and varieties of central activities, organizational complexity, technological sophistication, ideological elaboration, geographical dispersion, and so on. The published diagram gives a summary graphic representation

	Consequences For						
Pain Tasks	Illness trajectory	Life and death	Carrying on	Interaction	Ward work	Sentimental order	Personal identity
Diagnosing							
Preventing							
Minimizing							
Inflicting							
Relieving							
Enduring							
Expressing							

Figure 8. A balancing matrix.

of relationships among organizations located within the same or quite different worlds and subworlds.

Figure 8 is another visual device, a matrix or grid. It first appeared in a theoretical memo during the study of pain. (See the chapter on codes and coding.) Later the entire memo, pretty much in original form, was included in the ensuing publication on pain management (Fagerhaugh and Strauss 1977). The invisible memo transformed into its visible publication is Figure 8.

Dimensions, Matrix, and Contexts

The most basic questions about balancing are: What is balanced? Why? The answer to those questions, both in general and for particular interactions, can be approached by considering a grid (see Figure 8). This grid consists of *dimensions* listed vertically and horizontally. The vertical ones pertain to various pain tasks. The dimensions listed horizontally pertain to various matters that may be affected by or affect the pain tasks: the illness trajectory, the maintenance of a life, the interaction itself, the personnel's work, the sentimental order of the ward, and the identities of various interactants. If we think of these dimensions as *crosscutting*, or balancing each other (i.e., pain expression versus interactional disturbance), then the entire grid might usefully be referred to as a balancing *matrix*.

To know what is being balanced in any given pain drama, or in a scene from such a drama, we need to find out which dimensions are being balanced by each interactant. Suppose that, for convenience, we call that particular calculus the balancing *context*. For instance, a child is balancing crying against keeping outward poise in the face of pain inflicted by a catheterization which he's agreed to, while the physician is balancing the infliction of pain against needed diagnostic information. If the child decides the pain is too much and decides against further catheterization (i.e., the diagnostic test versus enduring the pain), then the staff may utilize coercive measures since their choice, unlike the child's, remains in favor of the diagnostic information versus infliction of pain. If the press of work is not too great, the physician may take the time (i.e., "work") to be more careful and gentle in the continued catheterization, thus minimizing the inflicted pain. Each of these contexts is different, and each can be diagrammed easily on the balancing matrix. This is done by putting a mark in the box (or boxes) where the vertical and horizontal dimensions cross. (Since the patient and the staff may not balance identical dimensions, use a different letter for each relevant interactant: *p* for patient, *n* for nurse, *d* for doctor.)

A very important condition for any given balancing context is the array of organizational properties bearing upon it. By this we have in mind such properties as the ward's ratio of staff to patients, the experience of the personnel, the pain ideologies of the staff, and so on through the large number of properties implicitly or explicitly discussed in the preceding chapters. When an organizational property changes (the staff at night may be less experienced than the daytime staff, or there may be fewer personnel, or an inexperienced resident may take the place of an experienced one), this change may profoundly affect a balancing context. For example, recall what happens when an empathetic, psychologically oriented nurse attending a mother in labor goes off duty, and her place is taken by a brusque, no-nonsense, medically oriented nurse. To designate the most relevant organizational properties bearing on balancing, we suggest the term *organizational context*. Balancing never takes place in an organizational vacuum.

It is probable that researchers are prone to overuse the more common devices. This too-frequent usage should be guarded against. Why? To employ them too often must surely reduce the potential flexibility of thought processes, and so of the analytic process. A corollary point is that different kinds of visualizations are useful for different kinds of problems: for visualizing multiperspectives of actors, for portraying temporal events like phases or stages or processes, for visualizing the relationships of categories, and so on. If the researcher restricts herself or himself in use of types of devices, whether personal or discipline-derived, the analyses and presentations will be less stimulating than if the visual means were more imaginative, more freely engaged with the data themselves.

Also, one other caution: Often, as Howard Becker has noted (personal communication) there is some danger that a certain amount of valuable data can be lost in the transfer into graphic representation. This is an additional reason why such representations cannot function in analysis as the sole carriers of information, but must be supplemented by theoretical memos and their sorting.

Graphic teaching devices as memos

In research seminars and consultations, visual elements can further analysis and function as a type of theoretical memo. Indeed, after the sessions the students and consultees often write memos that incorporate or improve on those operational visualizations. In the teaching and advisory sessions, often I find myself using graphic representations, especially diagrams, to clarify for myself what can be done analytically with the presented data. In the seminars, for example, I sketch diagrams of what seems salient in each presenting student's data, usually elaborating or modifying the diagram when classroom discussion brings out additional features. Sometimes at the conclusion of the session, I put my diagram on the blackboard, intending it to function as a summary of the total discussion; or to show some steps that have been missed in the discussion. Sometimes I chalk up the diagram earlier, when the discussion is floundering, and ask whether that helps to give a more useful direction to the discussion. At other times, the students are immediately curious about my penciled diagram and ask to see it: Then it can function to set the initial direction of the seminar discussion. And occasionally, midway or so during the two-hour session, I put a diagram on the blackboard to show the students that they have concentrated only on certain features of the data, asking them: "What would you like next to explore?" Or, "This relationship is left unexplored, so wouldn't you like to tackle that?" Or, "You have developed terms for these concepts now, but *this* one is just a name, you have blackboxed its contents – how about focusing on that next?"

In short, the teaching diagrams (true diagrams, or matrixes, or whatever) function as operational ones, since they serve to move the collective analysis along. Usually the student who has presented data in the class uses the diagram afterward for further stimulation, or when analyzing both previously gathered and new data suggested by the diagram and discussion. These teaching visual devices can also be used effectively in research consultations with students, even with project associates who happen to be puzzled by some features of their data, dissatisfied with their current analyses, or blocked in their analytic efforts. Listening carefully to the consultee, one senses what the specific problem is (sometimes personal but more often analytic), then figures out what kind of diagram might help, or quite literally works it out in the joint session. In my experience, these teaching devices can work wonders in moving inexperienced or less experienced researchers along faster or getting them in motion when they are blocked.

Rules of thumb

To summarize: The *rules of thumb* pertaining to this combination of memo writing and graphic representation are:

1. Since these representations can measurably help analytic operations, researchers should attempt to develop skills in using them. This may require considerable practice. The following steps are recommended. First, begin with standard means: fourfold tables, simple graphs, uncomplicated matrixes. Second, chart your way, using diagrams or more complex tables and matrixes, through some of your past analyses. This will enable you to see graphically what you have accomplished previously. It may also cause you some chagrin, if you now see something important you had omitted! Third, get into the habit of making graphic some of your current analyses. By using such means, you can see if that captures what you have just done, or you can tackle a difficult analytic tangle.
2. Incorporate these visualizations into your theoretical memos, permitting them to affect your later thinking, when reviewing and sorting the memos, as well as to inform your current thinking.
3. Try to make some of these diagrams, tables, matrixes, etc. cumulative, so that they will contribute to the total integration of your analysis. In short, incorporate their information into successive integrative diagrams.
4. In the last stages of your analytic integration, however, you must utilize the most recent integrative diagram *and* additional sortings of your memos. Each can contribute its core clarity and core density. Of course, the final integration comes only during the actual writing of papers or a monograph.
5. The final integrative diagram is probably too complex for inclusion in your publication, except with much additional explanation. Some of the operational visualizations, however, scattered throughout the memos, can be incorporated directly into the writing, and sometimes, with adaptation, so can at least portions of the accompanying memos.
6. If you are teaching or consulting about research, then let your visual skills enter into your sessions, so that others can benefit and afterward write their memos based on both the visual and verbal interchanges.

7 Excerpts that illustrate common problems

In this chapter, we reproduce fragments of seminar sessions, and in one instance the summary of a student–teacher consultation. Their presentation here is designed to illuminate, beyond the materials in preceding chapters, *common problems* encountered in learning–or doing–grounded theory analysis, whether by students in a seminar or even by experienced researchers when confronting the problem of analysis during a new project, perhaps with new kinds of data and data-collecting experiences. (Some of these common problems are found in any style of qualitative analysis, not only in the grounded theory mode.) The teacher's strategies for helping to surmount these problems are also reflected in the cases given below.

Among the stumbling blocks to effective analysis, at various points in the learning process or during the evolving research project, are the following:

1. learning to persist, line by line
2. learning to move from description to analysis via dimensionalizing
3. breaking through to analytic focus when flooded with experiential data
4. asking for too much data
5. illustrating the connections between macro- and microconditions and consequences
6. determining the central issue in the study
7. filling a hole in the integrative diagram.

Case 1

A rudimentary line-by-line analysis

As noted earlier, an initial step in teaching grounded theory analysis is a close examination of the interview, fieldnote, or other document, done quite literally line by line, even by a word-by-word scrutiny of the opening lines. (See Case 1, chapter on coding, for an example.) This gives the fledgling analyst a vivid sense of what can be gotten from patient, detailed scrutiny. Watching the instructor do this is one thing, doing it by oneself is another.

The memo reproduced below was written by a graduate student who was not yet skilled in this kind of microscopic inspection of data. Nevertheless, it illustrates how a student can be stimulated to make detailed commentaries and raise astute questions by utilizing this procedure. However, she had not yet gained any facility in coining terms, for the categories implicit in her commentaries and questions (for example, the in vivo code, "coming to terms"), that would allow for the next steps of dimensionalizing the categories, as well as lead quickly toward more explicit if provisional answering of questions and issues raised in the memo. She had written the memo not long after a seminar had discussed with her one of her first interviews, done with physically disabled persons who were living independent, relatively normal lives, aided by good equipment and paid attendants. Afterward, the instructor had asked her to closely scrutinize a paragraph or two of an interview, since he suspected she was having difficulty getting off the ground with her analyses.

12/17/82 MacCready
I am not sure how to do this; I am going to try it phrase by phrase and see how it goes.
Coding – from D's[1] Interview

	"Our society is so locked into the physical . . . when you've got a major deviation you've got to come to terms with it, what's meaningful, what's bullshit. While someone else may be dealing with feeling too fat, for example, if your whole body's different, you've got to come to terms with it. It can be a liberating experience, a time for reassessment."
"Our society"	"Our society is so locked into the physical . . . ." . . . this implies a notion of some big "Society," with a capital S, as opposed, perhaps, to what she or some others might think. Anyway, I think it implies that there is another way of looking at things than "our society" does. "Our society" is a powerful, impersonal, abstract entity. Normative and impersonal.
"is locked into"	*Locked* – this is a very strong word. It suggests a strong connection, not just "partial to" or "accustomed to" but "locked into." Sounds permanent, restricting, involuntary, certain.
"*the* physical"	Physical – as opposed to psychological, emotional, spiritual. Notions of beauty, perfect bodies, magazine ads. Not various standards of beauty, but one. Does not imply being understanding and tolerant of physical differentness, but being locked to one standard of physical perfection. Judging people physically.

[1] D is a young quadraplegic with limited use of her hands and no mobility below the waist. She is also a graduate student in psychology.

"when you've got a major deviation"	A deviation from the physical standard "society" sets. Deviation implies the existence of a standard, a norm, a mainstream. But she doesn't say "when you deviate," it's "got a deviation." Being deviant versus a deviation.
"you've got to come to terms with it"	Learn to think of it in another way? Face up to what the "deviation" means to others, to "society" and to yourself. Evaluate it, face it. Reach a truce with it. She does not say, "society has to lump it," or "others have to come to terms," but *you* have to. A sense of personal responsibility for coming to terms with your own body in the face of society's opposing vision.
"what's meaningful, what's bullshit"	Having to evaluate "society's" notions on your own terms. Having to decide which of "society's" notions of the physical you choose to accept and which you do not. You stand back and distance yourself from what this "society" might think of your "major deviation" and come to your own stance. Independence.
"while someone else may be dealing with"	Here again, the language implies self as opposed to others, self as dealing with things that are different from what others may be dealing with. But also, that other people have to deal with things, too.
"feeling too fat, for example; if your whole body's different"	Contrast between a minor deviation – too fat – as opposed to a "major deviation." Whole body's different and yet it isn't – not like a Martian or a little green man with eight legs and a giant eye in his stomach; it's only not moving her arms and legs and sitting in wheelchair. But still, the notion of degrees of different ranges of deviation. The disabled person's difference as being of a higher degree, or even different quality than something like feeling (not necessarily even being) too fat. The sense of dealing with something more extreme than other normal or "merely fat" people do. Strong sense of difference.
"you've got to come to terms with it"	Emphasis. This is important to her, she's said it twice. It's not "want to" or "helps to" or a "good idea to," but "you've *got* to come to terms." Necessity as opposed to choice or volition. What if one didn't come to terms, I wonder? What would happen then?
"It can be a liberating experience"	Not is, but *can* be. A sense of possibility rather than inevitability or necessity. Other alternatives – what are they? "A liberating experience" – connotations of freeing oneself, of releasing oneself from the "physical" that "society" is "locked" into. Prisoners are locked in; she is liberated. Like in women's liberation. Part of it is freeing oneself from conventional notions of how men and women should be. Coming to a new view which one believes is superior to "society's" old, locked-in view.

	Release from restrictive norms and images. Very positive connotations. A surprising way to look at things: disability as liberating. Associations with social movements as liberating from old ideas and expectations.
"a time for reassessment"	"Experience," "time," – both these words suggest not a revelation but a process over time. Standing back and looking at things freshly – an *opportunity* to not be "locked in" but to decide oneself what one does and does not believe. Coming to a new view, a new way of looking at things, a redefinition of the situation. Fits in with "coming to terms," with "liberating" oneself from old ideas, with challenging "locked-into-physical"-ness. An opportunity for change. A transition. A positive step to a more thoughtful stance. A considered rather than a "locked-in" view. Reflection, growth, transition. Acceptance of a new ideology.

What this student needs to learn next about coding is not only how to isolate and give names to categories but also how to dimensionalize them and discover their conditions, consequences, and associated interactions and strategies. She has only reached the first steps of learning what is involved in line-by-line analysis.

Case 2

Dimensionalizing

The next excerpt is taken from a seminar session whose participants had worked on the American Indian reservation materials. In a later session, this class was invited by the presenting student (Katarin) to address the issue of how Indians "construct" the concept of medical care. The selection from the class discussion centers around: first, some answers given by Indian respondents to a questionnaire; and second, a short section of an interview with an Indian woman. The entire session was a lesson – directed strongly by the instructor – in dimensionalizing the Indians' conceptions of medical care and relating these to conditions and consequences.

At the beginning of the session, the instructor underlined some terms, potentially relevant for dimensionalizing, on his copies of the questionnaire responses and the interview, then asked the students to try dimensionalizing through the analysis of the questionnaire answers. They managed to do that with a little initial guidance. Then they turned to the interview selection: Again, the discussion turns around dimensionalizing the properties of perceived medical care, with attention also to some conditions and consequences. Some of this class discussion, the interview itself, and a few questionnaire responses are reproduced below.

Questionnaire responses

What would be nice is if those people would *send more* doctors and if those doctors would take *time to examine* people and *listen* to the people's problems and that would be good for the people.

Those field doctors and nurses could be more *kinder* to the Indian people.

We don't have specialists. *They don't send anybody* good. The doctors we have don't spend time *examining* people and they don't know how to *listen.*

They never visit in our homes. They won't come near us. *If they don't see how we live* how can they understand the problems?

They don't *care* about us. You can tell because they *hurry through* every examination, don't know *what they're* doing, and, most of all, don't know how to *listen* when people talk to them.

They don't take a *personal interest* in us. They don't *understand* our ways or our means of livelihood.

They *should speak plain simple language* so people can understand.

Those doctors should get on the ball. *We don't even have real* doctors. What we get are interns. Ones who need lots of practice material.

They make us wait while they drink coffee with the nurses, *laughing* and *joking* around while the waiting room is filled with really sick people, some in pain. We wait and wait and wait. Then if you're ever seen, they hurry up and poke here and there, *throw some medicine* at you and send you back home without even *listening* to what you have to say.

We need doctors that really *care* about the health of the American Indian not just to see how many patients he can put through his door. From some of the doctors I've received the impression they *detest caring* for a *dumb* Indian.

Why can't the government give us a bus or something to *help the people get to the hospital*? Here we are, 90 miles (or 125, if you take the better roads) away from the hospital and if it's a day the local clinic is closed, you have to get yourself in to see the doctor. Many don't have cars so they have to pay a friend to drive them, sometimes $20.00 just for gas.

They should send a doctor to live out in the districts. *See how the people live. Learn* about the people's ways. Learn about our life and customs. Someone who could learn and could help us. *We'd be willing to work with that person and help him.*

An interview

Last summer I took my baby who was four months old to the hospital here with a fever of 102° and a cough. The doctor gave me Tylenol and told me to take her home and force fluids and have her rest. He said she had a cold and would get better in a few days. A week went by and she got worse, even though I gave her Tylenol as directed, forced fluids, and rest, and sponged her to bring down the fever. I took her back to the doctor the next week and he told me that she was getting better. I told him she was getting worse – she was more congested and wasn't sleeping well at all. He said to continue with the regimen of Tylenol and fluids and rest. I tried this for a few more days. Then she got real sick, fever of 103.6° and breathing difficulties. I took her to the off-reservation hospital (60 miles away), where the doctor took an x-ray and told me she had a very bad case of pneumonia and should be hospitalized immediately. He got real mad at the reservation doctor and called him on the phone to find out what was going on. He then told me to take the x-ray back to the reservation hospital and ask them "if their x-ray machine was on strike?" and have the baby hospitalized. I did that and they did hospitalize her immediately and began antibiotic treatment to cure the pneumonia. I had to pay $90.00 for the doctor's fee but my baby's life was worth it. I'd do it again, too.

You *just can't get good care here*. The doctors are *young*, have only *book learning*, none of them have children themselves and don't know a damn thing about kids, are scared of them and *just don't know* what to do other than a well-baby check-up. *They can't even assist* at a normal delivery. We are just lucky to have our midwife. She's the only nice thing in that hospital. And they tried to get rid of her too. That's 'cause she *cares about her patients* and treats them *like human beings*. She *spends time* with us and *listens*. If she doesn't know, she says so. She *doesn't try to fake* you out, like she knows it all. Those doctors are *so ignorant* and *insecure*. And we have to pay for it. They are treated like little gods, act like they are doing us a favor being here, putting up with reservation life, and then treat us like *less than humans*. Then, if you go elsewhere, *they act* like you have committed a sin. They told me I wasn't taking care of my baby right, and that's why she got sicker. Now that just isn't so. I didn't sleep but a few hours each night for a whole week because I was up rocking her and trying to force fluids and help her sleep. But *they don't believe* you when you tell them that. They just laugh you off and look away and say something ugly to the nurse about how "they know" when a baby gets good mothering. Somehow only they can tell. It is a very painful experience, being Indian and being really sick and needing some care. It probably will never change . . . . Sometimes I think that they are just trying to get rid of us, that it is *just a big plot to exterminate the Indians*.

You know sometimes we get rare doctors who care, *who really care*. Then he gets *overworked* since everyone wants to see him. He gets *burned out, fast*. Or if he tries to help us get better care with other doctors, he gets *bad-mouthed* by his associates. Once there was a doctor who would even make home visits. It *created such a problem*, since the others were not willing to do the same, that the rest petitioned the administrator to make home visits prohibited. They voted on it – the majority won. Now, that's real democratic.

Things have gotten so bad that you are afraid to take your sick child to the hospital. The *first thing the doctor suspects* is that you have *neglected or abused* your child, that it is *your* fault the child is sick, and they want to *turn you in* to state welfare as unfit.

Seminar discussion

K.: The mother gets a real different assessment from the other doctor. He also has a different treatment. Both reinforce her assessment monitoring, so that she gets some kind of reinforcement from the second doctor. She gets some support for her judgment of incompetence of the first.

A.: (the instructor) Sure, he even says sardonically, "Ask them if their x-ray machine is on strike" . . . . Go on to, "They just can't give good care here, the doctors are young, etc."

K.: Are we going back to dimensionalizing, then? Is she identifying for me the dimensions, as she sees it, of bad medical care?

A.: Yes, and the dimensions all become conditions or explanations of why she got poor care. They are rather general. She is not talking only about this particular doctor: She's talking rather generally. Which is interesting, because she could have said, "He's a resident and without experience," or something like that.

P.: So she doesn't see this particular doctor as being incompetent, but all of them as incompetent, of which he is just a representative.

A.: Interesting, since then it wouldn't matter whom you saw, they were all incompetent. So she begins to draw out the dimensions for you. However, on the fourth line from the bottom she tells you something else.

K.: There is someone there whom she does like, can trust.

A.: So she is making an exception and telling you why. She thus gives you the condition. The "because" signifies the condition.

K.: Yes, she "cares" and "treats us like humans," "spends the time" and "listens." She also brings in the interesting parts that the hospital tried to get rid of that particular professional. And that is a very salient point. The institution acts to impede the acquisition of some of the qualities that the people feel are important for good medical care. It is a very real problem for the professionals.

A.: But that's a consequence of how *they* look at good medicine.

K.: The administration; it reflects the administrative definition and boundaries to good medical care. Perhaps not the doctors, separate from the administration.

A.: She's built in a comparison for who gives good care and who doesn't, along these dimensions, these sets of conditions. She has also added for you, in a very nice way, what happens when they do give good care, in her terms of good care.

K.: It all seems so tight. They are each bound to their own perceptions and particular definitions and visions. Neither side can look at nor understand the other's perspective. It's like they can't pass from one camp to the other.

A.: But that's just a general statement. If you ask yourself, "Why can't they pass?" (under what conditions) or, "Why can they pass?" you will immediately see that there is a set under which some can/do pass. Or you have to spell out the conditions under which they might, though in reality they may never exist there.

S.: In some ways, the nurse crosses that boundary, as she says. So some people sometimes do cross.

A.: So your initial statement which was so general – upon closer examination you begin to specify under what conditions certain things do and do not occur.

K.: And when you begin to understand particularly the connections that this issue has on the broader, more macro level in terms of institutional and governmental health care to these people some of the . . . .

A.: Well, that's fine, you just built in another level, and are making connections . . . . But what we're saying with this instance is that there is a woman who did pass the boundary because she did care. OK? So there is one big condition under which passing the boundary occurs – caring. And you have in your head some of the macro reasons why passing does not occur, where professionals cannot cross the boundaries . . . . All right, keep going.

S.: She helps you identify some more of the conditions of bad medicine. Trying to "fake you out." But acknowledging deficits, she aligns with good medicine. And these doctors don't have that.

P.: But isn't that part competence?

K.: I'm not sure.

N.: Competence is closer to knowledge. Acknowledging lack of knowledge is not necessarily competence or incompetence. Like this nurse can be OK and still not know. But the deficit of not knowing is excused because she is willing to say she doesn't know. But the doctors are perceived as not knowing, but trying to camouflage that, or purposely circumvent and distort that.

K.: So there is a kind of "bluffing behavior."

A.: Look, dimensionalize it. Doctors know or they don't know. If they don't know, how do you break that down?

K.: Under the don't-know category – either they bluff and say they know when they don't, or they "'fess up."

A.: Yes, and she's telling you that the nurse 'fesses up and the doctors don't.

N.: She also gives you a clue as to why they don't – they're insecure.

A.: How do the patients know whether they bluff or not?

K.: Through experiences like this woman is telling us about – when the patients' assessing and monitoring and intervention are as dichotomous as these two, and they have to go elsewhere to get some resolution and "good care."

N.: I may be reading this in, but she says that the nurse midwife does admit she doesn't know – so, perhaps by the absence of such acknowledgment by doctors, they come to an assessment of "faking us out." The nurse, being able to do this, may point out that it is lacking in the doctors.

P.: Also their experience with a "more competent" physician off reservation leads them to know that the doctors don't know. It is sort of by implication.

K.: I think it is more than just trying to find a doctor who agrees with you. I would attribute it to the continued experience of not having relatively meaningful and successful, and mutually agreed upon assessment and treatment experiences. It is also knowing that people don't always know everything, but these doctors act like they do. They know people aren't perfect, but the doctors continue to act as though they are.

A.: We don't have too much time left today, so let's move rather rapidly down the sentences. "Those doctors are so ignorant" – (we know that one already) – "and insecure" – ah, there's a new one. So security and insecurity have something to do with knowing and competency.

N.: Security and insecurity have to do with the ability to say you don't know.

A.: The next sentence, "And we have to pay for it" – meaning not money, but human costs. And the next sentence?

K.: I thought that was deferential and demeaning behavior. For the doctors, the reservation life is just not good enough for them and, by inference, neither are the Indians.

A.: So, there's nothing particularly new here. Now the next sentence: "Then, if you go elsewhere, they act like you have committed a sin." It's practically a scientific proposition! "If . . . then." They are giving you the condition of going elsewhere, and the consequence – punishment. Now your experiential data can tell you that there are various kinds of possibilities. If you don't like one doctor, you go to another. You can leave one doctor for another without blame. They may just forget you or think you are crazy. So there are a whole series of possible consequences of looking at comparison relationships . . . . The next sentence is what?

K.: When she says, "they also told me I wasn't taking care of my baby," she is explaining what else they do. They blame her, as part of the punishment for going elsewhere, for the entire sickness of her child.
A.: When you say that she gets blamed, this is a *negligence* accusation. That is different than an accusation of incompetence. She's actually making an incompetence accusation of the doctor, for not making the right assessment. But he's saying he doesn't think *she* has any competence at all, and is making an accusation of negligence.
N.: Which is related to how he sees her competency as a mother.
A.: So now you have a couple of new categories to add in there also.
N.: And the competency accusations go back and forth.
A.: And one consequence of being negligent is that she gets sicker . . . . OK. Now, the phrase, "That just isn't so!" What is going on there?
A.: The mother has made her own assessment, about her mothering. Her assessment is no, they are wrong.
A.: She's answering the accusation, right? And then she's telling you how she did the monitoring. But "they don't believe you" when you tell them about your monitoring. That's the interaction between them.
K.: Sure, and she describes it: "they laugh you off" and "look away" and say snide things within earshot.
A.: Yes, that's all interaction: how the doctors handle her . . . . All right, let me move you on, as our time is almost over. "Sometimes I think that they are just trying to get rid of us. That it is just a plot to exterminate the Indians." What does that mean? It's the first time we hear that issue.
S.: Are you talking about that a consequence of bad medicine is extermination of the Indians?
A.: I don't think she means that.
P.: She's saying it is deliberate.
A.: Sure. She is saying, they are not trying to cure, they are trying to kill. Who is "they?" Not the physician. So another condition of poor care – that is, noncare – is an extermination policy. Interesting, because now we are not just talking about care, but also noncare. And noncare may kill you. A condition of noncare is an extraordinary policy. When you think through all that, then you can see now that you have sort of ordered what is implicit in what she is saying. You have a translation of what she is saying (excuse me for being so dogmatic and directive here). The translation is: "Given the condition of an extermination policy, they give us no care at all." If you want a kinder interpretation: "They give us lousy care, but it is really no care at all."
P.: So that's her underlying understanding.
A.: Sure, she's saying: "Look, that guy gave me terrible care, but behind it all that operates because of the governmental extermination policy. That's why they give us incompetent doctors; that's why there are no resources; that's why they send us people who don't understand us, don't listen to us, and so on." It is the larger macrostructure that lies behind all this action. (It's the first time we've hit this. K., do many people believe that?)
K.: A good number. It is also part of the explanation given for the economic conditions.
A.: Then you have to ask yourself, under what conditions would some people believe it and some not, and some others wonder. That may not be relevant to

your final writing, but it belongs in the analysis, doesn't it? If some people believe it, they do one thing; if they don't, they do another; if some people are dubious, they do another. We have time for just one more paragraph. "You know sometimes we get doctors who care, who really care." That's your comparison. That's the person who crosses the boundaries. She doesn't tell you why, which is fascinating. Then she tells you something else: He gets overworked because everyone wants to see him. Let's look at the first part of the sentence: and "then he gets overworked." What does that mean?

K.: He doesn't have time to see patients and give them good care.

A.: Right. You can't give good care without time to do it. OK. But there is a reason for why he doesn't have time. So that sentence gives you a double condition. A condition for a condition. We don't have to do the rest of the interview; you can see what that is all about. All right, K., what have we done? Let's see if we can summarize. Start from the beginning. What did we do today?

K.: I think we set up the dimensions of good medical care from the patient perspective, and kind of outlined many of them. Then, we looked at the interview to see under what conditions these things come into play in peoples' experiences in trying to get medical care. And we attempted to set some of the specific incidences in the context of the larger system. That's what I think we have done.

A.: Now see if you can make a diagram that will pull these things that we did today and the last time together. That may give you additional visualization . . . for these integrative diagrams can be very, very useful. If it doesn't come easily don't break your head over it.

In summary, the questionnaire and interview excerpts illustrated the technique of underlining key words and phrases in questionnaire and interview responses, in order to suggest quickly possible dimensions in those responses. This class session shows how the analysis is carried forward by utilizing the dimensions, linking them with conditions, consequences, and with respondents' interactions and strategies. Thus, the analysis moves from an appreciation of the respondents' perspectives to an analytic perspective, enabling the researcher both to utilize and yet spring free from being captured by the respondents' views of events, institutions, and other actors. A general rule of thumb is, then, to look for in vivo categories, examining them not as themes, as is often done by qualitative researchers, but in terms of dimensions – then to create hypotheses bearing on possibly relevant conditions and consequences, strategies and interactions.

Case 3

Flooded with data

The next selection is taken from a research-seminar session and, like the preceding, is designed to indicate a common problem faced by researchers, especially when they are doing research in areas in which they have worked, or are working, as practitioners: social workers, nurses, educators, and so on. The problem is that they are flooded with personal experiences: They know too much, in a genuine sense, about the areas and events they wish to study. What to do with all that experiential data? Where to cut into it analytically? Indeed, what to study, specifically? In this seminar session, what is illustrated is this issue and the attack on it for both the class and for the individual researcher.

A. is a social worker, who has worked for some years in a public school, where she deals with the children defined by faculty as "difficult." She would like to write a doctoral thesis on some aspects of how these children and their parents are handled by the school staff. She does not know, however, how to focus that broad interest nor how to formulate issues precisely. She has had a conference with the instructor, who suggested she present a couple of anecdotes to the seminar for its reading at the outset of the seminar discussion. She does this. Then she says to the seminar participants:

A.: I must say that one of the big things for me, starting this way, is I feel I have *so* much information from my work in my head that I'm interested in working on, that I don't know how to approach it. So, I wrote this down for you to read. I don't have any rationale for having presented the material this way, but it's a beginning, and I was hoping I would get help from you in terms of not only how to focus issues, but how to present the data. You know. To work on my fieldnotes . . . .

The instructor: In this kind of situation, what happens is you get someone like A., working in the situation, and so get all that marvelous data – none, or little of it, is written down. Given all the hazards of analyzing that kind of material, the issue really is, What does she do with all that data? How does she handle them? The problem is not how much she knows, but what the issues are. Does she know them? I think she does, if we *push* her. See now if you can't get her to recognize explicitly what's in her head. After all, she's been living with all that data.

A. (immediately afterward): It's kind of frightening. (as I blurted out to you, Anselm, yesterday) for me, because I don't know how to approach it. It's all in my head; it's also in my fingertips, because I'm doing it all the time. And as I'm working, I generate all kinds of hypotheses about what's going on – and there seem to be so many – you know, they mushroom. A listing of them, and I can't seem to handle them conceptually to get the kind of focus I need.

Instructor: "You know, you can't write about everything. You need a central generative question . . . . Somebody, begin the discussion."

For the next half hour or so, there is a lively discussion during which many substantive questions are asked, and substantive suggestions are made, on the basis of the discussants' personal experiences. A. answers the questions with additional experiential data, and so on.

The instructor then stops this discussion to draw out one major issue: that of the many different perspectives of school staff about the difficult children, and their corresponding definitions of the childrens' behavior. He asks A. if *that* is central to her thinking? She responds, "Yes." The instructor requests the class to target their questions to that issue – but only *that* issue and no other. The remainder of the session produces a number of useful questions. The participants also focus on whether A. already has answers to those questions, or perhaps needs to look more closely at "what happens" by gathering specific observational and interview data.

A general *rule of thumb*, then, when one is flooded with experiential data, is to get distance from them by raising theoretically oriented questions about items in the data, possibly even selecting one such question and then focusing the usual kind of analysis around it. One looks at it, then, in terms of categories, thinking in terms of hypotheses about possibly relevant conditions, consequences, etc.

Case 4

Getting information rather than getting down to analysis

Young researchers share a common tendency to become entranced by data, as well as being hesitant to take analytic steps without "more" data. In the early sessions of a research seminar, this tendency is evinced repeatedly, as the participants press the presenting student for additional substantive details – which are usually given with some relief, since the presenter feels relatively secure with those, but relatively insecure, or barely started, with his or her own analysis. The instructor is sometimes inclined to allow the discussion to flow for a while, even to let it flow "all over the lot" before intervening, either to give the discussion analytic focus or to request that the class itself do that. Sometimes, also, the instructor may point out why the discussion is getting nowhere analytically and why, therefore, some participants are getting impatient. Even presenting students may grow impatient, being eager to get help with their analyses. Relatively quickly, students learn to blow the whistle themselves, recognizing the difference between knowing enough to raise intelligent analytic queries, or to move sensibly into a microscopic analysis of the data. Yet, if the presented material is especially fascinating, one sees the students begin to get captured once again by their entrancement with it. When that happens, the instructor may again have to warn against that tendency, or call it to their attention after allowing the discussion to flow for a while.

A striking instance of this occurred once when a young visiting professor, a clinical professor, who had been attending the class for two quarters and was well integrated into it, chose to present some data from a large project on which she and several colleagues had been working. She began by specifically stating what she wished from the class discussion: namely, alternative sociological perspectives on her data, which were already well analyzed from psychological perspectives. So, she clearly issued an invitation: "I want *new* ways of approaching the materials." It was not that she was dissatisfied with the old, but that if she were to get anything from the discussion, she explained carefully, it would be something to supplement her current perspectives. She did not need specific analyses – she could do that for herself – but fresh analytic questions about the materials.

The class discussion began immediately, as it often does, with questions asked of the presenter about substantive descriptive detail, and since the topic itself was fascinating, and the students were curious anyhow about "findings," they fell into the trap of asking more and more substantive questions. They also narrated comparable incidents from their own experiences. Since this particular class was many months along in its training, the instructor was, at first, annoyed, but restrained his impatience to see how the class would resolve what was turning out to be an endless – if descriptively interesting – session. He waited about thirty minutes before reminding the class, in a mild tone, of the presenter's original request. The students nodded, but shortly thereafter returned to their substantive questioning. The presenter patiently answered all queries, while the instructor grew increasingly impatient both with her politeness and with the participants' wandering in this sterile desert. But he waited another half-hour in order to underline the lesson about to be drawn. Finally, he cut abruptly into the collective conversation, pointing out that G. and her colleagues had spent two or three years collecting data and analyzing them, and she was filled to the brim with information – of course they could go on "forever" getting more and more descriptive information about the data from her if they continued in this vein. But, he reminded them again, though you are all enjoying this session, you are wasting her time unconscionably, and you are *not* answering her request for new, alternative *analyses* of her data. "Tell me: What do you think you are doing? What has been going on for the last hour?" The students quickly realized what had been happening, laughed at themselves, and turned to the real work of the day.

The students were violating a *rule of thumb* in the grounded theory mode of analysis: Namely, don't waste valuable time and effort in gathering *too* much data before beginning the analysis. Analysis can begin with very little data as long as the researcher takes the analysis as provisional – to be checked out.

Case 5

Connecting macroscopic conditions and microscopic data

As noted earlier, making connections between larger structural conditions and interactional consequences is a skill that needs to be developed. The following seminar session illustrates some first steps in directing students' attention to the necessity of making these connections and in suggesting how they might begin that operation.

This particular group of seminar students had been analyzing data given by one of them (who was also an audiologist) about her work and occupation –

about the sales and selection of hearing aids, and so on. At the beginning of this particular session the participants moved into a discussion of the manufacturing and marketing of hearing aids, since some data were also available immediately about that. The instructor asked if they wished to focus on the hearing-aid industry? They did. And so the class elicited rich, experiential, professionally derived data from the audiologist in their midst, while they themselves quite effectively analyzed the data.

During the closing minutes of the class, the instructor put two lists on the blackboard: to the left, the properties of the industry and market; to the right, the option processes. He noted that the left list included structural properties of industry/market (macro, in common parlance); while the option processes might be referred to by some people as microconsequences. Then he posed a question and gave an assignment. "In writing your memos for next week's session, I want you to connect up the two lists written on the blackboard. How will you go about doing that?" The students came up with several possibilities.

1. Do a memo connecting up many of the items on one list with items on the other list.
2. Take one or two structural conditions and see, in some detail, how they affect the option processes.
3. Take one or two processes and link them up with the market/industry conditions – and besides, the processes could be conditions affecting the structural items.
4. Perhaps we have blackboxed intervening conditions that lie between macro and micro items; so try to open up the black box and connect it with both lists.

The instructor then asked the students to divide up these tasks and write their memos for next week's presentation.

The *rule of thumb* for linking the structural and interactional was suggested in the chapter on coding: First, think both structurally *and* interactionally; but second, collect and scrutinize data bearing on both aspects, looking carefully for *specific* connections between them. These will be in the usual form of conditions and consequences, as well as associated interactions and strategies. The seminar session described above is just beginning these necessary operations, under the guidance of the instructor.

Case 6

Determining the central issue in the study

This graduate student had never attended the seminar on qualitative analysis because of the exigencies of his full-time position as a highly skilled social worker at a large medical center. There, he has for some years counseled the families of very ill children, many of whom have been operated on successfully for inherited cardiac anomalies but whose illnesses and illness experiences impinge forcefully on both them and their families. This researcher, therefore, has an enormous amount of experiential (professional) data, as well as other data gathered recently for his thesis research. His major problem is having too much data, as well as having very little training in grounded theory analytic procedures. In this regard, he is much like many other researchers, faced with mountains of data and not knowing how to make the most out of their examination of these data.

The instructor's strategy is to get him (much as if he's an anthropologist's key informant) to articulate what he senses is *the* major issue; then to get him focused analytically on it through a series of specific research questions. (The second section of Chapter 2 reproduced a short interview and analysis of it, done some weeks later.) After the session, the student wrote up the following précis.

A.S.: The new director of the Social Work Department, where I work, is very supportive of me finishing my degree as soon as possible. He will allow me time away from work to concentrate on it. I'm basically in need of some structure and direction in terms of how to organize both my thoughts, ideas, and data. Will there be another course, offered soon, on analysis? I would like to enroll in it. Could we negotiate some time when I could come in to see you more often?

A.L.S.: Maybe something in the fall. However, if you can call from work, we can discuss some things by phone, that would save money and time. Also, last year's cohort group established an interesting network. Some tapes of classes are made: Maybe you can contact them and listen to some of the tapes. It would be good, also, to meet with them, share some ideas and data with them, and get feedback. That's always helpful. You have a lot of data in your head that needs to be coded. You could sit down and have a conversation with self and record what emerges.

A.S.: I'm needing all of that. I want to finish, and bring this experience to closure by getting my degree.

A.L.S.: Let's start somewhere. I want you to think for a bit and tell us, What are the major issues involved in the situation that you're studying? What do people argue about?

A.S.: (After some deliberation and thought) Could one issue be the reality of the suddenness of the situation? The immediate impact the child's cardiac condition has on the family?

A.L.S.: Well, it could be. However, what else could be a real issue that's specific to the situation you work in?

A.S.: Could the matter of the decision to do or not to do surgery be an issue? There are occasions when heated discussions develop about the family's abilities to handle the surgery, yet the child may need it.

A.L.S.: Could be. Go on. I'm moving you along for a reason, which we'll talk about later.

A.S.: Sometimes defining the patient is an issue. Some physicians and others will only work with the patient, excluding the family altogether. Others will treat the child as well as the family. So I think one issue could be who decides who will be treated and how. Another issue could be the nonmonetary costs to the family. Is it really worth all of the effort to treat a child who will never be completely well or perfect?

A.L.S.: Continue.
A.S.: I think that information control is an issue. What are parents told? How do they find out what they know?
A.L.S.: Information control is a real issue. However, much of what you've talked about are elements of the situation or conditions, and not necessarily issues . . . . Let us look at this from another angle. Let us play an unreal if perhaps psychologically difficult game. Suppose you were told that you had cancer and had only six months to live. What message would you want to leave for physicians and others?
A.S.: (Thinks for about one minute) Tell them that one must provide caregiving that goes beyond taking care of. Got to understand better what factors and interactions that converge and have impact upon the overall outcome.
A.L.S.: What outcome?
A.S.: Whether the children survive or not: I'm convinced that with only medical care they would not survive. The family's efforts are essential.
A.L.S.: *That's* your main issue . . . . There is a book by the Massies you should read. (Goes out and returns with *Journey* by Suzanne and Robert Massie.) This couple writes about their son, who has hemophilia. It's somewhat different from the illness you're studying, but in a sense it's related. The book is journalistic, but worthwhile. Are there children whom you work with where death is a potential reality, but you never know when it's going to occur?
A.S.: Yes. Quite frequently, there are children whose status is so tenuous that it's always hanging over their heads and their parents'.
A.L.S.: If you were to write a monograph of your experiences with children you work with, how would it differ from what the Massies have done?
A.S.: (After some pondering) I'm not sure I know.
A.L.S.: Your reference base was developed by thirteen years of experiences with hundreds of children and families; theirs was based on living with one child, as valid as that tends to be. I think you would be able to collect a great deal of descriptive data and treat it analytically, focus it . . . . I would think that your primary issues are, How do families keep the child alive, and manage to hang on?; and of course, ultimately, the two are related.
How do the children survive physically and live with minimal emotional scarring? This is done largely by nonmedical means.
You are convinced that nonmedical aspects are central and paramount.
What goes on biographically with families and their kids which affects the survival of the child? – which medical people don't see.
What you will be doing sociologically is bringing news to people that they already vaguely understand. However, if they really understood, they would act differently. Physicians know the importance of the family, but do they really *know* it?
Your job is, How to be more convincing?, and a monograph would be dramatic.

One of the basic questions for parents: How "normal" can they keep this situation? The whole normal–abnormal spectrum: What does it do to the family that affects the child's survival? You should think through issues, and ask: How does it fit in with keeping the child alive and moving into adulthood? This couldn't be done with medicine alone. When it's done otherwise, there's failure, with devastating impact. Think about how this fits into the main issue of keeping the child alive with minimal emotional scarring.

How to pull it together:

Think about the numerous experiences you've had with families over a thirteen-year period.
Write memos to yourself.
Reach some conclusions about connections and write them down.

Look at the data:

Ask what does it have to do with *survival issues.* Take each incident, each paragraph, and see where it fits.
If it doesn't, discard and forget.
Parents must work at balancing risks – present-moment impact on future identity building.
How to draw on data – look at the issues as they emerge and fight it out with yourself.
The reality issue of *work* – the red thread that runs through the fabric: They work *hard* at the child's survival.
What is the nature of the work, and of work that has to be done, Where, and how?

In short, the instructor was giving this beginning researcher a series of *guidelines*. First, how to have confidence that, indeed, he knew what his central theoretical issue was, and how to muster the courage to commit himself to it – to the extent at least of asking theoretically informed questions addressed to it. Second, by taking a series of steps to further the analysis (the usual ones), but always by keeping an eye on the central theoretical issue, doing selective coding, then, rather than merely doing open coding.

Case 7

Finding the hole in the diagram

The next case illustrates how a student's integrative diagram, which had a crucial hole in it, was corrected during a lengthy seminar discussion of the diagram. The presenting student had finished virtually all of his data collection and analysis and, at this session, was seeking validation from the seminar participants and anticipating questions that might fill in analytic details. Otherwise, he was quite pleased with his analysis and diagram. He was a clinical psychologist, well trained in a tradition consonant with what he had learned about grounded theory method from attending this seminar. He was making a study of the socialization of young physicians who had chosen the specialty of family practice and who were now doing the first year of their residencies at a family-practice clinic. What the researcher had discovered was that these young practitioners had entered their residencies with considerable idealism and then become increasingly frustrated, as they struggled with the realities of "too many patients," "not enough time," "not enough resources," and so on. They found they were learning far less than they expected because of these impeding contingencies. The researcher had elaborately diagrammed his analysis, and presented this diagram at the beginning of the seminar session. It covered three pages and consisted of numerous structural conditions and categories diagrammed within boxes, various arrows indicating linkages among the diverse boxes, etc.

After scanning this diagram, the seminar participants peppered its author with a series of questions, trying to understand what he meant by various boxes, the arrows between them, etc. They also made both substantive and analytic comments, following on his confident and, on the whole, quite clear responses. The instructor remained relatively silent, allowing this probing discussion to flow along. Finally, as the discussion reached the last set of boxes, the students "pushed" on the issue of what the residents' work overload and resulting frustration did to their idealism, notions of good medicine, and lives: That is, the discussional focus had moved more to the biographical than to the work aspects of the residents' lives.

The instructor then entered into the discussion, recognizing this new focus was essential to the diagram and its integrative function. The researcher evinced some confusion, saying that maybe the biographical was causal. One student pointedly said; "*That* isn't in the diagram." So the instructor confronted the presenting student with the blunt question: "During this crisis of identity of which you are speaking, during this period, are they reconstructing their lives or just figuring out strategies for handling their educational work more effectively? If you can answer that, you can ask pertinent empirical questions." The researcher thought briefly and replied: "Strategies for handling their educational work?" Instructor: "That is the heart of the socialization matter," and noted that it wasn't in the diagram, but would have to be spelled out analytically. The researcher laughed, the students laughed, and the discussion moved further along those lines. By the following week, the researcher had made much progress in closing off this central part of his analysis.

A general *rule of thumb* illustrated by this "filling in of the hole" is essentially identical with that applied to the last case, except that here it is applied to a researcher who is much further along in his work. The rule pertains to discovering the central theoretical issue to which all further analyses will relate. And the rule itself: Force the researcher (who may be oneself) to confront alternatively considered central issues, making him or her choose the one that seems to encompass analytically the greatest part of the data. This choice will indeed yield the core category (survival, in the previous case, and here, educational work). Researchers are often loath to make those choices, because of uncertainty about "the" right one, because so many options still look exciting, or because they fear that choosing one will force them to abandon seemingly relevant other ones, and so on. But when the researcher is as far along as in this particular case, he or she must make that choice at least provisionally, to see both what it encompasses and what it might leave out. (See the section on pruning, in the chapter about writing for publication.) In short, the rule of thumb furthers the process of integrating the entire study when the central issue and core category fail to "emerge" unambiguously. The instructor sensed the incomplete integration in this researcher's analysis, first when looking at the diagram, then from the students' questions about it, and finally when noticing the hole (literally) in the diagram.

Summary

As remarked at the outset of this chapter, some common problems encountered in doing qualitative analysis can be illustrated by students' floundering over them, yet even quite experienced researchers may encounter these problems when engaged in new projects. Especially is this so when the research terrain and its data are unfamiliar. Then an experienced researcher may falter under the weight of too much data, gathered because of the circumstances of the field itself or because of anxiety, or because for the first time he or she has collected historical or masses of other documentary data. The researcher may also have difficulty in determining the central theoretical issue, may be flooded with experiential data, may even encounter more problems than usual in linking structural and interactional aspects because of too much focus on only one when collecting data. Such common problems are potentially lurking in the wings not only for grounded theorists but for researchers engaged in other modes of qualitative analysis.

8 Integrative diagrams and integrative sessions

Undoubtedly, *the* most difficult skill to learn is "how to make everything come together" – how to integrate one's separate, if cumulative, analyses. If the final product is an integrated theory, then *integrating* is the accurate term for this complex process. (See also Chapter 1.) This is why the inexperienced researcher will never feel secure in how to complete an entire integration until he or she has struggled with the process, beginning early and ending only with the final write-up. Perhaps the integration is more difficult for grounded theorists because they cannot integrate their research by opting for "story lines," resting only on a conceptual framework, or on several themes or on a few concepts, or on concepts that are not carefully related to each other in the total analysis.

Correspondingly, the most difficult to convey feature of memo writing pertains to the integrative features; including, how the important categories are kept doggedly in analytic focus, and how that focus is embodied in a sequence of memos. These memos will be sorted from time to time, and from that summary, memos will be written, such as the one on safety in the previous chapter. Along with the sorting, the memo sequences and a succession of operational integrative diagrams, together, can help to keep the cumulative analysis much more orderly – and more clear, in the researcher's head.

Given the difficulty of teaching, let alone learning, how to integrate the complex analyses involved in grounded theory studies, we shall present in each of the next two chapters a fairly lengthy case. These will illustrate integrative steps. There are also, in this chapter, commentaries written, after three consultative sessions, by the recipients. For getting maximum benefit from the materials, you should probably do the following. First, scan each case to grasp the general ideas they illustrate. Second, re-read the cases later, more carefully, to see how the integration actually occurred. Third, when actually faced with integrative problems of your own, study each case intently *and* the rules of thumb given at the ends of this and the following chapter. By then, the illustrated integrative steps or phases will make additional sense, linking with your own struggles to achieve analytic completeness and unity. This mode of reading should be helpful, although one cannot get a genuine sense of the steps and the intense work involved in the entire integrative process until one has actually been through it at least once.

Steps in integrative diagramming: a work session

This case consists of a conference between a predoctoral student, Leigh Star, and Anselm Strauss. The session itself and the analytic commentary on it illustrate (1) *how* integration begins to build, (2) the kind of *work* it takes to do this building, and (3) the several *functions of operational diagrams* in the integrating process. What are these functions? They include, at least:

helping to pull together what you think you already know;
thereby contributing to analytic and psychological security;
stimulating you to follow through with the implications of the diagram;
clarifying what you do not know (i.e., gaps in your knowledge or understanding), and so stimulating the next steps in filling gaps;
acting as a touchstone that allows you to relate new analytic advances to the main line of your previous analysis.

Turning now to the session itself: The student was far along in collecting and analyzing her data; has a fine analytic mind; and will be seen interacting essentially on equal footing with her "official thesis advisor." The student is deep into the sociology of science and into the substantive materials of her research (on the development of brain localization work and associated debates during the late nineteenth century); whereas the instructor knows relatively little about these materials, something more about the sociology of science, and much about the sociology of work. Readers of this conference summary may have difficulty with its substantive detail (as did the instructor) but should readily understand the main flow of the discussion. The main issue is: How did a particular view of brain functioning get widespread and persistent acceptance (it still persists) among neurophysiologists, despite its very shaky scientific basis? The study had proposed going from a more general focus to targeting this now seemingly salient issue.

The student will be seen, here, feeding into the discussion – especially data collected recently during a visit to English hospitals to scrutinize medical records and other historical documents – but doing so in conjunction with specific analytic points as they arise. Many of those points are raised by her: some directly suggested by the discussion, and others deriving from her earlier reflection on the data. The instructor moves along in his analytic thinking in response to her choosing of areas for discussion, but also raises a host of specific questions and issues (this is his main procedural tactic) about each area during the evolving session. The purpose of this session as enunciated by the student was to map out salient areas in the data, to get an overview, before going into greater depth later and by herself.

During the conference, the analysis turns largely around the unforeseen evolution of an integrative diagram. Its initial version emerges quickly, during the first minutes of the session, and gradually becomes revised and more elaborate as the analysis fills in possible relationships among the main diagrammatic elements, and adds new elements and relationships. The diagram provided visual stimulation, too, which helped visualization of some of those possible relationships. All that amounts to saying is that the total analysis got systematically furthered, that integrative steps were taken, and categories were rendered more precise and analytically powerful. The session is notable also for the speed and cumulative development of its analytic evolution. Of course, this first productive integrative session was followed by the student's further analytic struggle, leading to new diagrams throughout the course of her investigation.

The session will be presented now, divided into phases, with the instructor's analytic commentaries inserted after each phase. This kind of commentary was directed at seminar materials in Chapter 5, and many of the points made then are similar to those made below; however, here the emphasis is also on the integrative process rather than on just the coding.

Phase 1

1. L.: What I'd like to do in this session is to map out salient areas in the data and try to get a sense of what the territory looks like overall, then go back to individual things in greater depth later.

2. One of the first things I noticed, probably the most striking thing, is that these guys didn't change all that much over time. I saw these notebooks that spanned from 1873 to 1890 or so, and Ferrier had the same ideas, the same concepts, that he used right from the beginning. It was almost like he was just filling in a grid.

3. A.: Ferrier?
4. L.: Ferrier was one of the guys at the National Hospital. He did monkey experiments and was also a physician . . . I looked at his notebooks which were left from this period.
5. A.: So he spent eighteen years essentially filling in this map?
6. L.: Yes. It got more packaged up over time . . . . After the first ten years, for instance, they started having mimeographed forms of the brain, that they could just shade in the areas. Which was really silly because brains are very very different from one another, and he was trying to achieve a precision of up to one-sixteenth of a millimeter.
7. A.: So there was a kind of fitting process that went on, using the information from the monkey experiments and putting it into this map? But there was information that didn't fit, what happened with that?
8. L.: Well, that was a really interesting thing, too. One of the other areas that I wanted to talk about is standardization. It seems like these guys were more interested in coming up with standard maps than with verificational work that would get the right function mapped onto the right place in the brain. For instance, I read a referee's report for a paper that Ferrier submitted to the Royal Society in 1876. The referee, Rolleston, notes a discrepancy between the places labeled by Ferrier in a publication of his in 1873 and the paper submitted in 1876. It's like in the 1873 paper he had the following diagram (Figure 9) and it was 1 = leg, 2 = arm, 3 = eye. Well, by 1876 he had something like 3 = leg, 2 = arm, and 1 = eye. Now these are places that are supposed to be very precise indicators of the motor areas in the brain and so forth. But what Rolleston says is that Ferrier should re-label the diagram to make it fit with his earlier diagram. Not re-do the data to check out this anomaly! . . . So there's some kind of tension or conflict that I'm seeing between the mapping work and the verification work.
9. A.: What's important here is the perceived priority in terms of the work that they were doing. Obviously, the mapping work was perceived as the main enterprise, and other kinds of work either were used to further that concern or somehow dumped. Would it be possible for us to make a list of the kinds of work that were going on, and what the priorities were?
10. L.: Well, there was the standardization-mapping work. And then there was verification work, which was underneath it, not so much of a priority. And then there was clinical work, too (Figure 10).
11. A.: What about the inical work? Was that used to further the mapping work?
12. L.: Yes. No. Well, they claimed that it did. But I'm not really sure if it did.
13. A.: Did the mapping work feed the clinical work?
14. L.: Well, it legitimated it, I guess.

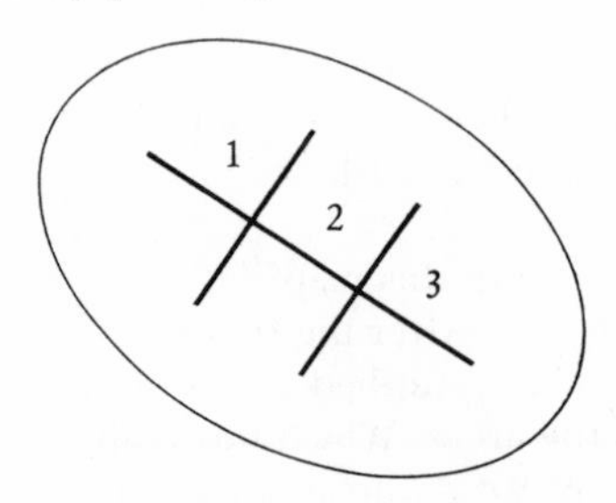

Figure 9. Toward integration: first operational diagram.

Standardization-mapping

verification

clinical

Figure 10. Toward integration: second operational diagram.

15. A.: But did they use it in the clinical work?
16. L.: They claimed to. That's the funny thing, it was like they were trying to use it to make the clinical work so routine, and to use the clinical work to help legitimate the mapping work. But I'm not sure how much of it really went on.

Phase 1, Commentary

1.		Student states purpose of session.
2.	(2)	Student gives data and insight.
	(3–6)	Filling in of information.
	(7–8)	Question: the information that doesn't fit?
	(8)	Answer stimulates talk about standardization, standard maps, giving of specific data, and prompts first version of diagram.
	(9)	Question: Can we list types of work and priorities?
	(10)	Three types of work are diagrammed.
	(11–16)	A. S. directs question at clinical work and mapping relationships . . . question and answer, and insight from the data.

Phase 2

1. A.: Let's put that aside for a minute, black box that relationship, and go on and talk about this verification work and what was involved

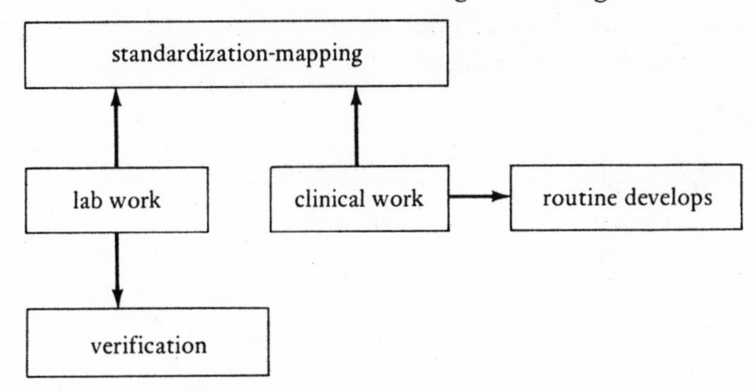

Figure 11. Toward integration: third operational diagram.

in the other kinds of work. What was the verification work, and did it feed the mapping enterprise? If so, How? So, first of all, Where is the verification work done and by whom?

2. L.: Well, it's done in the laboratories. Ferrier, for instance. But not a lot of verification work.
3. A.: Are there other kinds of work that go on in the laboratory?
4. L.: Sure, there's this mapping . . . the surgery . . . the electrical stuff.
5. A.: So the lab work is used to feed into the mapping, and there's other kinds of work that go on there, too?
6. L.: Yes.
7. A.: Ok, so we have both lab work and clinical work going into the main enterprise, which is producing this map. And there's also verification work going on at the laboratory. Is there also verification work going on in the clinical situation?
8. L.: Well, not really. That is, the press in the clinical situation seems to be toward developing flags that will allow you to look for the usual. To develop a routine. This was another thing on the list that I wanted to talk about. There's some kind of tension between this trying to get the usual, the routine packaged up, and looking for the extraordinary case to write up to present to further legitimate the mapping process. The diagnosis stuff did get increasingly routine. Just like the physio guys had mimeographed brains to fill in, they developed forms to fill in for epileptic fits, and for brain tumor symptoms, localizing symptoms. In 1860, I get long narratives in the records about epileptic fits, for instance. By 1880 they have "fit sheets" that you can just check off the movements of the hand or jerks of muscles, etc. Same thing for the brain tumor symptoms.
9. A.: So it was pretty much routine procedure by this time.
10. L.: Well, it seemed to be. They wanted it to be. (Change diagram – Figure 11.)
11. A.: Did the routinization serve to further the mapping process?
12. L.: The routinization claims did. In fact it wasn't so routine at all. A lot of times they would just let the tumor patients die because they

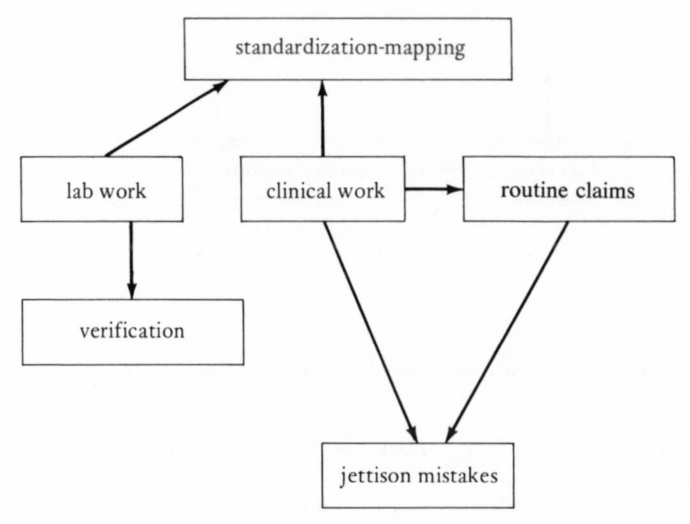

Figure 12. Toward integration: fourth operational diagram.

didn't have any idea what to do or what the diagnosis was, even after the operation had been invented.

13. There was also a historical accident, which was the introduction of potassium bromide for the treatment of epilepsy. It was the first really effective drug to control seizures, and it was introduced some time in the 1870s, I think. They used, literally, tons of it at the National Hospital. So you had this effectiveness that somehow could be parlayed into legitimating the maps, I think.

14. A.: OK, but what about the tumor patients who died? These clinical people were, literally, burying their mistakes. Why? Why were the mistakes ignored?

15. L.: Well, that's it. It's not routine at all, it claims to routineness. And the claims are used to legitimate the mapping enterprise . . . anything that jeopardizes the claims to routineness is jettisoned. Like this (Figure 12).

Phase 2, Commentary

1. (1) A. S.: wait (black box that): go on with "work". . . questions about verification work.
   (2–5) Questions, data feeding: further analytic understanding.
   (6) An elaboration of the diagram.
2. (7) Summary point; next question: relating verification work to clinical work.
   (8–10) Insight and data: the issue of routine and "usual".
   (11) Question: routine and mapping?

(12) Data re patients dying.
(13) Analytic questions about that.
3. (15) The answer leads to further elaboration of the diagram.

During the next four phases of this work session, more questions and issues were raised, more data were reviewed and pinned down provisionally, and a third diagram was drawn. During the sixth phase, an additional important linkage of two boxes was added to the previous diagram. In the eighth and final phase, a summary diagram was worked out that covered all relationships built into it by this consultatory discussion. One would have to be conversant with all details of this discussion to know exactly what was meant, but one can immediately see the great advance in detail over the intial two diagrams. It looks like this:

Phase 8

1. A.: Good. So now let's look at what we've got in terms of a diagram (Figure 13).
2. L.: The only thing that's missing is a list of the clinical verification work and the way it relates to routine claims. I noticed in England that the blank on the patient records for "diagnosis" was often, especially in the case of a brain tumor, filled in after the postmortem was done! I could tell 'cause the P.M. was written in different ink. Isn't that amazing . . . it wasn't a diagnosis at all. But it was presented that way.
3. A.: So part of the clinical verification work involved reconstruction.
4. L.: Yes, a deletion of uncertainty retrospectively.
5. A.: So there's a connection there between both burying the mistakes and claims to routineness. And both feed up into the mapping thing.
6. You need to go back to the data and look at that one more carefully and trace it out.
7. L.: So what I'm getting out of this is breaking down the idea of "theory as commodity" into the various types of work that go into the mapping enterprise. The inertia I picked up is structured in multiple ways into the various parts of the work and their relationships.

Phase 8, Commentary

1. (1) Back to diagram to summarize all relationships now built into it by this conference discussion.

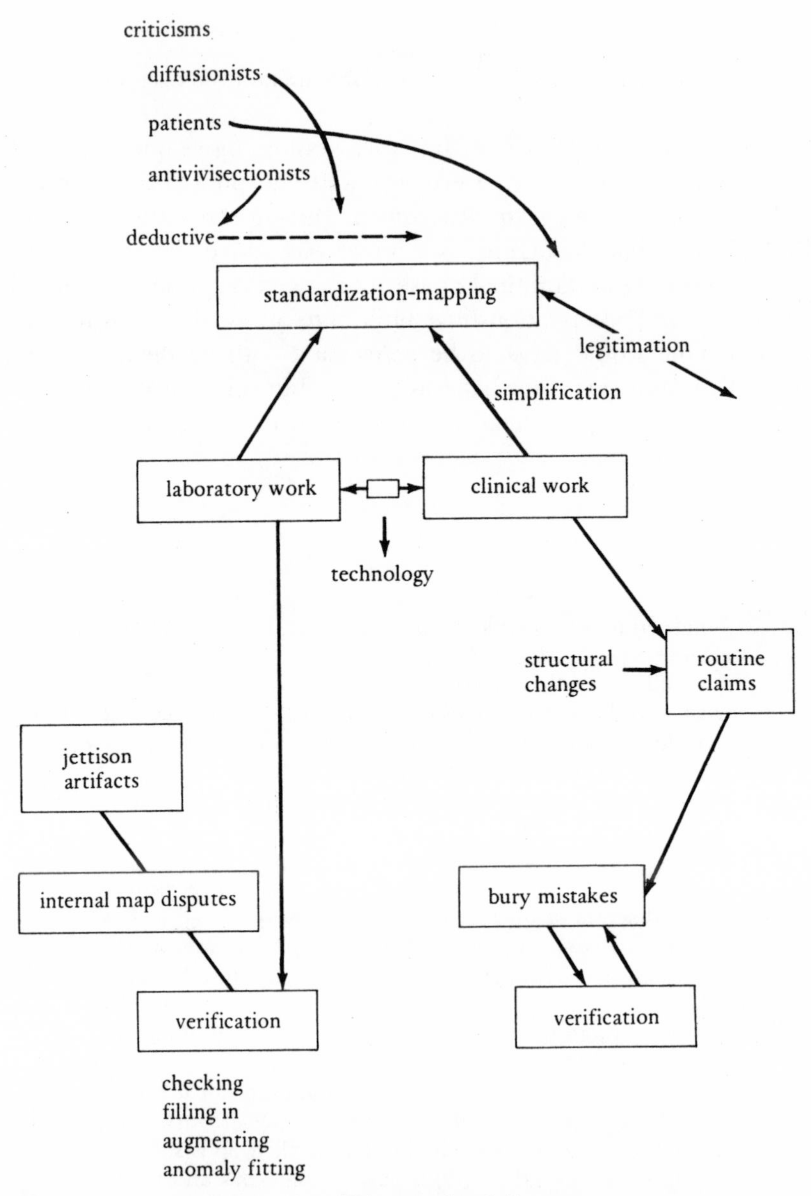

Figure 13. Toward integration: fifth operational diagram.

2. (2) Noting missing detail plus data.
   (3–5) Discussion of that.
   (6) Advice on a next step that should be taken.
   (7) L. S.'s summary of general line of attack taken in this conference; then one last relational connection is made by her.

Here now are Leigh Star's retrospective notes on the session, written three weeks later.

(Star, February 28, 1983): Retrospective notes on diagrams

Anselm asked me to write a page on how the diagramming process had been useful to me, looking back over the thesis process.

The diagrams have functioned in two ways: first, as sort of resting places in the process, places to tie up loose ends, take a deep breath, and feel (at least for a while) that some order had been brought to the chaos in the mountains of data; second, as ways of ferreting out unseen connections, unthought-of relationships. It's this second function I want to talk about here.

The diagramming process would begin with a phrase of single code, perhaps even a hunch about what was important in the analysis at that point in time. Several kinds of questions would come from Anselm, Elihu, or from students at that point: "And then what happened?" "Who else was involved?" "How does that relate to the point you raised last month?" "Doesn't that contradict what we usually think about in relation to this point?" "Did it always happen like that, or were there exceptions? What were they?"

These kinds of questions formed the "tendrils" out from the original idea – arrows and boxes showing connections of temporal progression. Some of them were intended to analyze the material over time; some, intended to get at variance, or the completeness of the diagrammatic representation; some to show logical, historical, or political connections between events or processes. After some time of sketching in these arrows and boxes, one or several gaps would usually appear in the diagram, and these were critical. "Now what would happen if you connected these two together?" someone would ask, often Anselm. I'd realize that that was a connection I'd never thought of, a gap in my knowledge of the data, or a blind spot in my usual coding procedures. Filling in that gap, or at least questioning intensely why it should be a gap, was often the next step in the analytic process. A couple of times it became my work for the next few weeks.

Looking back over the diagrams we've made over the course of this research (about two years now), they appear as records of questions, blind spots, and gaps, as well as increasingly complex syntheses of the data. This visual "story" of the thesis process is a useful organizational tool as I begin the final write-up of this material. It helps keep me close to the data, and to remember that the codes and concepts I'm writing about grew from mistakes and collective work, not from logical imperatives!

Two postconsultation comments

It may also be useful to reproduce two additional statements written respectively by a student and a junior co-researcher after their consultatory sessions. These illustrate, again, the feelings of closure that come after successful integrative sessions. The first statement is by Barbara

Bowers, who had been struggling to make a final determination of the major integrative features of her thesis research. The instructor focused on getting her to make specific commitments about relationships implied in her most recent integrative diagram. Afterward, the student described vividly her experiences before, during, and after the consultation. Her frank account underlines the difficulties involved in achieving integration when the young researcher has not yet gone through the entire integrative process, and so still lacks the requisite experience and confidence.

Memo to Anselm

I spent about an hour with Anselm yesterday. After we finished, he asked me to write a memo on something we had talked about. Anselm wanted me to write a memo on how I spent the better part of three months wallowing, suffering, agonizing over my data. I can remember clearly, from about three months ago to fairly recently, looking at my data and wondering what it was all about, where any of it was going, what I could possibly do with it, if I had much of anything there to work with, whether I was going to be able to write a dissertation at all!

That was what I was thinking. What I was feeling was even more distressing. There is a period of time when I can only describe the experience as very lonely. It was a feeling of being out there alone with my data, drowning in the data. So many ideas and fragments of insights kept flying around, but none of them seemed to be very connected. There were some moments that felt almost hopeless. I should add that there were times during that three-month period when some of those insights and ideas really caught my attention and I was delighted by them. That was fun and it felt great. There are definite highs and lows in this process. The highs are terrific. The lows just need to be recognized as a necessary part of the creative process and used to advantage. Those times are more likely, for me, to occur when I'm being flooded with data and can't keep ahead of it.

So what did I do with that time? I coded, read (articles, novels, whatever), interviewed, coded, read, interviewed. I stacked up piles of memos on categories, dimensions, fragments of possible relationships. If someone had asked me to explain 'what my data was about' or what the 'story line' was, I wouldn't have been able to tell them. There was no central issue or consistent thread.

Then, a few weeks ago, something happened. I remembered going to the dissertation seminar and saying, although I hadn't planned to, "I think I'm coming up out of the mud." I don't really understand what work I had been doing "down in the mud," but I'm sure it was necessary. Something really exciting was starting to happen. I made an appointment to talk with Lenny Schatzman. The first thing he asked me was, "Tell me what your dissertation is about." That dreaded question! To my astonishment I told him. It made sense to both of us.

That evening I went home and took out my stacks of memos, spread them all over the floor, and began to sort. I spent the next three days sorting, writing memos on new relationships I was discovering while I was sorting, resorting, and writing more memos. The relationships were emerging.

Still being a bit reluctant to trust myself, I labeled my memos in pencil (so they could be easily erased and relabeled). Next, I began drawing diagrams of categories, the way I had watched Anselm do when I would talk to him about data. By the time I went to see Anselm, a week later, I had several pages of experimental diagrams. None of these had arrows to indicate the direction of relationships, only lines to indicate that a relationship existed. I wasn't finding the diagram terribly helpful. I was also worried about where or how to integrate the policy questions that I was trying to integrate. I felt like they look glued on, like I had parts of two dissertations uncomfortably stuck together.

Anselm asked me to tell him about my dissertation. As I talked, he drew a diagram. Then he asked questions about the direction of relationships, which I hadn't done. We drew in the arrows and added some things I hadn't included in my earlier diagrams. The diagram fit! It felt good. Then, when Anselm asked me to show him where the policy work fit in, it was obvious. There was no problem placing it. It no longer felt forced.

I could see the relationships clearly. (I went home and put together a new diagram with a few additional categories and several new relationships.)

The second commentary was written after a conference between Juliet Corbin and Anselm Strauss, meeting specifically to handle the analytic integration of their joint study of chronic illness and the work of spouses. (See the memo sequence in the next chapter, also memos in Chapter 5.) Corbin had for some weeks been asking, "When is all this going to come together? We have so much now!" Strauss had deliberately been delaying a head-on attack on integration to keep their thinking more open, but asked her at the close of the previous team meeting to spend the next week doing two things. First, to go through all the codes, but not the memos (of course, she had the gist of many memos in her head), to see what would happen integratively if she did that. Second, to make an integrative diagram when she had finished with the codes. The session that followed turned around discussion of these completed tasks, and led to the further linking of categories and the development of the original diagram into a more elaborate and exciting version. (Three later ones were done over the next year.)

Reflecting upon this experience, I see it as having many different aspects. After the session, there was a feeling of excitement about the quality and extent of work that had been accomplished, but most of all excitement about the product of our work, the integration of our major concept into an overall theoretical scheme. There was the joy that accompanies discovery, not only regarding the overall scheme and relationships, but also from the discovery of how this piece of research contributed to the unfolding of the term trajectory. Coupled with this was a feeling of pride and accomplishment at having played a part in that development.

The session also led to an increased awareness of the complexity of the scheme as it finally evolved. As each new insight was added to the original diagram, I could see how complicated illness management, all its related work, and the impact it has really are. At the same time, there was the realization that there still remains a great deal to be discovered by us and others about trajectory, not only as it relates to illness but extends to other types of work situations. There was a great deal of learning that took place on at that stage of the analytic process, and on how to use diagrams and code outlines as powerful analytic tools. There was relief from the anxiety that came from wondering how we were ever going to make sense from all this data and a feeling of, "We did it!" when the relationships were finally firmly established and the scheme outlined. There was a sense of direction for the remaining work to be done on this project. Since we now know what our major and minor concepts are and how they relate, we can theoretically sample to test the hypothesized relationships under various conditions and to increase the density of the relationships. Finally, there was fatigue. It was a long, hard working session.

This researcher's commentary on the integrative session relates to the sequence of memos and to the tying-up, last, integrative diagram on this research project, which are reproduced in the next chapter.

Rules of thumb

Here are a few rules of thumb for doing this kind of work on integrative diagrams and the associated memoing. These may be scanned now or studied later when doing your integrative analyses. They apply to the solo researcher as well as researchers who are so fortunate as to have cooperative instructors or teammates with whom to work. The rules of thumb are:

1. Choose "next issues" in your data.
2. After examining the data, code and think about the codes, then make your first operational diagram.
3. Continue thinking about the codes and also about the diagram, meanwhile asking questions such as the following: How does the diagram suggest reexamining the data? Where does it fail to cover the data? What should be added to it? In short, play your questions and your data against each other repeatedly.
4. Keep raising questions about the codes, diagram, and data. All the while, you should be using the usual coding procedures, if only informally, since the conversation – with others or with yourself – will be taking precedence over strict coding during the session itself.

5. Then make a second diagram, incorporating the answers to your questions. The incorporation will involve changing the previous diagram so as to add boxes, make the boxes more complex, fill in missing relationships to the diagram as a whole.
6. Repeat, then, steps 2, 3, 4, and 5. Continue that entire process until the end of the session, perhaps even for additional ones if they follow through on the same codes and issues.
7. The last diagram in the series may find its place directly or indirectly in an overall integrative diagram, which in turn will be successively – if only occasionally – elaborated in much the same manner as these operational diagrams.

9 Integrative mechanisms: diagrams, memo sequences, writing

This chapter will address other means for integrating the entire final analysis: namely, integrative diagrams, memo sequences, and writing itself. First, however, three preliminary points need to be made. To begin with, the operational diagrams, and perhaps other operational graphic devices, help directly to integrate *clusters of analyses*, but only indirectly the *final analysis*. These diagrams, however, may contribute to filling out the more general integrative diagrams drawn from time to time.

The sorting of memos likewise will usually contribute directly to the integration of analytic clusters but, especially near the close of the research project, may also contribute quite directly to the total analysis. Memo sorting does this latter by clarifying the current integrative diagram, whether early or late in the project, and by clarifying for the researcher what the total analysis is and ought to be.

It can do the latter even with the use of integrative diagrams. As for coding: This makes a contribution to integrating both analytic clusters and the total analysis. Coding results are incorporated into the memos, and besides there is a recoding of old data along with coding of new data from time to time, as those procedures are deemed necessary because holes in the current analysis become apparent. In addition, the coding contributes to conceptual density, which in it itself is a part of the final total integration. It is true that integration can be made without much conceptual density (the multiple linkage of many categories, all linked with a core category or categories), but then recollect that this would leave the analysis very thin. (See Chapter 1.)

Integrative diagrams: rules of thumb

The functioning of a succession of integrative diagrams was touched on in Chapter 1. To supplement that discussion, here are additional points which essentially constitute *rules of thumb* for constructing this very important type of diagram.

1. An integrative diagram helps to give a clearer picture of where you have come from in the research after all that data collecting, coding, and memoing. It puts together into a larger pattern, however provisional, a lot of otherwise scattered materials – or scattered sense of those materials – into a sense that this project "has really gone somewhere." It also gives added assurance that, "We really have something here that makes the total study important or at least interesting."
2. An integrative diagram also gives direction to the forward thrust of the research. It does this not only for psychological reasons but also for analytic reasons. Examined carefully, but sometimes even casually, the diagram helps you to see what is lacking in your previous data collecting, coding, and memoing. Just as with the operational diagrams, black boxes will need to be opened up, relationships between them specified, clarified, and supplemented.
3. In addition, integrative diagrams need to include the separate analytic clusters provided by particular operational diagrams and memo sortings. A review of both in relation to each successive integrative diagram is strongly recommended.
4. There should not be just one integrative diagram but a succession of them over the course of the project. Each diagram should incorporate not only the preceding one, but also the new analyses done since the latter was drawn.
5. The number of such diagrams should not be numerous: You must not be obsessive about "keeping the analysis all together" every minute or at every point in the project. From time to time, however, the current integrative diagram should be looked at and you should ask of it: What does this fail to incorporate that I now know? When the additional knowledge becomes great enough, then it is time to draw another diagram.

Before leaving the topic of integrative diagrams, we shall reproduce one example of it near the close of this chapter. It is the last of three such diagrams done during the research project on chronically ill persons and their spouses. (For other materials on this project, see the memos in Chapter 5 and the integrative session commentary in Chapter 8.) Not all of the categories in this particular diagram may be understandable without reading the memo sequence taken from this project, reproduced in the next section, but the general idea of what such an integrative diagram looks like, and its considerable complexity, should be useful to see.

Integration through memo sequences and sets

Another major mechanism that contributes to the final integrated analysis is a progression of memos. This progression occurs quite naturally, as the researcher's attention is drawn to one or another theoretical issue during the research. The memos then tend to fall into sets of memos, for the researcher tends not to continue working on a given issue for very long before shifting to another. Sometimes, however, the "kick" may last for quite a number of weeks or months. It is useful to refer to the product of this longer period of attention as a *memo sequence*. Memo sequences include a number of memos and are characterized by their intensity of analysis as well as by their cumulative results. There are several useful rules of thumb that apply to these memo sequences and sets, but they will not be given until the end of this section, after readers have gotten a better sense of what a sequence might actually look like and what some of its main features are.

To that end, we reproduce next some memos from a long series written during the chronic illness–spouse project. Considerable space will be given to this series because it is important, not only to see how the progression of memos becomes cumulative but also for some other characteristics, touched on in pages introducting the sequence. In our experience, students and other inexperienced researchers learn quite quickly to write on-target memos, while their occasional and final sorting of memos permits them to finish a project with a fair amount of conceptual richness, density, and integration. But they get there much more slowly and with more diversions along the way than when – and if – they learn to juggle all the necessary analytic operations, in conjunction with the patient and persistent elaboration of thought exemplified by the sequence of memos reproduced below. Although the memos below are written by two researchers to each other, there is no reason whatever why a solo researcher cannot learn from their exchange.

The series of memos was written over several months by two researchers (J.C. and A.S.), working closely together on a study of chronically ill persons and their spouses. (For additional memos, see Chapter 5, and the postconsultation comment about integration in Chapter 6.) In the preceeding year, their attention had been focused on married couples' illness-trajectory work – work done in the service of managing a spouse's illness. However, the researchers had also coined terms like *identity work* and *biographical work*, based on analysis of many interviews. Yet, they had unknowingly blackboxed these terms, not yet realizing that these kinds of phenomena needed intense scrutiny.

The particular memos presented here were based principally on the analysis of a "cache of data" (Glaser and Strauss 1967, pp. 167–8), consisting of the writings of highly articulate people. Some of them were professional writers, who had detailed their own or their spouse's or children's experiences when "coming back" from acutely severe phases of chronic illnesses or when dying from fatal ones. (Some of their accounts were based explicitly on diaries or other chronologically ordered notes.) While the researchers were recording their cumulative analytic work in memos, they were also coding the contents of these books, which coding reflected a deepening and integrating of their analysis. We shall not reproduce this coding, but only the sequence of some of the analytic memos that incorporates it.

This sequence begins with the final memo of a preceding sequence, developed over many months (cf. Case 2 of Chapter 5). The next memo, written by the theoretically and socially sensitive researcher (J.C.), opens up the path to what turns out to be a necessary series of analyses of body, time, biography, and their respective relationships. Note how long it takes to do the exploration; also, how this analytic chase moves steadily along. The exploration is breaking new theoretical ground not only for this particular inquiry, whose focus is on illness trajectory management, but more generally for theory pertaining to body–time–biography and work. (See Corbin and Strauss 1986, forthcoming. See also Chapter 10, Case 1.)

Several other points are deserving of our attention about this sequence of cumulative memos. *First* of all: It consists of several sets of memos, because a given category (or relation among categories) comes into the forefront of examination and persists there for a time.

Second: There are also shiftings of foci when the exploration of one phenomenon is dropped while another is examined. The conditions which produce this shifting of focus include:

1. temporary exhaustion of the line of analysis;
2. coding calls attention to the possibilities of another line of analysis;
3. or further development of one category opens questions about aspects of another and so calls for further development; and
4. when there are two or more researchers on a project they are very likely – even when working closely together – occasionally to shift each other's focus as one of them becomes struck by "something," often because of collecting or scrutinizing different pieces of data. However, another condition for a team member shifting the focus is that one of them (as this particular sequence of memos will illustrate) grows dissatisfied with the team's current understanding of a phenomenon. He or she may mention that dissatisfaction from time to time without getting any answering response, and so may eventually pressure for a hearing verbally or through writing a memo. This can touch off a new line of analytic development which will, of course, include a series of cumulative memos about aspects of phenomena either not perceived before or unwittingly blackboxed.

Third: The memos sometimes come in quick sequence, but sometimes there are many days between the beginning of another series of memos, or even between one memo and another in the same series. This can be due to the researchers' attention turning to further coding: as they sense the need for stimulation of their capacities to theorize; or are a bit tired of memo writing and need coding as a contrast; or quite without design, more data are stumbled upon – in this instance, new biographies – and so they are coded.

Fourth: In these memos, theoretical sampling was only partly designed. ("Let's look at aphasia, where there is a speech disability, rather than a purely bodily disability," or "Let's look at a child's illness because it should highlight *x*.") Partly, however, the sampling was fortuitous, insofar as different books were simply drawn from library shelves. Like any other set of documents or interviews, these provided theoretical samples just because they turned out to emphasize new phenomena or to highlight others already analyzed. Of course, the more advanced the theoretical formulation becomes, the more sensitized is the researcher to such fortuitous theoretical samples.

Fifth: Occasional references to technical literature, either for their stimulation or contrast, are also notable in the memo sequence. On the whole, however, this strategy is not much resorted to at this phase of the project by these particular researchers. Yet, sensitivity to biographical and temporal issues was affected by background knowledge, doubtless drawn on implicitly (cf., Mead's conceptions of time and of self, 1934).

Sixth: It is easy to see, upon reading the memos, that conceptual density is increasing throughout the sequence, and so is integration. The researchers, of course, understood this; but their drawing of integrative diagrams at two points in the study, and occasional re-reading of codes and memos, helped to locate, "Where are we now?" as well as to initiate further integrative efforts.

Seventh: It should be emphasized that the whole process of formulating such *cumulative* memos is not always accomplished without fumbling, going up blind alleys, or some measure of intellectual doubt. This is especially true of earlier phases of a research project, or when researchers are less experienced than the senior researcher on this particular study. So there was more rapid and obvious snowballing of conceptual density and integration, as reflected in the memo sequence, than would often be characteristic of such memoing.

This reproduction of a memo sequence does omit a number of memos which were repetitive. The whole sequence represents only a set of provisional formulations; later, these were thought through further, aided by more data collecting and coding. The latter, as always, contributes to additional conceptual density, which of course is only schematically outlined in these memos. The author of any particular memo is noted, and the recipient's initials are put in parentheses – for example, JC(AS). Portions of memos omitted here are indicated by ellipses: . . . .

As with the memos reproduced in earlier chapters, you may not entirely understand the substance of each memo but, scanning the memo sequence, you should get a vivid sense of how each memo more or less follows through with some preceding ones and how the analyses cumulate. Later, a careful re-reading of this progression may help, as we shall note below, to further your own integrative efforts, especially when you are actively engaged in doing that aspect of the research. To get the fullest from the sequence may take quite close study, especially as the substantive materials may be foreign to your own experience. Also, pay special attention not only to the general points made below and the specific commentary in each memo, but to how the researchers are putting clusters of analysis together, as they examine new data, rethink the old, and reexamine older categories and their relationships with newly emerging ones.

The key developments to watch are biography, biographical time, and body conceptions, in relation to trajectory work: For trajectory is the core category in this study. Recollect that trajectory pertains to the course of an illness, and to the work done to control it. Additionally, it involves the effect on workers' relations (in this instance, mainly the ill person and the spouse), and how that in turn affects the trajectory work.

The sequence begins essentially with a discussion of the trajectory projects (the forecasting of the trajectory's course) and trajectory schemes (the anticipated immediate trajectory work). The first memo is focused on these two categories. The next begins what will develop into a new line of inquiry, involving the central topics mentioned. As they move along, these analyses become integrated with each other and with preceding ones (which had focused on various kinds of trajectory work and the associated division of labor between spouses). The immediately next memos reflect the researchers' struggles with relationships of the spouses' biographies to the trajectory work, including visualization of that work through trajectory schemes and projections. The potential instability of those illness trajectories leads the researchers to focus on biographical reviews (6/23). Later they are struck by the many and subtle temporal references in their data, so begin to relate these references of *biographical time* both to trajectories and to spouses' lives (7/22, 7/23, and 7/30).

Reading and coding a book by Cornelius Ryan and his wife about his three-year fight against cancer, while he was closing off his productive life, leads the researchers next to explore the concept of *closure* in relation to biographical and trajectory phasing (7/27); also to the "accepting" or "not accepting" of the limitations of an ultimately fatal disease (8/17). The 8/22 memo begins a closer linking of closures, biographical reviews, and biographical time; also it suggests the differential impact of biographical projection on each spouse in relation to their respective biographical phasing, as well as to trajectory phasing and work. And the 8/23 memo looks again at the Ryan book in terms of linked biographical and trajectory phasing and related biographical time and biographical review. Some weeks later (10/2), the researchers are thinking more abstractly about *types of time*, attempting to compare biographical time to these types.

Next begins an intense focus on questions pertaining to the body and its failure during illness. These questions were precipitated (at last! for it is rather late in the project) by the coding of more autobiographical and biographical accounts, particularly one by a handsome young woman, a model, who had undergone amputation of her cancerous leg. In the resulting memos, body issues are related quickly to biographical and trajectory considerations. The ramifications of what the researchers now term *biographical body conceptions* (BBCs, 11/15) are developed further over the next months (11/2, 11/11, 11/16, 11/21, 12/7, 1/25, 2/10, 2/17).

Near the end of this memo sequence, the researchers begin to think about types of trajectories (comeback, stabilized, downward), coding and memoing for them for the next two months. Into those memos are incorporated many of the conceptualizations developed in the preceding months. One of the first memos in this new series – on comeback – is dated 1/21/83. (See also the comeback trajectory analysis in Chapter 11.) We might add, in case readers are surprised at the many pauses between memos and the length of time covered by the entire sequence, that these occurred partly because of continued coding and partly because neither researcher was working full time on this study.

We suggest that there are at least five ways of profiting from reading this memo sequence.

1. It is probably best first to *scan the headings* of the entire sequence to catch:

its scope,
the repetition of its topics,
its persistent topics.

2. That picture can be filled in by next, similarly, *scanning the subheadings and italicized terms* which refer to categories and/or their relationships.
3. Then *read* the entire sequence *quickly*.
4. If interested in a particular topic (time, biography, etc.), then *read that intermittent series quickly again*.
5. Careful *study of the entire sequence or any of its subparts* can be especially helpful if you are embarking on such a sequence in your own research. However, remember that the sequence reproduced here is only one exemplification of the integrating process, not the only mode or model.

The memo sequence

JC(AS) – Late spring, 1982

Some additional thoughts on trajectory

A. Trajectory projection
   1. Defining the trajectory:
      (a) crystallized – defined image of future
      (b) not crystallized – undefined image of future
      (c) crosscutting are formulations and discovery of limitations by means of surgery or beginning therapy or rehabilitation, or trying and being able/or not able
   2. Actual trajectory (a) may work out as anticipated (b) unanticipated turn of events . . . .

B. Biographical work components of trajectory
   1. Identity recapturing – thinking about and coming to realize what has been lost – status, money – children, physical disability, activities, etc.
   2. Identity reconstruction – working out a new identity on basis of new trajectory projection
      (a) definition and redefining potential limitations and sometimes new opportunities
      (b) values reorientation – establishing new priorities about what is important in life
      (c) direction refocusing – taking new paths in light of new identity
      (d) adjustment comes – when not dwelling on loss but getting satisfaction from what one can do – coming to terms with . . . .

JC(AS) – 6/1/82

Trajectory projection and time

The following thoughts might spark you or we can explore further. I don't have a full grasp on it yet, but find the concept intriguing.

A. *Trajectory projection* includes not only present and future time but past time as well. In doing the interviews I found people not only wanted to talk about what is going on now but often gave an entire medical as well as biographical history. They take me through the house, show me their pictures and other mementos. Somehow I felt this was significant, but in what way I was not sure. Now it is beginning to make some sense. Then I began to read *A Private Battle* by Kathryn and Cornelius Ryan; it seems to jell in my mind even further, though I still feel that perhaps the logic is not quite right or that there are some missing pieces.

B. The trajectory projection involves *contextualizing the illness into the biography*. This contextualizing has two dimensions.

   1. *A type of illness-related work* which leads to visualization of its possible course. Will I ever be pain free? How long do I have? This involves at least

      a. acquiring, interpreting, and incorporating *information* about the illness

      b. *historical sequencing*, which is an ordering and relating of past events, symptoms, experiences with this illness up to the present, and at times other illnesses

      trying to make some sense out of past illness events in light of present events and knowledge.

   2. *A type of biographical work* – which leads to forecasting of the future of one's life in light of the illness. This involves:

      a. A *biographical review* – who was I in terms of appearance, relationships, social life, etc.

      Physicians, spouses, as well as the ill person do review.

New options may be opened up, others closed. A *new trajectory-projection* process takes place, a *recrystallization*, and a new scheme developed, incorporating illness and biographical components, *each* in terms of the other.

Where I am at now is looking at the reviewal points.
What brings them about?
For whom? physician, spouse, patient.
What new projection is formed? What do they include for each?
Degree of biographical or illness knowledge.
How do these projections compare for each party?
What new schemes are devised? What kind of work involved?
How are they carried out? Division of labor?
What sources are involved?
What are the consequences?

      b. An *imagery of present and future biography*, what I will look like, my ability to work, my relationships are changed and will never be the same, etc.

      The trajectory projection may be *crystallized* or *not yet crystallized* or *uncrystallized*. A crystallized projection is one in which one can say with a certain degree of certainty (until the next change in status of illness or biographical contingencies) that

I will be able to work this much
I will have so much pain or other symptoms
I will have to undergo so much treatment
I will be able to go here and not there, etc.

Once the trajectory projection is crystallized, a management scheme for the illness and one's life can be worked out, in terms of one another.

Sometimes the trajectory projection becomes *decrystallized*. That is, some change occurs either in the biography or with the illness (increase or decrease in symptoms, crises, new treatment, etc.) so that I no longer know how much pain will I have, how much work can I do, what will happen to my relationships, etc. With the appearance of new symptoms, or decreased ones, changes in the biography, etc., a trajectory *reviewal* takes place.

AS(JC) – 6/23/82

Trajectory projection and biographical reviews

(See preceding 6/82 memos)

1. There may be more than one biographical review: i.e., conditions for this being changes of illness, crises, nonmedical contingenices, etc. So think of *review #1, 2, 3, n.*
2. CRYSTALLIZATION AND BIOG REVIEW: *Decrystallization* of the trajectory projection requires precipitating conditions (illness and nonillness). These could be necessary conditions, but not sufficient. But they must occur in order that decrystallization occurs.

   If the reviewal occurs, it may not necessarily lead to decrystallization, then. The review may lead to conclusion that nothing much has changed; or to rearrangement of resources, division of labor, etc., to ensure that "nothing much has changed."

   In other words some kinds of trajectory projection will decrystallize, others don't. But you need a review for decrystallization to occur.

   MATRIX OF REVIEWAL: the set of conditions affecting the reviewal.
3. Iceboxing crystallization: When the trajectory projection gets decrystallized, it may not yet RECRYSTALLIZE because the sick person (and spouse) may ICEBOX the situation of their reviewal. Why?
   a. if a re-reversal is still possible, or
   b. extent of reversal is still unknown.

   But this *decrystallization is unstable* because it's neither recrystallized nor crystallized. Hence, "icebox". . . .

.
.
.

5. Kinds of bio revs: Julie, you are supposed to work on that! And from the data!
6. Variations in who and how and when reviewed: Husband's and wife's reviews can be different not only in kind, but when they occur (and IF they occur).

   These reviews may include not only "my" trajectory, but how it impacts on the other – or may not. And whether the other weighs heavily in imagery itself as important to it. AND "his trajectory" and how . . . on me, etc.

   Combined trajectories (both ill) involve: (a) husband's trajectory (as viewed by each); (b) wife's trajectory (as viewed by each); also (c) their combined trajectory as viewed – by each, and *together*. . . .

AS(JC) – 7/22/82

Biographical time and types of trajectory comeback

Some preliminary thoughts –

Reference: David Knox, *Portrait of Aphasia*. See also Fred Davis, *Passage through Crisis* (the polio families).

Comeback trajectories have a series of questions – which people involved re-ask all along the trajectory. These questions are convertible analytically into biographical time and often are couched that way by the people themselves.

1. Given present medical status after initial crisis (heart, stroke, whatnot), how reversible is this illness, how far back will–can I go?

THE FUTURE *may* be again like the PAST, *if* reversible, and so NOT like the PRESENT.

2. But how far back will (can) I come? All, much, some, a little? And with regard to which of my activities (since some may come back all the way, some not at all, etc.)?

This question of degree, again, has to do with the future being like the past, being simply a variant answer. In other words: How much like the past will the future be? Or if certain skills can't return, this may mean (say skills like playing piano or high diving) that the past returned to in the future is relearning of those particular skills. Thus, after my MI, I could not play fast piano for a while but could play very slowly at first, then faster. (Though that gets us into later questions: as below.)

3. How fast will I come back – whether partway, most of the way, all the way?

How quickly will the future be identical to my past? Does the *future-like-the-present* stay this way for a long time, medium, short time? (Before the future is like the past.)

This is also a PRESENT-DURATION question: How long will I be in this present condition (more or less)?

4. How long will I REMAIN ON THIS PLATEAU before moving?

This is the STUCK PRESENT: I am stuck here at this current level of functioning. So how long? Before I move, in the future, to a new level? In a certain sense the QUESTIONABLE FUTURE is a part of the STUCK PRESENT, collapsed into it; until one moves to the next level.

5. The second part of the previous question is really: Is this a permanent plateau, will I never progress any more?

Will my PRESENT PRESENT BE PERMANENT? So that I *have no future* (in terms of more reversibility, or "progress"), *other than my present present*?

Indeed, when a central activity gets killed (like a pianist's fingers or a lost arm) then "I have no future," as a pianist, is clearly said. If it is more total, as with small-comeback strokes or aphasics, then they may say the same. Knox at early phase of comeback trajectory kept saying she wanted to die.

6. Setbacks, when "progress" seems to reverse itself (i.e., another stroke, another heart attack).

LOST TIME is one conception of the person, now he must retrace steps – in more or less distance – thus the whole reversible process will take more time. "So I have lost time."

7. A total setback is a finishing off of the progressive steps toward making future like past. "This is it!"

This is a clear definition that #5 condition has arrived. *No future other than the present present* – at whatever "level" of past functioning is now present. Since new attacks don't necessarily thrust you all the way back – but they may, or even further (thus further into the past is possible).

8. A key question for actor is whether his/her own actions can affect the comeback: in degree or rate? Or is it all just dependent on fate? And more specifically, what actions will affect (exercise, following regimen, praying, etc.)? Also minimize setback probabilities.
9. Even when there is a total comeback, is it permanent? How long? How long before I know? How far back will I go if not permanent? (This is simply a variant of question–answer earlier.)
10. Fluctuations (hourly, daily, weekly) in actual functioning. For instance, no matter how far I came back – at whatever level of functioning – I would fluctuate (still do) in capacity. How far back will vary from slightly to enormously. But it is temporary: minutes, hours, a day or two or three.

TEMPORARY PAST IN THE PRESENT, with experiential knowledge that THE SOON FUTURE will be similar to JUST PAST.

Some subsidiary temporal issues in these fluctuations. How often? For how long? How far back will THIS one be? How shorten? Minimize effects? Maybe prevent this one? Waiting it out (lost time) – WAITING TIME. LOST TIME MAYBE, TOO . . . .

The spouse's biograph time picture shares some or all of the above, but probably involves other issues too, temporally. I won't bother to think that through now.

Temporality in the Knox book

The whole book is about slowly moving into the future with some measure of progress, with plateaus, one big setback, questions about how far reversible, and final pages, where it's only partly reversible to date and probably forever not much more reversible.

And the way such a book is written (present looking back, with open if cloudy future), it's really a narrated sequential past up to the present moment, but "as if" author and reader were moving through presents sliding into future step by step. The vivid sense of moving into the future – it all feels too much like he is looking back into the sequential-stepped pasts, in the pasts.

AS(JC) – 7/23/82

Biographical time: paradigm notes

First a word about Wolfram Fischer's excellent study. He was limited to interviews with relatively inarticulate people, who talked retrospectively about the past, and he studied only two illnesses. This means that he did not get much of a trajectory feeling into the study. And his coding doesn't, probably because of the above, catch many of the structural conditions for the sick people's specific time constructions, nor very much of specific consequences (except when he is analyzing particular cases). Anyhow, his coding language is relatively sparse compared with, say, our memo of 7/19 on bio time coding of the McGrail book.

That memo of 7/19 (and see the 7/22 memo too) is very rich in terms for past, present, future, and showing lots of "overlapping" and "interpenetrating" of past–present, present–future, etc.

Both Wolfram's analyses and our memos indicate that G. H. Mead was on the right track, but terribly abstract and just touching the surface of the richness. He was, after all, not concerned either with biography, except abstractly, or with time in relation to trajectory time. (A big hole, and peculiar, considering that Mead's whole book has to do with "time.")

Paradigm notes

1. *Trajectory* is the place to start, and we have done that.

As the McGrail and other books reflect, the biographical time conceptions *appear during trajectory phases*. They don't just float in abstract space, but can be temporally located by analysts.

2. *Structural conditions*, therefore can be specified for why a specific time construction appeared in a given phase (or miniphase – like just before going into an operation, a flashback).
3. *Consequences* can also frequently be specified, are clear from the running narratives in these books. Consequences for immediate or later action, for marital work, together, for different kinds of work, for mood, for trajectory shaping in general. For marital relations, too, no doubt.

Integration issues

Biographical time conceptions obviously link with trajectory shaping, trajectory projection, the self continuity–discontinuity, biograph review/recasting, etc. Part of our job, presumably, is to make the linkages clear, at least in the more general sense, and perhaps somewhat specifically too. Then, Julie, you will want to do this with other central concepts. It isn't necessary to do this with all, though, as that would be overkill.

But, as an instance, take closure work. This work looks different at different phases of the trajectory. We would have said that what's done by phase depends on trajectory projection before, but now can link it up with biograph time conceptions, too. Maybe I will try this later.

AS(JC) – 7/27/82

Closures: multiple and phased along trajectory
(from the Cornelius Ryan book)

1. Begins with his tape recording (despite future still being open, if problematic, but he is taking no chances).
2. And his decision to go ahead with his BRIDGE book, for money for family and own identity.
3. Unless I missed next closure steps, there is no next closure until p. 199, before his crucial surgery and crystallization: "then will, papers, assignments, and the like" had been put in order. But (199b–200) he can't even close psychologically with his young daughter.
4. With wife, last time before his operation (206b), but complicated because of their mutual front of optimism. (See also 211.)
5. And inviting the priest in, that same night (206b–207t).
6. His last taped words FOR HIS WIFE just in case he dies, same night (208).
7. Going on outings with family so children after his death would look back at a good and interested father (294).
8. The Xmas party for all his friends, with injunction to wife afterward to send out Xmas cards after his death with his photo on the cards (296 onward, and especially 301).
9. Making the book outline for his wife and secretary, J.
17. (440) And more with his wife: "My poor little girl," he said. "My Katie." She was looking down at him and he was watching her as he said that. (442) Geoff, the son, was in the room just before he died. There's a picture of a fisherman casting on the wall. Calls Geoff. "The wrist," Connie whispered. "Remember to use the wrist. . . ." Son says he will. Connie smiles.

Julie: I think the above *closure sequence,* linked as it is with the discernible phases of C.'s trajectory, make the point about how *self-continuity is maintained in some part through these closure activites/gestures.* When my favorite aunt – the one who helped raise me – died, I lingered at the grave, the last, and threw a rose to her.

AS(JC) – 7/20/82 (from Connie Ryan book)

Biographical time conceptions
Don't be wearied by this listing. I have reason for getting this down to look at.

"That I am dying" (18b) Present moving into a finite future and future
Present objects have a pregnant past (biograph flashbacks) 19
The future will always be present (i.e. cancer) 19b
Linked past/present/future re future was in past and now present 20(1)
The limited terminal future 20(1)
Pushing away the terminal future 21(5)
"Grief." Future always in the present 32(m/b)
Symptom review (and biogr detail with it). Going back and carrying forward to the present the past 38

The terrifying present–collapsed future 68(m)

Present completely focused on future 84
Planned sequence of future steps 89(m)
Pushing off the future by will (i.e., prolonging life) 102
The shortened potential future 102(b par)
Present storage against future (looking back then) 108(t)
Possibly "last time" 111(m). . . .

AS(JC) – 8/7/82

"Not accepting": an integrative memo

J: In answer to your challenge and beginning notes on "not accepting" in the Limey stroke book: See how the categories crisscross!

Issue: this man does not "accept" his limitations, etc. What does this mean sociologically?

1. Three dimensions of nonacceptance: illness itself, symptoms, activity limitations. This man doesn't accept any of it, but the symptoms primarily, coming from nonreversible illness, impact on his life and activities drastically.
2. I translate this into BIOGRAPHICAL TIME CONCEPTIONS. There's a radical discontinuity with the past, and the future/present will be forever the same (or worst future). Some people can "adapt" – they can't live with or transcend (the self). This man can neither transcend nor live with.
3. This really means that his past is too discontinuous with his present–future.
4. Or in "self terms," his *identity work is a failure.* He cannot close the gap between past and present–future (discontinuous) self.
5. His trajectory projection has not only crystallized but there is no foreseeing that it can become any other crystallization – except worst projection of future.
6. This is all the more striking because people who know they are dying for sure can still accept by doing appropriate *biograph reviews* and *closures* (*review work, closure work*) . . . .

Closure can be one-sided, and *not involve interaction* with others or another. Or, if it involves interaction, the other may not know his/her part in the closure drama. But it may involve interaction (see below).

Closure can be *one-sided.* It can be *mutual.* It can be *two-sided* (*wife/husband*) *but not mutual.*

Mutual closures involve reciprocal gestures, "closure work" together.
One-sided may involve other's interaction, even help.
Two persons may be getting closure (dying husband and wife) but doing this separately.

Awareness context insofar as closure gestures/actions may be kept secret; but may be open or suspected.

Note that the *pacing of closure* for husband–wife may be perfect, come together. Or it may be out of phase.

Mutual closure. How do they do that? Invitation and acceptance, both. Then doing closure work together. Involves, if not talk, then clearcut nonverbal gestures. I think in the biographical books you see both sides *offering closure possibilities* to the other, and when they are accepted, then both do that work together. No?

Unsuccessful Closures around interaction, conditions of: (a) proffered invitation or request, (or signaled help) – or unnoted – or seen but rejected because of AC considerations – or misunderstood as meaning something else. Note: There are conditions for all of these.

(b) accepted, but unsatisfactory interaction. (Condition then for successful presumably might be signals which are then noted, communication that might "improve" responding reaction, etc.).

Re pacing, there is also the issue of *tactics*: How to get the other to get closure, how to get closure for oneself, how to ensure that both get closure separately or mutually?

Consequences of unsuccessful closure (a few). Different than when closed or open AC. Concerning open AC: apology, sadness, grief . . . possibly later realization then move to more successful closure.

AS(JC) – 8/22/82

Biographical reviews

Perhaps you should check the Connie book again (also, probably, the Wertenbacher book) just for biographical reviews, to flesh out my impressions. In the Kavinoky book (8/20 memo) I found:

simple flashbacks
flashes forward (i.e. images)
memory flashbacks
whole scenes
reviews of past
accountings focused on past to present
biographical forecasts.

In *Goodnight, Mr. Christopher*, p. 116, there's . . . end of life *biographical closing off through reviewal.*

I think it's important to get the classification of these pinned down, so that we can catch the entire scope of the phenomena. And to be able to say something about how they do – or do not – affect present biographical passing and future biographical work and impact. *Yes*? . . . .

JC(AS) – 10/2/82

Biographical phasing/trajectory phase/b. time reviews

Different phases of the trajectory may lead to or be associated with changes in biographical phasing, synchronous, dissynchronous to varying degrees. These changes may be major or minor. The change may occur along any dimension of the self (and many times along various dimensions of the self), body image, spiritual, social, identity, sexual. These changes may be occurring at the same time or at different times. With each biographical phasing there is some continuation of the old as well as evolution of the new. In each, therefore, there are elements of past, present, and future. This biog. phasing is not sudden in response to the traj. phasing, but rather, evolutionary; that is, evolves over time. Entering into this changed biog. phasing, or serving as a condition for, are *reviews*, which are often triggered by phasing along the illness traj. (perhaps the link between illness phasing and biog. phasing).

Another condition affecting biographical phasing is time. Trajectory phasing affects *time perceptions* as well as management and it is this time perception, such as, "I only have so much time left," or time taken up in illness management, or buying time, etc., that then enters in as a condition affecting biographical phasing.

Of course, also entering into biog. phasing are other conditions micro and macro, the expected and unexpected contingencies of life, interactions with others, spouse's biog. phasing, etc.

What is important for us is how Biographical Time and Trajectory intersect . . . .

AS/JC session on 10/7/82 summary

Types of time: There is *clock time* – which is characterized by regularity.

There is *everyday time* which is chopped up, scheduled, routinized, disorganized, ordered. It can be stretched or constricted, or adjusted to.

There are *time perceptions*, or *perceptions of time* – time becomes an adjective of experience.

There is *biographical time* (see below).

Perceptions of time: Has to do with activity, attention, mood, energy, quality of event. It is *personal, situational,* and has *structural elements* in the explanation of it. The same person can have different perceptions of situation and therefore of time at different times. (Though the situation may be structured essentially the same. Or as Mead says, when looking at the same situation from different presents, one interprets it differently.) Time may be perceived differently under conditions of repetition such as having taught the same class so many times that time seems boring, or never have taught the class at all, in which time seems stretched out, endless.

Biographical time: When speaking of biographical time we are talking of a breaking into (the total stream of something we call trajectory, which in this case is a chronic illness) an individual's life. Preferred time patterns become disjointed and misaligned, that is, patterns of living one's life. Examples of this would be the army which is disrupting or breaks into one's life. There is a mismatching of normal lines or activity. During a vacation or time off, normal life is bracketed for a period, the flow of activity over time is interrupted, in this case voluntarily.

Time related to biography is not therefore just mood, the handling and organization of, or related to the experience such as with perceived time. Rather life is structured around time, whether that be conscious or not.

Where is the time in time????

Where is the time in clock time?: The time in clock time has to do with the articulation of work, sequencing of task, duration of task – how long, rate of task. All take articulation, juggling, coordinating. Time is built into it as time lost, or used up, or to be used or managed.

Where is the time in time perception?: In time perceptions, time is central, time becomes a salient adjective in describing an event experientially. Time is an indissoluble part of an event. For example, when one is having fun, one might describe time as having flown by, or describe an event as a momentous occasion. An event or experience is described directly or indirectly in terms of time (or can be conceptualized by the analyst in terms of time).

Where is the time in biographical time?: In biographical time: mood, the situational event, activity are really linked with the projection and its phasing. It is also linked in a structural sense with the phasing of the trajectory itself. Whereas perceptions of time are more momentary and fragmented, though may also be linked to phasing of the illness, its treatment as well as to biog. events; the time in biog. time is more related to long-term (or past, future) conceptions of time. These time conceptions become an underlying cause or condition for action. *Where is the time in biog. time then*??? Biog. evolves over time; biog. is processual, implies movement, change. Time is at the core of biog. What do time conceptions have to do with biog.? Time conceptions are integral to biog. (Biog. is thought of in relation to past, present, and future, thinking back, thinking ahead and interpreting in light of the present situation – which in the case of chronic illness is the trajectory itself.)

Time conceptions have to do with evolving time past, present, future – from simple to complex. Biographical time (because of its bringing together of the past and future within the present) is more than just situational, just related to mood, the experience of something, etc. Any description of it must include both elements of time (past, present, and future) and the individual's conception of that time (in light of interpretation of past, present, and future).

Therefore the answer to where is time in biog. time is that *time is totally within it*. It is not adjectiving or schedule time. Time is central to biog.: one can't view one without the other. Time can't be split from it. There are conceptions of time in relation to biog. and perceptions of biog. in relation to time. Time and biography can't be separated one from the other: time as central attribute of biography. (As Mead says, the past and the future come together in the present with the past acting as a condition to action in the future and present.)

JC(AS) – 11/10/82

Some thoughts on body

We project an image of ourselves through our body. And the image we have of our body feeds into our image of ourselves . . . . We need to look at images of body and how they evolve, and strategies for managing. We also have to look at how they relate to images of self . . . .

A concept worth mentioning because it has come up in just about every book I have read is the one of *wholeness*. They all talk about no longer being whole and then something or other making one feel whole again. Whole is more than just physical; it has to do with one's perception or image of self. This perception may be broken down into inner and outer perception. The inner being, How does one appear to the self? The outer being, How does one appear to others? And what is the relationship between the two?

AS/JC Meeting – 11/11/82

Body-self conception

Body carries with it projection and time. Change in body impacts self-conception and change in self-conception impacts body. When one changes it can cause or lead to change in the other.

In body failure, the individual must discover the bodily part or system that has failed; has to explore, experiment, and find substitutes for the nonworking or failed part.

The biographical projections related to body are profoundly affected by where individuals put the salience, whether body or mind, and if body, which parts. For instance, a pianist with arthritis would be more affected than a writer with heart disease.

Important to look at is how individuals distinguish mind from body; how do they balance, if they make the distinction? How does the focus shift, that is, under what conditions and with what difference does it make? When is the focus on and when is it off bodily or mental activity? When do they focus on one or the other?

What is the salience of body, when does it come into play? When doesn't it? Body obviously controls actions. When is body shoved out, quelled, put aside, transcended? When does it intrude on itself?

What metaphors do people use to describe body?

With mind, what activities are people referring to? When are they using it as a metaphor and why?

Can't do what normal person can do given the limitations. Crippled head can't do what he eventually can do.

What does body function mean in terms of salience to that person? What activities? Dominant themes in the person's life? When body fails you are in trouble in terms of time and energy. What impact does it have on the self? The most profound question is, *What does body failure hit at*???? This is how it relates to the self.

Image of self becomes shattered like glass. A new and different self emerges to which we seek validation through interaction with others.

Wholeness and crystallization.

How does body relate to work, time, biog. processes? We need a language first, and then to bring out the connections.

Body image – Body self-conceptions – Biog. projections Trajectory projections

Includes: sensations perceptions at the moment, present-time activities

AS(JC) – 11/14/82

Body–mind dichotomy

This distinction appears in paraplegic, stroke, and in my own "case." See the various books on this. As we discussed it the other day, it comes down to this:

MIND is the equivalent of (is a metaphor or shorthand code for) various activities, other than the sheer "bodily" ones that are now not possible or are limited.

These activities are mostly, but not necessarily entirely, "mental" and "psychological." That is, the capacity to think, work at thinking, write/compose; but also *bodily things that are mental* like responding to smells/sound/stimulation from environment, etc.

What happens is that when bodily functions fail, then the return to mental/psychological activities – or discovery of what one can do now in terms of those activities – has to be a process of coming to terms with instead of the lost bodily activities. In Connie Ryan's case we see how, also, though body still there, he anticipates losing those functions while still working on book. And in my case, while still not walking, I was "working" on book/papers, etc.

This leads to the *balancing process*: balancing mind against body. When that can be done in some stable, permanent manner, then *COMING TO TERMS with the side of limitations equation* has been accomplished.

And this is tied in with *body/mind salience.* Under certain conditions (see your codes on the books), body or mind comes into focus. And the other fades to background partly or totally. One condition for getting body out of focus is to concentrate on mental activities. This is not an *integrative balance.* It is, as they write about it, dissociation.

Under what conditions do our people use the mind metaphor? Perhaps a useful question.

"Wholeness" (their term) or integration after crystallization, coming to terms: means bringing mind/body to *durable balance.* Come to terms with body limitations and with what activities remain or are discovered.

Body, biography, trajectory (Figure 14)

AS(JC) – 11/16/82

Biographical body conception (BBC)

This is a key concept for us.

It is the *conception the person (patient or other) has of body in terms of biography.* So the body image embedded in it (crippled body), and often phrased as a metaphor, is not simply an image (I look good today) but related to the whole biographical span. And in our people, to their biographical projections (see the autobiographies).

The *BBC has biographical time* built into it (as the language in the books shows clearly).

BBC is to be distinguished from the following: bodily sensations – hurt, burning, dizzy; bodily perceptions – feel good today, feet look swollen, one shoulder lower than the other. (These seem present focused, and may have immediate consequences for next action or actions in proximate future.)

In biograph reviews can recollect past sensations and perceptions, of course. And when doing that (see Joni, p. 111) it is related to biography, past tense; and as in Joni, past tense in relation to present/future body failure. So this is now biog. time related, just as is the BBC.

Body image is a wastepaper basket term, covers too much (unless literature when we look at it is more specific). I see it as entering into sensation, perception,

AS(JC) – 11/15/82

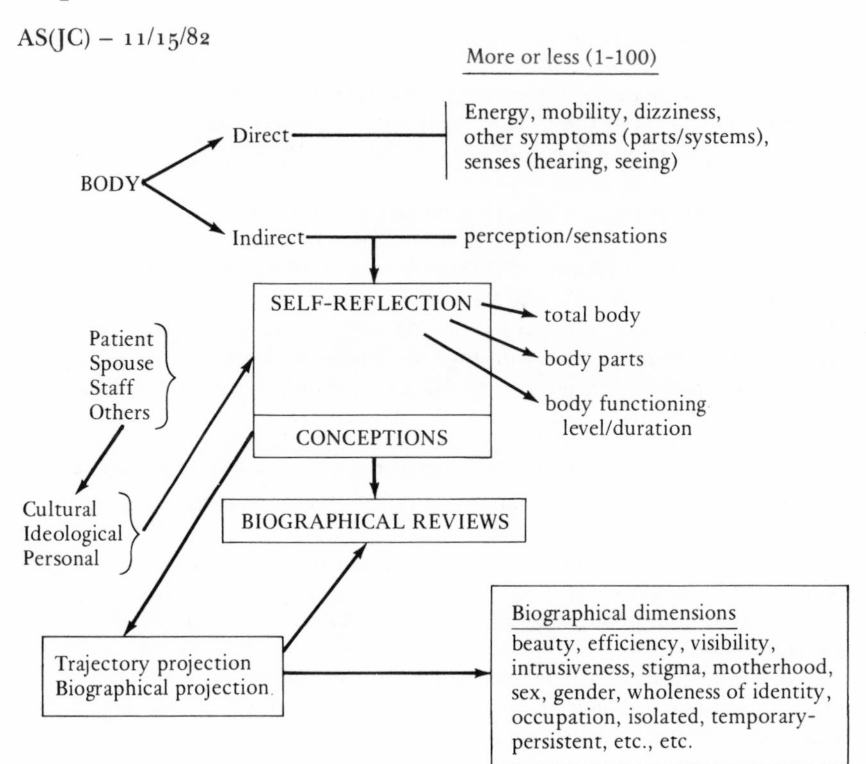

Figure 14. Body, biography, trajectory.

remembrance of perception/sensation, AND BBC. But images do that in different ways, as can be seen when think of, say, "my entrapping body" that is this way now and will continue that way forever.

Joni, p. 45 has a gorgeous paragraph that ties together image, BBC, TP, BP, and Bio scheme. In the following steps: 1 – others react to their perceptions of her body – transmuted into version of her BBC. 2 – Later, Joni looks into mirror and see what they see – and gets the same version of BBC. 3 – This translates into a hopeless trj. proj. and bio. proj. 4 – The latter then leading to begging nurse to help commit suicide (since she can't even do that by herself). 5 – After, tried to help look good to others; id. work to neutralize body perception and concept (somewhat). So this kind of paragraph, and writing elsewhere, shows how perceptions feed into conceptions . . . .

As trajectory moves along (whether comeback, terminal) BBC changes. And in tandem biog. proj., bio. time, bio. scheme, etc. I don't see the BBC staying stable until the traj. proj. stabilizes. In the autobiog. where movement is the name of the game, we see the BBC changing, of course.

AS(JC) – 11/21/82
Memo from Helen Wulf, *Aphasia, my world alone*

Mind–Body Add to 11/10 memo that one can also (and see Agnes De M.) make the distinction between mind and brain and body. Failed brain means failed body parts and functioning (including speech). But mind means mental activities–unimpaired. (Agnes gets panicked, or at least much upset, when thinks memory is affected occasionally) . . . .

AS(JC) – 12/7/82

Comment on the book by the ex-model – (lost leg) and BBC

This book's main theme, aside from comeback trajectory, can be phased as follows. And "as follows" is nice because it gives us a case where *a* lost body part (but not so many functions really) is central.

A (only one) body part is lost – leading to BBC*a*, then *b*, then *c*, then *n* . . . . BBCs feed into bio. projections – where comeback is all about a normal functioning and feeling self despite the loss of the body part – with proof to self by activity that she was like she was (sex, driving, dancing, cooking) in terms of leg's activity – with crucial role of others in validating that she's still herself (their active part in this, or acting right to her by setting up situations) – crucial issue of articulating BBCs and identity (cf., p. 33).

The body loss, especially later with an artificial leg substituting, allows for most bodily–leg functions to take place anyhow, or substitute ones (horseback rather than tennis) to be explored and done.

The body loss, symbolically re femininity, is handled with general body style, clothing, "willpower," and then with the searching out and getting a beautiful leg *and* using it in the way she wants. That femininity, as you clearly code it, is central: And the legs are central to that centrality!

So she was handling both the symbolic-femininity business AND the functioning normally business.

Note my turned down pages of your coding for a few other comments, and noting of your good concepts. Like body-identity work; comeback projection, BBC projection, physical limitations management, BBC confirmation, salient aspect of past recaptured . . . .

JC(AS) – 1/18/83

A summary session on coming to terms

Mobilization/Crystallization/Transcending

The first phase of coming to terms comes with crystallization, or a realization of what the situation implies, derived from a combination of a traj. proj. of eventual death and/or being disfigured, ill, crippled, etc. the remainder of one's life, along with a biog. projection of an unfilled future as a result. This in turn leads to a change in one's perspective of identity along various aspects of self (especially if those aspects of self lost are salient) in relationship to one's world. (I am no longer who I was or thought I was.)

Crystallization?

(Who am I now?) B.P. identity change: shattered biog. – task or work is putting biog. back together again in order to gain a new perspective on who I am or, on one's identity (I may be changed in these aspects of self but I am still the same along these dimensions) and achieve some degree of harmony about this new self. To gain this new perspective, the work involves mobilizing oneself to some degree in order to get the mobilization mechanisms going. In turn, the mechanisms feed back into the mobilization process. The level of where you were before affects the *mobilization mechanisms* and how good the mechanisms will lead back to how well you mobilize . . . .

To move forward toward coming to terms or to grips with a changed lost biog., and changed self, two processes must occur. One is a feeling process, or *recrystallization,* and the other is the action process, or *mobilization.* In moving back or failing to come to terms there is decrystallization without the recrystallization or mobilization necessary to move forward (even if at a lower level of functioning, having accepted increased limitations). As one continues to move forward through confrontations, decrystallization and recrystallization, and mobilization, one eventually comes to know and accept this new self (though acceptance is never static).

Though one may wish one never had to die, or be ill, in coming to accept and know this new self, one sometimes finds that this new self transcends the old self, is a better self in many ways. Degrees of this transcendence occur along the way and manifest themselves as increased sensitivity to others, acuity of vision; that is, the ability to see and experience old or different situations in new ways, with new appreciation, or realization perhaps that one has indeed fulfilled or has the potential to fulfill that biog. but in a better, or if not better at least different, way.

This process of coming to terms involves a lot of letting go, with a lot of consequent grieving, anger, disbelief, shock, bargaining, feelings of alienation, anguish, apprehension, fear, hope, despair, self-pity, etc.; a lot of fact finding, sorting; a lot of mustering up of inner and outer resources; a lot of different kinds of reviews; battling illness or struggling at comeback or to maintain the status quo, closure work, etc. Feedback, working back and forth, as conditions and consequences.

JC(AS) – 1/25/83

Biographical body conception (BBC) chain (Figure 15)

The means by which we *take in* and *give off* knowledge about the world, objects, self, others. For the most part this is an unconscious process which takes place through sensations, sight, sound, smell, touch, taste, and the perceptions we form from these. It is also the means by which *we play out our biog.*

During illness that results in a failed body, this chain is interrupted (*BBC Chain*). It is interrupted because the failed body alters or interferes with the process by which we take in and give off knowledge about the world and through which we play out our biog. The problem becomes one of working around this failed body with its limitations to mend the chain or put the chain back together so that the cycle can continue the inner reconstruction through coming to terms.

That is why in comeback we see each comeback indicator acting as an identity booster, and this in turn feeding back to the person to act as a condition to

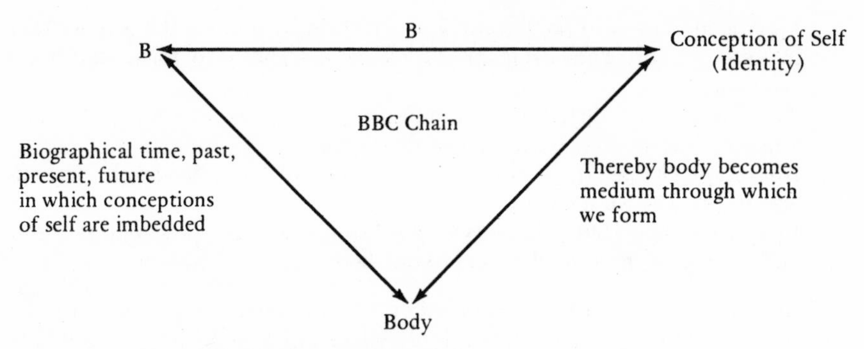

Figure 15. Biographical body conception chain.

continue working at comeback. It is tied into trajectory because the status of the illness and type of trajectory, and its phasing, are tied into degrees of body failure. You can see how it would change with traj. phasing. As A.S. says in memo dated 11/17/82 on BBC: "As trajectory moves along (whether comeback, terminal, etc.) the BBC changes. And in the tandem biog. proj., biog. time, biog. scheme, work, etc."

This memo ties in with other memos on body and with the memo Biog., Identity, and the Self.

JC(AS) – 2/17/83

More on BBC

Mead's focus was on the development of the Self. He was not interested in body; nor does he dwell on what happens to this Self when there is a salient loss of one or more of those multiple selves through loss such as divorce, loss of a job, or body failure. Our interest and emphasis is upon body and the role it plays in the development of the Self because of the interrelationship we have seen between the two in our data. This memo summarizes and clarifies what was written in the BBC memos.

Not only do we need social experience to develop a Self (an ever-evolving Self), but we also need an intact body:

1. senses, sight, hearing, smell, taste, touch, means by which we take in and give off knowledge of the world, means by which we build our shared significant symbols
2. means by which we communicate and engage in cooperative and adjustive activity
3. the encasement of our Self which enables us to play out our biog. through the various aspects of Self – the multiple selves
4. perceptions of body feed directly into our identity; (of course, so do things like our professions, being loved, or not loved, etc.) by means of taking the role of others toward ourselves . . . .

In illness, where there is a failed body:

1. May be alteration in body sensations which in turn alters the way we take in and give off knowledge of the world/our ability to form significant symbols.
2. The means by which we communicate and engage in cooperative and adjustive activity may be altered or disrupted.
3. Alteration, change, disruption of our present and future biograph; perhaps also in the way we interpret past.
4. Body blows may lead to identity blows through changed perceptions of our body – taking the role of others to our body.

Body failure can lead to disintegration or loss of wholeness of the Self through loss or change in any one of the multiple selves which make up the total Self. (I.e., it can result in alterations or change in the relationships of the various selves which make up that whole.) The degree of alteration is determined by the severity of body failure – which leads to the number and degree of limitations – which leads to the salience of the loss of one or more aspects of selves – which leads to the disruption or change in relationship of the multiple selves – which make up the whole. The individual now has to reconstruct his Self, or put the changed or new selves back into relationship with his old or unaltered selves to form a whole once more.

What is interesting is that the loss of body wholeness through body failure can lead to loss of Self wholeness, but the individual can reconstruct a new identity, a new Self, and sense of wholeness, though his body remains unwhole, by coming to terms with his limitations or transcending them, be they time limitations, body limitations, etc.

Mead has given us a good start, but he has left us a lot of room to take off and work from, like how we reintegrate or reconstruct the Self after significant loss.

Not long after this sequence of memos was completed, a third and last diagram summarizing the project's work was done (Figure 16). Even after reading this chapter, readers may not be able to understand all of its concepts, but the diagram should be useful for suggesting several points: its overall design, sketching of relationships, processual flow, potential for further expansion, and of course its complexity. Only the portion of the diagram that pertains most directly to the memo sequence is reproduced here.

From memo sequence to writing

To illustrate how some of these memos and a portion of the integrative diagram get converted into writing for publication, we reproduce next a short segment on *coming to terms*. This material was written for a forthcoming monograph.

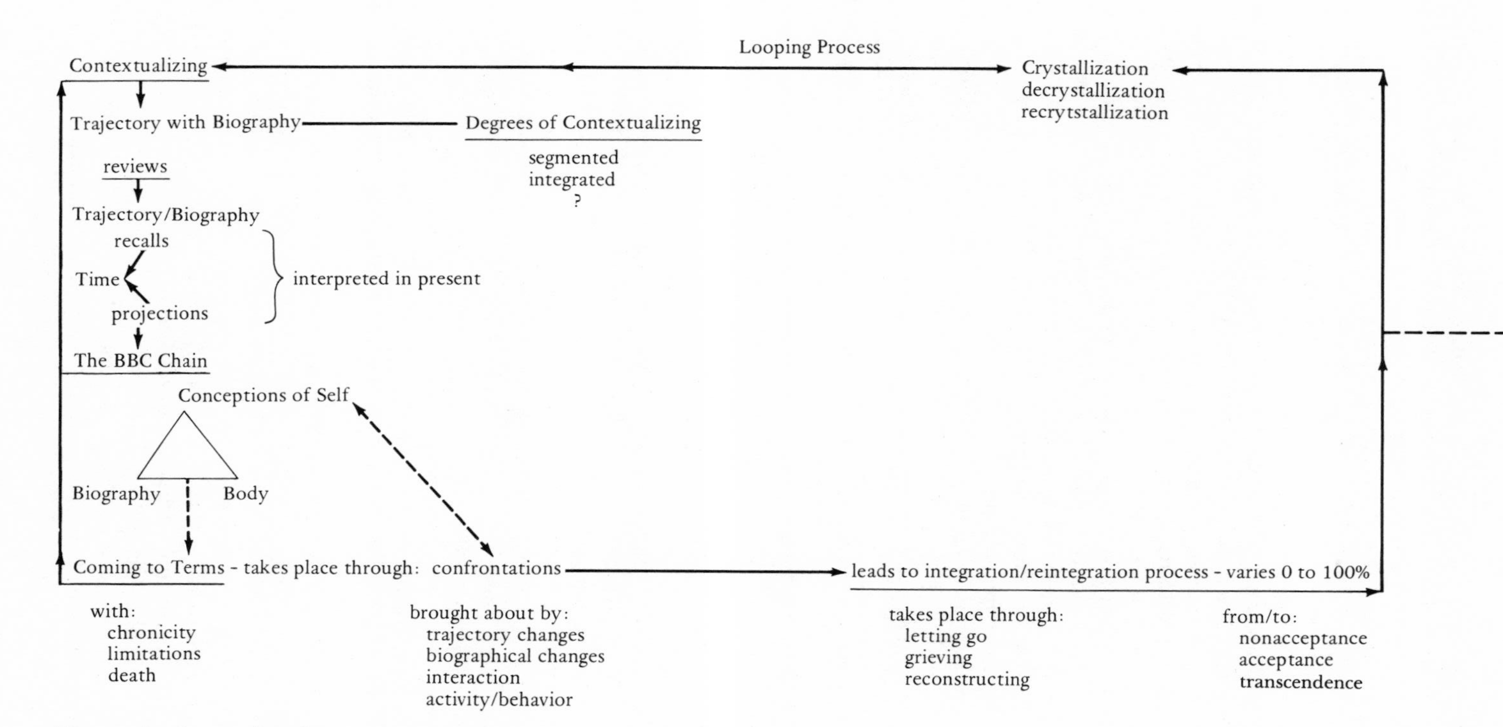

Figure 16. Integrating the memo sequence.

The process of coming to terms involves the following, although the specifics vary according to trajectory type and degree of body failure. First, there is a confrontation of the potential for lost performances. This is followed by one or more reviews that may lead to denial, anger, and in some cases bargaining, as the person tries to hold onto salient and meaningful aspects of self. Gradually, through further confrontations and reviews, the ill person realizes that these aspects of self are gone, no longer possible, and in doing so begins to relinquish the past through a series of closure acts involving self and others. This relinquishing involves grieving for what was lost.

With grieving may come depression to varying degrees. Eventually, the person begins to realize that he or she can no longer live in the past, but must begin to look toward the future. In embracing the future, acceptance begins, for one cannot accept unless there is hope for a better future. Without hope there is no incentive to move from letting go toward some degree of acceptance. Hope, here, we translate to mean the perception of an "exit" – a way out of the present situation. The future will be better. "Perhaps they can prolong my life until a cure is found." Or, "Maybe this medication will work." Or, hope may be seen as freedom from fatigue, suffering, and pain, as in the release of death; or it may mean that there is a life after this one. Or, "I will live to see my child graduate from high school." Or, that limitations will decrease, as when one makes a comeback. A state of no hope exists when there is despair, arising from a perception of no exit from the present situation being possible (as in the play with the same name by Sartre). There is no way to ease this suffering, "even death eludes me."

Acceptance, of course, does not mean a state of happiness. Nor does it mean liking the situation. Acceptance here means that a person has found a way of biographically accommodating to an illness through altered or changed performances and, in doing so, a way to give meaning to life despite ongoing and progressive body failure. Some ill people not only reach the state of acceptance but go on to an even higher level, a state that we call *transcendence*. Transcendence occurs when persons have found a way to overcome bodies in such a way that they are able to find real joy in living, or even dying, although their performances may now be severely limited. Life has taken on a new meaning and is in some ways better than before. Some, like Mrs. Bayh, are able to transcend their bodies through beliefs in an afterlife. Others, like ex-Senator Javits find the impetus to go beyond their limitations through the challenge of work and the contributions that they are still able to make to society. Still others find that, because of their limitations, they are now able to see the world with new eyes. For the first time they can keenly appreciate the beauty in nature; or the value of some persons whom they had previously overlooked. Agnes de Mille has quite beautifully captured the essence of transcending experiences. Reviewing her previous life and finding it "stale and used up," and regarding now her poststroke life as "a fresh fight," she was able to experience "new delights and none of the old constraints" and to continue growing and learning. "It was a feeling of freedom such as I haven't known since I was, in chronological time, five years old" (de Mille, p. 205).

Of course, some of the ill are not able to come to terms, to accept their limitations, whether of the illness itself, the symptoms or of performance.

Nonacceptance essentially means there is a radical biographical discontinuity with the past, and that the future and present will be always the same, or even worse. In self terms, this means that biographical work is a failure. As in the case of Limey and some of our interviewees, the ill person cannot close the gap between past and present–future self. There is no future imagery of progress or of better times ahead to pull one through the rough periods.

Rules of thumb for memo sequencing

The introduction to this chapter as well as the memo sequence suggest general rules of thumb for carrying out this kind of integrative work. They are:

1. Possibly the most important procedural rule is to think – and go on thinking constantly – of your memos as potentially cumulative. They become memo sets, which eventually become linked integratively.
2. Sort memos occasionally, in order not only to help your memory but to give a sense of continuity in your memoing.
3. Follow through on any memos, either after sorting or from your memo, that stimulate you. In fact, be ready to be stimulated. The stimulation can come because an interesting path has just opened up, because you are dissatisfied with an analysis or set of analyses, because you just had an additional ideal or two to add to a memo set, and so on.
4. Coding will call attention to the possibilities of another line of analysis, and if so then you should memo it for an immediate or later follow-through. Coming unexpectedly upon suggestive new data can also lead to a new line of analysis: Follow immediately if you have an urge to do so, while you are excited about it.
5. Do not be afraid to abandon work on one set of ideas (a memo set) if you happen to get started along another track of ideas. It is very important not to get compulsive about finishing off one ideational track before moving on. Trust yourself, indeed trust your subliminal thought processes as well as your memory and later sorting, to come back to your older ideas when the time is ripe. They all integrate better that way. If you are working in a team, and someone else hits a potentially interesting new vein, allow both yourself to be stimulated, and the option to carry the cooperative work in that direction.
6. When you get bored or tired with working on one ideational track, that should clang a bell of warning that you should get off that track and onto another. If other teammates are getting bored, they should be allowed to redirect the immediate analytic attack that will produce memos along different ideational lines.
7. In formulating these memos, remember that their integrative cumulation does not necessarily occur, and probably will not, without complete sureness about what you are doing and without chasing occasional phantoms. Uncertainty is very likely to be a part of this integrative game. The distress caused by this uncertainty is often counterbalanced by the wonderful

breakthroughs, as was illustrated of course by the last statements in the preceding chapter.

8. Dovetail technical reading into the flow of your memos, either in terms of data or their concepts or hypotheses.
9. Do not be compulsive about keeping the memos flowing in regular order. There must be time for coding, time for thinking – and time for recreation and for the business of daily life.
10. Occasionally draw either an operational diagram to further integration of clusters of memos, or a major integrative diagram to build on prior successive ones. (Of course, other graphic means can be used for these purposes.)
11. Do the later segments of your memo sequence with a careful eye kept on your core category or categories. This attentiveness will help the final integration which is achieved only when you begin to write up your materials for publication.

Writing as an integrative mechanism

Ideally all of the integration, or at least its major features, should have been accomplished by the time that actual writing for publication takes place. Yet even when a first draft of a monograph (or the initial articles) is written, understandably, researchers always find themselves discovering something that tightens up or extends the total analysis. When that first draft is reviewed and revised one or more times – sometimes many more – then additional integrative details may be added. Even entire integrative steps may be taken, since further data collecting or at least further coding may be deemed necessary, not only to add more detail but to add to the final integration of the analysis.

That is the ideal and its qualified reality. In fact, however, under certain conditions a very great deal of integration may continue *while* the researcher is doing the writing. He or she may literally be sweating out not merely the writing but the equally difficult task of bringing an incomplete analysis to satisfactory completion. There are at least two situations where that kind of incompleteness will occur. One is when the researcher (or research team) decides to write an additional manuscript, whether a monograph or paper, and therefore has not coded or memoed in nearly enough detail to sustain the analysis to be presented in this additional publication. For that reason, a great deal of recoding of old data and coding of new data needs to be done. Otherwise the analytic presentation will be thin, or at least thinner than it might be. And this, despite perhaps a fair amount of coding and memoing having been done on the phenomenon under consideration.

A second situation under which considerable integrative work must be done while writing is when the researcher discovers that despite much memoing and coding a great many connections were left unspecified and that some black boxes were left unopened. In short, there are many holes and ambiguities in the extant analyses. Then the author must grit teeth and painstakingly do the necessary integrative work. Probably this is the most usual pattern of doing the final analysis, rather than the ideal of accomplishing virtually all of it before writing ever takes place.

So think of there being a continuum running from more or less integrative work that will accompany your actual writing for publication. Here are the *rules of thumb* for this kind of integrative effort:

1. Be aware that integration may be incomplete, and get into the frame of mind that welcomes this rather than resents it.
2. As you construct each chapter, each section, indeed each point, keep questioning: "Is this construction complete? Does it hang together? Are there holes in it? Are there details missing that weaken the integrative structure here?" This dictum does not require you to become obsessive about or fearful over getting yes answers to those questions. The more yes answers, in fact, the better your final integration will become, despite the time and effort and anguish that this work may cost you. Just remember, it is not additional work, but expectable work, expectable in general if not anticipatable in all its details.
3. Wherever you get yes answers, then you must think through what that means. Thinking through may involve better specification or elaboration. That means drawing on your memories of data, codes and memos for that specification or elaboration. Alternatively, it may require recoding old data or new coding of data gathered expressly for these purposes. On the other hand, thinking through the analytic deficiencies may even entail a major reconstruction of your original ideas about how the analysis finally would hang together. Alas, then you would have to face up to that more extensive task. Of course, this is much more likely to happen if, for one reason or another, relatively little integrative work had been done previous to the write-up period.
4. The overall integration done during the writing of a first draft must be carefully reviewed before embarking on a second draft. You cannot expect only to do some editing on the first draft and then blessedly that will be the end of your work. You must expect at least some additional integrative labor, even if only a filling in of details, a tightening of a few bolts here and there. If you have done a careful job on the first draft, probably your additional work will be minimal. But keep yourself in the frame of mind that it may *not* be minimal. Besides, reviewers of your manuscript, after it reaches a publisher, are likely to come up with things that either you have overlooked or that you may wish to add for the kind of audiences they represent, and who might otherwise miss those details. So you must fight

the temptation to be finished, to wash your hands of this project once and for all. You are finished, when finished! On the other hand, do not go to the other extreme, as many authors have done, and spend years tinkering with the product long after it could have withstood public scrutiny, and perhaps long after its effective impact might have been made.

A last note

After reading the material in this chapter in an earlier draft, a British sociologist friend, Paul Atkinson, wrote the following commentary. It expresses exactly what all qualitative researchers confront as they seek to bring their studies to an integrative close.

> This aspect – making it all come together – is one of the most difficult things of all, isn't it? Quite apart from actually achieving it, it is hard to inject the right mix of (a) *faith* that it can and will be achieved; (b) recognition that it has to be *worked* at, and isn't based on romantic inspiration; (c) that it isn't like the solution to a puzzle or math problem, but has to be *created*; (d) that you can't always pack *everything* into one version, and that any one project could yield several different ways of bringing it together.

10 Presenting case materials: data and interpretations

As anyone who is familiar with qualitative research studies knows, their publications almost always include "real live" data. Monographs, reports, papers, even speeches based on qualitative research are replete with illustrations: quotes from respondents, fieldnote excerpts, chunks of historical material, short case accounts, chapter-size biographies or career stories, even entire books consisting of a case history narrative or a case study of an organization. Such illustrative data are used in all of the social sciences and in professions such as social work, education, and nursing.

The reasons for that have been extensively laid out in the literature. For example, Diesing (1971), an informed philosopher of science, has discussed what he calls the *case method* or *holistic method* of doing and presenting research, as contrasted with other variants like experimentation and survey research.[1] He has systematically explored the assumptions that lie behind this holistic style of thought and action, as well as how methods of research follow from these. There is no point, however, in reviewing here why data illustrations are used so copiously. Rather we shall discuss first, and briefly, some issues attending the use of illustrative data, then note some rules of thumb that can usefully guide the constructions of case studies, also of long case histories, whether they are published separately as monographs or as parts of them.

Illustrative data

Since qualitative researchers do not generally use or present much in the way of statistics, they face an interesting set of options. They can keep the presentation very abstract; or they can give very little theoretical

[1] But, for an interesting recent discussion of the possibilities of integrating sample surveys with case studies, see McClintock et al. (1983, pp. 149–77).

commentary but give a great deal of data, allowing it to speak for itself. They can seek some balance between those two extremes. Which option they choose for a particular publication depends not only on personal predilection but on the purpose of the presentation and the anticipated nature of the audience. Traditionally, it is understood that illustrative data can be used to give a sense of reality to the account, so that readers will feel they are there, in some sense, as the researcher had been there. Or illustrative data can be used deliberately to convey the viewpoint of actors, giving so-called *verstehen*, especially when their viewpoints are far removed from that of the readers. Or the data can be used to lend added credence to the author's theoretical commentary – or argument – in short, giving evidence. The included data are meant to function sometimes as information that helps the readers to understand what is going on, at the site studied and in the phenomena being analyzed, otherwise what is being presented would be more or less incomprehensible.

These traditional grounds of data utilization point to the general issue alluded to earlier: What should be the ratio of interpretation to data? The general issue, however, is really constituted by a set of subissues. Among them are the following:

1. What data should I include?
2. What form should they take?
3. Where should I put them?
4. How do I combine my interpretation and my data?

Doubtless, a great many researchers do not think through these issues, they just do what feels right, given their interpretation of the data, the data themselves, and the audiences to whom they are directing their messages. Clearly, others expend much thought on these issues when constructing their manuscripts, whether they are veterans at it (cf. Becker 1982) or stepping out into relatively uncharted waters in their disciplines (cf. Polsky 1983).

Grounded theorists face the same issues and often respond to them – and to the requirements of credence, *verstehen*, and so forth – in practice, much as do most other qualitative researchers. However, the features of this mode of analysis affect (or at least should!) their practice. Among the features of course are: the considerable open and theoretical coding, the emphasis on core categories, the insistence on conceptual density, the use of theoretical sampling and of constant comparative analysis. Understandably too, the concerted attention to presenting theory rather than, as it is sometimes termed, low-level description or pure description precludes incorporating large gobs of raw data with little analytic commentary. In general, there is much more reliance, as we shall see below, on an interweaving of discursive propositions – utilizing the results of coding and memoing – with carefully selected pieces of data. The latter may just be quoted phrases in combination with the theoretical points being made, or very short quotations or fieldnote items following on some systematically made theoretical point. Or an actual analysis is built into a descriptive précis constructed from a number of fieldnotes recording field observations, interviews, or other documents. Here are two examples of those respective uses.

> Staff members, again especially if they are inexperienced, must guard against displaying those of their private reactions to him and to his impending death as might arouse the patient's suspicions of his terminality. For instance, young nurses are sometimes affected by terminal patients of their own age whose deaths become standing reminders of their own potential ("I found . . . that the patients who concerned me most when they died were women of my own age . . . "). Identification of this kind is quite common, and makes more difficult the staff members' control of their behavioral cues (Glaser and Strauss 1965, p. 38).

> In addition to all these devices, other techniques reduce cues that might arouse the patient's suspicion. Space is carefully managed, so that talk about him occurs away from his presence. If a nurse believes her involvement with, or sadness about, the patient might give the secret away, she may move quickly outside his visual range. She may even request assignment away from him. Possibly revealing cues are reduced by decreasing the time spent with the patient. Personnel who fear that they may unwittingly disclose something may remain with the patient very little, or choose to work on his body rather than talk much with him. They may keep tabs on his physical condition by popping in and out of his room, but thereby keep conversation at a minimum . . . . If the patient becomes genuinely comatose, nurses or aides can again circulate freely in the patient's room (Glaser and Strauss 1965, p. 37).

That kind of presentation, unless supplemented with large slices of quotations from interviews and fieldnotes, sometimes seems to disappoint readers brought up in the traditions of "just let the actor speak," or "let the data speak for themselves." Properly done, however, this style of rather tightly interwoven theoretical interpretation and descriptive data meets all of the classical requirements of *verstehen*, credence, sense of reality, and reader comprehension. In addition, it builds in the specificity and variation that should be the concern of anyone who is really serious about generating and presenting theory. (Apropos of "sense of reality," as mentioned in Chapter 1, a well-written monograph is likely to be read by those who were studied, other laypersons, and even by some social scientists, as description – "It really happens that way" – rather than as a theory accompanied by theory-informed data.

To quote an illustrative phrase from a review of one of our monographs: "Why and how this happens is described in much detail with much insight" (Sharfstein 1985).)

Case histories and case studies

Among the persistent uses of illustrative data are life histories and biographies, narratives gathered through interviews from respondents about their lives. These are, however, only one form of case history. A chief feature of all case histories is that they cover some temporal span or interlude in social life – a biography, an occupational career, a project, an illness, a disaster, a ceremony. Also, the case history involves a story about one social unit – a person, group, organization, relationship. It is useful to distinguish *case histories* from *case studies*. In constructing the latter, the researcher is focused on analytic abstractions for purposes of presenting theory at some level or another. We shall discuss case study construction first, before turning to similar issues with case histories. (See also Strauss and Glaser 1970, where some of the materials of this and the next section first appeared.)

Case study construction

A great many publications by qualitative researchers are written in the form of case studies, as in the analytic depiction of various kinds of groups, organizations, and cultures. There is no attempt, as with case histories, to tell a story as such, for temporality is not the ordering principle. Assuredly, however, many stories about evolving events, people, their careers, and so on are embedded in the longer case studies. Indeed, something of the evolution of the social unit under study may be presented, but temporality is not a main organizing feature of the analysis. Indeed, one researcher, Fritz Schuetze, who is studying biographies as a general phenomenon, writes general theory about this, but necessarily uses a combination of theory and temporality as his organizing principle for presenting his materials (Schuetze 1981, 1985).

In the grounded theory style of analytic presentation, case studies are constructed not very differently than by most qualitative researchers. The principal difference from many of theirs is the density of conceptual analysis and the tightness with which the presentation hangs together. In general, however, because this analytic mode uses theoretical sampling and constant comparison so extensively, its practitioners tend less to write case studies, joining the ranks of many others who not do this either. They write about phenomena more generally rather than about one hospital, one trade union, one science laboratory.

Some case studies take the form of short descriptions which are included as cases within papers and sometimes even as separate chapters or sections within monographs. Or the author may briefly contrast two or more cases. The construction of those cases is relatively simple, since it consists mainly of highly selected descriptive detail put together as a more or less coherent whole, to illustrate one or more theoretical points. Usually the latter are introduced before the case presentation, and perhaps restated or elaborated afterward. The same form may often be used even with case studies that constitute a book-length monograph, except then the theoretical commentary generally is more elaborate, and may appear at intervals throughout the descriptive account or after every internal section, as the focus shifts from one aspect to another of the social unit under study. When book-length case studies are published, their descriptive materials are organized in close conjunction with the theoretical points being made throughout the publication. From time to time, smaller cases supplement and illustrate the theoretical points.

The construction of this kind of commingled theory and data is not difficult once it is learned, but unquestionably it takes some writing practice to do it effectively, and for presentation to different audiences. Where one or another emphasis (credence, comprehensibility, *verstehen*) is deemed necessary, then the usual modes of illustrative presentation will be utilized from time to time: such as snippets from fieldnotes, quotations of varying length from the people studied, also shorter case studies and histories. (For example, cases about one or more departments of a total organization, a business firm or a scientific laboratory perhaps, studied by you.) These materials are framed, however, within the context of specific theoretical propositions or points.

Useful *rules of thumb for constructing case studies are*:

1. Collect data and analyze it, building theory around a core category or categories as usual. Data may include, of course, brief or lengthy case-study and -history documents which can contribute to building your theory.
2. Then, construct a working model of your case study, with major attention not to the illustrative material yet, but to your theory. Pay careful attention to the necessity for clearly specifying all of the theoretical elements and their connections with each other.
3. Afterward, build in illustrative data, but selected according to the salience of your requirements (*verstehen*, credence, comprehensibility, reality) either

overall or in particular sections of your manuscript. However, very carefully choose these data to bring out precisely the many theoretical aspects that need supplementation by illustration. The tendency sometimes is to overload the case with too much descriptive material because it is so colorful or interesting – at least to the author. Remember that these data should function mainly in the service of your theory.

Sometimes researchers collect a number of cases, but in their publications are concerned not so much with presenting case studies as such as with reaching some general conclusions about the phenomenon under study. They will tend to construct first for each case (for example, of Turkish immigrant families living in Germany) an overall descriptive picture of its organization, including both the relationships among its members and the relationships to the external world beyond the social unit principally under study. Commonly, the second step is to analyze each case separately, although comparative knowledge may enter informally, or implicitly, since the researcher or researchers after the first case have considerable data and knowledge about two or more other cases. Then, a third step is to draw general conclusions about all these cases. If a decision is reached, however, to present an actual case for publication, or included in short version within the more-abstract discussion, then it is important for us to note what takes place. Then the original analysis of any given case, and even the first-step description, no longer can serve without considerable alterations. Why? Because they would only be informed by partial theory and not by the final theory (or, "general conclusions"). The task of constructing a theory-informed case study must still be managed. In that event, the rules of thumb suggested above would still useful and, of course, much affect the selection of illustrative data offered in the final version of the case study.

Case history construction

As for case histories, they can be very useful when wedded to theory. After all, their basic ordering principle turns around presenting theory along a temporal line, thus "applying" to "the case." Probably more case histories should be written, and the rules of their construction should receive more attention, rather than be done haphazardly or in traditional modes, and done only occasionally. In fact, during research projects, a great deal of case history data are probably collected and then used as basic analytic grounds for the eventual publication of some sort of publication, but not as a case history as such. Yet a case history can be very useful if brought into very close conjunction with a grounded theory. Through it, the researcher can depict a type, an average, an extreme, or an exemplary case. The case history provides a readable and lively vehicle, full of vivid imagery, often in the actors' own words or in the on-the-spot words of the field observer – but *can* also provide that imagery in the context and service of a theoretical account of the descriptive materials.

The purpose of a theoretical commentary in interpreting the case history is to give a broadened picture of the particular case. The theory puts the case within a more general context of understanding what could have happened under varying conditions, therefore why this case happened in this particular way. The issue of variance is therefore of importance in thinking about these case histories and, as we shall see, in selecting a particular case or cases for separate publication or inclusion in another.

Usually, theory is applied to a case history in several commentaries which explain and interpret parts of its content. The commmentaries commonly appear after each section of the case history. Sometimes they only appear as introduction and conclusion to the story, or may supplement the episodic commentaries. Some case histories are published, however, with virtually no theoretical commentary, when the author assumes that readers are so familiar with a current theory that they can interpret the case history without guidance. Whether the theory in the presented case is to be extensive or not is only one relevant dimension to be decided on by the researcher. Another decision pertains to the level of the included theory's generality: low-level, middle-range, formal theory? (See chapter 12, on formal theory.) A third is the source of the theory: generated by the author? formulated in a series of studies by others? If it is a formal theory, then probably it is part of the traditional heritage of the discipline. Two additional decisions pertain to the degree of systematization of the presented theory, and the density with which it is formulated. (All of these decisions also have to be made with reference to case studies.)

For grounded theory researchers, these decisions should not be difficult since their style of work leads fairly directly to the answers. Their theories are generally, though not necessarily, generated from the research itself, and are mostly of middle range, and intensive (dense) and extensive in character. Hopefully, also, they are systematic. Therefore, all of these features should enter into the organization of a case history presentation.

Rules of thumb for guiding construction of case histories include:

1. A case history meant for publication is *not* to be constructed before the theory is well formulated. Of course, in preliminary form a case history or story may have been collected as data and written up in the form of an interview or fieldnotes: a life history or a running chronology of a sequence of related events, such as a public ceremony.
2. For a maximally informative case history, the elements of your theory must be clearly formulated and specifically related to each other.
3. Next there is a decision to be made about what aspects of your theory you wish to illustrate with a case history, given both the theory and your visualized audience or audiences. Of course, if you have only one or two case histories, the choice is easy. Or if you have been deliberately collecting data for case histories along the course of the project, utilitizing theoretical sampling, then you must still decide which of these to use. If the case history will be used as a chapter or section of a monograph, then more aspects can be emphasized; but, in a paper, what can be highlighted is much less. Anyhow, you must select the cases not just choose haphazardly or on the basis of human interest.
4. There may be several criteria for selection of a case for presentation. What salient feature or features of your theory do you wish to highlight, or perhaps can be highlighted, given only your pool of case history data? And should the case illustrate an evolution, development, or set of stages in a process, and so forth? Should it illustrate an exemplar, an extreme, or an average type? What audiences do you anticipate will read this case history? This consideration too may affect your selection, since readers will be more interested in or concerned with some of your case stories than others, some aspects of theory than others.
5. This entire selection procedure sometimes can be furthered and made easier if you are thinking ahead during the course of your research, and collecting cases with publication in mind. It is very useful to be collecting the data for them by using theoretical criteria. You may even seek them out using theoretical sampling, a procedure that we recommend.
6. In beginning of the case construction, it is useful first to outline the chronology of events that gives continuity to the story. You need to have the story line clearly in mind, even if you do not actually write it up yet. Sometimes that may be necessary for you to do, too.
7. Then it is necessary to do a preliminary analysis of the main theoretical elements that appear relevant to understanding the story, doing this step by step or phase by phase or stage by stage. You will need to work out those steps, phases, or stages in considerable detail.
8. A useful procedure is then to go through the case materials carefully, selecting out sentences, paragraphs, events, and segments of those documents that seem particularly relevant to what you have done in step 7.
9. The next step is to make detailed analyses of those passages, some line by line but others of course done far less intensively. While doing this, you keep in mind both the descriptive story line and the theoretical temporal line.

10. To add to the final, overall analysis, you should draw on your comparative knowledge of other cases, and indeed may wish to incorporate some of that when writing the actual case.
11. In the writing, keep your theoretical story line clear. Keep the core category or categories front and foremost in the narrative. Keep the subcategories properly subordinate but relate them clearly, as you move along the descriptive story line, to the core category or categories. These should also be related to each other, whether they appear sequentially or simultaneously along that descriptive story line.
12. Select with care the narrative chronology – its events, incidents, interactions, and actors' behaviors. Don't put in everything that seems interesting to you: Be highly selective. If you think you have included too much, then prune away unnecessary detail. Don't be seduced by each and every colorful detail; after all, your readers may become bored by too many details. Anyhow, they cannot love them as you do, since they have not been actually involved in either living through the events depicted in your account or in collecting data bearing on the associated events. Your entire task is rendered somewhat easier if you have followed the earlier guidelines listed above.
13. If the case history appears as a section or chapter of a monograph, then you may not need much theoretical introduction to the case, since readers will already be familiar with the theory. If the case history is published separately, then you will clearly need to introduce that theory and its major elements. There is no necessity, however, for overloading the theoretical introduction. Use only whatever pertains most specifically to the descriptive story about to be related, and will actually be utilized during the account. Again, there is a temptation to say too much in the introduction, perhaps flowing from anxiety that your readers may not otherwise understand the theory. Trust your interweaving of its relevant elements with the descriptive materials in your final account.
14. If you have properly done that interweaving, then the case history will not require much of a summary, if any. The entire construction will stand on its own.

Now in the following sections, we shall present two case histories to illustrate how theory and data can be fairly tightly interwoven, yet the whole case can be readable and vivid. These cases are designed also to illustrate the treatment of data that are quite different in scope or scale, as well as of theories that are conceptually very different and addressed to different substantive materials. In these relatively short cases, meant to be papers rather than to be included in monographs, there are fairly extensive theoretical introductions. Then the authors have chosen to closely weave theoretical elements along the story line. They could, of course, have simply commented theoretically at the beginning or end of each phase, or even placed theoretical commentaries systematically on the margins of particular passages. This style of closely interweaving theoretical commentary and descriptive story, however, combined sometimes with occasional stopping of the story for commentary, generally suits the grounded theory style of construction better. Only a part of each case history has been reproduced here, since our purpose is to illustrate their construction rather than their full content.

Case 1

The interplay between trajectory and biography: the process of comeback

The presentation (by Corbin and Strauss, see also Chapter 9) begins with general considerations and the laying out of related key categories. Then the case history is presented, step by significant step, with the analysis interwoven with the descriptive material. The latter is sometimes quoted, partly for filling out the more abstract meanings of the analytic interpretations, and occasionally just for sheer color and *verstehen* impact on the reader.

The phenomenon

An individual's life and all the experiences it encompasses may be thought of as a biography, which is made up of a past, present, and future. Each biography includes a self or identity, which in turn is made up of multiple miniselves. At some point along the lifetime line, an individual may develop a chronic illness, which adds still another dimension to biography – and therefore to the self – and results also in an illness trajectory. (Trajectory is defined as the physiologic course of a patient's disease, and the total organization of work to be done over that course, plus the impact on those involved with that work and its organization.) The illness trajectory is but another aspect of life now to be managed. The management problems created by the combination of the trajectory and the biography are influenced by their nature; the points at which they intersect; the ways they combine, become entangled, and branch out. The very adjective *chronic* denotes that this merger is long term or even lifelong.

When an illness such as a myocardial infarction or stroke occurs, or there is a spinal cord injury, or a mutilating and/or debilitating surgery, then usually immediate attention is focused on illness management, prevention of crises, and survival. The concerns of physician, family, and affected individual, if conscious, enter around the illness trajectory, its projected course, both for the immediate present and the near and distant future. That is, what course can this illness, injury, post-operative period be expected to take? What illness work is necessary to pull him or her through? What will be the residual effects, if any? The illness trajectory is in focus at this time, while biography is in a state of moratorium. Once the immediate crisis or post-operative period is over and the trajectory stabilized, biography comes into play. The moratorium is lifted, as the individual tries to conceptualize the illness into his or her life. He or she now begins to ask questions, such as: How did it happen? What does it mean in terms of my future? How will I manage to integrate an illness and all the work it involves into my life?

Contextualizing is accomplished through trajectory and biographical reviews, done from the present perspective and looking backward and forward in time. Reviews may take various forms. For example, there are accounting reviews in which one looks back upon the past, seen now as lost forever, and asking oneself questions such as: Did I accomplish all that I had set out to do? Did I do the best that I could? Was I the person I wanted to be? Why didn't I take the time and the opportunity to do this and that when I was able? There are also body reviews, in which the body as it is in the now present, is compared with the body of the past. "My legs, once so slim and straight, carried me wherever I wanted to go. Now, they are twisted and limp and I cannot stand, never mind walk." There are also symptom reviews, in which past symptoms such as heartburn, forgetfulness, twinges of pain, are given new meaning in light of present knowledge. There are flashbacks, in which the individual perhaps catches glimpses of the neighborhood cripple whom everyone felt sorry for, or the heart-attack victim who was never able to work again. There are trajectory projections, in which the possible future course of illness is visualized. There are biographical projections in which there is visualization of what future life will be like; that is, lived as a cripple, an amputee, etc.

Eventually, the ill person confronts reality head on and begins to ask himself or herself and those around them: What parts of me are lost forever? What can I expect to recover? Will I recover mobility? Memory? My speech? If so, how long can I expect it will take? How far back can I come? Though the questions center around physical ability and the use of body, what the questions are asking implicitly if, indeed, not explicitly, are biographical questions, such as: Will I ever work again? Will I always be dependent on someone for even my basic necessities of life? What will I look like to others? Will I be able to travel again, enjoy my hobbies, make love? The point of crystallization is reached when he or she conceptualizes the extent of the impact the illness will have, not only in a physical sense, but also in terms of his or her biography, for the present and into the future. It is crystallization – along with the desire to have a future that is more like the past than like the present – that mobilizes the individual to devise a plan or scheme to manage both the illness and biography, in order to gain some control over that future.

Though the illness trajectory becomes contextualized into the biography, it does not mean that each receives equal attention forever. As with everything else in life, one or the other may come into focus depending on contingencies that arise in the course of living. In planning for the future, some individuals may decide that the limitations are too great, life with them is not worth living, therefore why not let nature take its course? Others may decide that perhaps there is a chance, and fight the illness every inch along the way. Still others find themselves in a situation of having survived the medical crisis, only to be left with a long recuperative period and varying degrees of physical limitation. Like it or not, they are stuck in the situation, and somehow have to find a way to live with it. That is where *comeback* comes in.

Comeback is the uphill journey back to a satisfying workable life within the boundaries imposed by the physical and/or mental limitations. It involves the attempt to regain those salient aspects of Self lost because of illness or injury. Comeback may be partial or complete, depending upon the nature of the illness or injury and degree of body/mind failure, and the nature of the biography, too. It may be easy or difficult, quick or slow. It is marked by visible indicators of progress or their absence. It is characterized by periods of acceleration, reversal, setback, plateau, and variation in the boundaries of the limitations. These periods may be temporary or permanent. Comeback takes place in phases. It may move along a course until the ultimate comeback potential is reached, or be arrested at any phase along the way, or may reverse, improve, and then stabilize on this tentative or new plateau.

In medical language, comeback is commonly referred to as *rehabilitation*. This term is too narrow and constricting, for it fails to include that which is occurring biographically as well as medically. In order to gain a more accurate picture of the real struggle involved in comeback, at least the three following elements must be considered. They are: (1) *mending*, the process of healing, which broadly speaking means getting better; (2) *limitations stretching* – the rehabilitation aspect – which means stretching the body to push the boundaries of current limitations outward, thereby increasing physical ability; and (3) *reknitting*, or putting the biography back together again around the boundaries of the residual limitations. These three processes may occur simultaneously, or they may be staggered. One or the other may take the focus of attention at any time. One or the other may impinge upon or accelerate the progress of another. For instance, trajectory phasing and the development of complications or crises and subsequent recovery may affect for better or worse the comeback process. So may biographical phasing such as life stage, career phase, marital stage. Though, for analytical purposes, trajectory and biography are distinct categories, in reality once the illness is contextualized into biography, the two become so entwined that they cannot be separated except analytically.

Embarking upon and making continued progress along the comeback trail requires the existence of certain conditions. These act not only as precursors but continually come into play in various ways throughout the comeback cycle. The conditions include:

1. The part(s) of the Self that is/are lost have salient aspects of that Self and therefore are felt to be worth working to regain, be it in the same or different form.
2. Physical recovery is possible, though just how far one can come back physically is limited by the degree of injury.
3. There is crystallization or a clear realization of the future, followed by mobilization to provide the impetus to embark: and recrystallization and remobilization should decrystallization occur, to keep the process moving.
4. The presence of a comeback initiator (usually the physician but another figure may assume that role), who devises the initial medical scheme and sets the individual upon the comeback trail.
5. There is a tailored fit between the comeback scheme and the individual, both medically and biographically.
6. There is a comeback articulator, who coordinates both the various types of work involved in comeback and the workers' efforts.
7. There is a pool of resources, including people, finances, and objects, to draw upon.

8. Both the individual and other comeback workers act as a team, each one moving in and out of the comeback process and undertaking specific tasks, according to the trajectory and biographical phasing and type of work to be done.
9. There are realistic future goals to work toward.
10. There is confidence in the future, that is, projection of a future that is better than the present.
11. There is the ability to laugh at mistakes.
12. There is the ability to be flexible, to be able to compromise, devise, and use the imagination.
13. There are periodic indicators of progress.

Comeback can be and usually is hard work and peoples' associated behavior is complex, for it is influenced by a variety of internal and external conditions. As such, comeback cannot be explained away by examining only a couple of variables. To understand the comeback process a systematic framework is necessary. Therefore, keeping this overview in mind, this case history will be devoted to illustrating how these concepts can be applied in a systematic manner to analyze a real-life situation.

The process and the case

The comeback to be analyzed is that of the famous dancer and choreographer, Agnes De Mille, taken from her book, *Reprieve* (1980). This poignant segment of a life story clearly illustrates the phases of comeback, the dynamic interplay between trajectory and biography, and details the struggle Ms. De Mille underwent as she moved upwards toward a successful comeback. Our analysis does not spell out every event that occurred in her comeback, for it is meant to be only a general overview. Also, since her comeback was successful, the analysis does not examine arrested comeback – why some people stop short of their potential and perhaps even fail to embark on a comeback. With these thoughts in mind, let us look at the comeback process and think about the first question that comes to mind: When does comeback begin?

Comeback begins with an initial awareness that there is something wrong with the body. It no longer performs as it should and once did. Hence, our labeling of this first stage of comeback as the *phase of discovery*. The discovery may be sudden or gradual: In this case it was a sudden one. The story opens as Ms. De Mille is rehearsing a group of dancers for a gala opening night at the theater. In the midst of the rehearsal, on the afternoon preceding the performance, she suffers a stroke and is rushed to the hospital. Thus, while we see the trajectory unfolding, we also see her biography, at least the professional aspects of it, suddenly coming to a standstill as she is hospitalized. The rehearsal stops and the show does not go on. In this early phase of the trajectory, the physician's prognosis is grave. He is not sure she will survive the body assault. She related, however, that even from the very beginning she believed that she would live.

Once settled into the hospital, Ms. De Mille describes a series of trajectory and biographical reviews through which she comes to contextualize the illness into her biography. She describes comparing her present body state with her body before the stroke, and the changes in her self-conception that result from this comparison. She refers to the sane, excited woman of the morning as now "a depersonalized lump that could hardly babble her name and had begun to drool, an aged, crouched husk of a creature . . . . " She also describes a symptomatic review (p. 29) in which she gives light to her present situation by relating it back to her high blood pressure, and unheeded signs of impending trouble like having, in the past, lost track a couple of times of what she was saying or doing (p. 33). She projects forward in time and asks, Will I recover my speech, vision, mobility, memory?

As in all crisis situations, the peripheral aspects of life are trimmed, and so at first her life revolved around and was caught up in the basics of survival. She says: "I was taken up with the minutiae of living. Everything was so extraordinarily difficult and so new to perform. Every single act became a contest of skill; and games can be tiring" (p. 46). Eventually, however, the basics became routinized, and she began exploring the extent of her limitations, arriving at the realization or crystallization that her body had indeed failed her (p. 55). The consequences of this crystallization were feelings of anger and the belief that there was no exit from this situation.

While some people at this point may decide that the catastrophe is just too overwhelming and would give up, Ms. De Mille, facing reality and knowing death would not provide a way out, decides that she wants a future, a future that will be different from the present. And so she says: "I tackled my strange and maimed existence" (p. 64). Thus, she moves into the second phase of comeback, *embarking upon the comeback trail.*

Yet, not everyone wants to or does embark and make progress. Four important elements come into play at this critical point. First of all, there must be an initiator, someone such as the physician who initiates the medical and rehabilitative plan and then sets it in motion. Second: At the same time, the affected individual must have a biographical plan or goals to work for, and accept the medical and rehabilitative scheme as necessary to achieve those goals, before getting underway. For all the rehabilitation in the world will not bring back a person who for one reason or another refuses to cooperate and do the necessary assigned work. Third, he or she must be mobilized. Mobilization comes as the result of having accomplished the necessary prerequisite work of confronting the situation and coming to terms with it, at least provisionally; that is, accepting limitations but believing that they can be stretched, and therefore having confidence in a better future. Ms. De Mille, explaining why she and those in situations similar to hers were willing to do the work required of them, says: "The work requires iron discipline, because one hopes to be useful and effective: Because although the heart of life lies behind and one faces a diminishing time with waning strength, one just prays and hopes to be less of a burden" (p. 68). Fourth, the plan must be right: right medically, rehabilitatively, and biographically. If the plan is found wanting, then it is necessary to revise it until there is a fit between it and oneself.

All of these conditions being met, comeback still requires another ingredient, if it is to be successful. That is, comeback requires teamwork. Not only does it require long, tedious, and at times superhuman effort on the part of the affected person, it also requires the coordinated efforts of other workers who share in the division of labor according to their respective specialties and abilities. Each moves in and out of the process according to the phasing of trajectory and biography and their associated contingencies. For instance, in Ms. De Mille's case, the team consisted of a variety of members (physicans, nurses, occupational therapists, physical therapists, speech therapists, spouse, son, many friends and associates). Each person in his or her own way contributed to her comeback.

At first, her efforts were directed at body mending and the stretching of limitations. Her biographical time and energy were quite used up in the performance of those tasks. Eventually, this work became routinized and insufficient to meet her growing need to move beyond where she had been. At this point, reknitting was added to mending and stretching. Reknitting, as a general process, begins when the person reaches the point of wanting to get on with his or her life. It may begin at different points along the comeback process, depending upon the degree of body assault sustained, energy levels, and the salience of these activities in the individual's life. The nature of this early reknitting varies according to the specific biography. For some persons it may take the form of seeing specific friends; others may begin to call business colleagues; and still others may call home and begin to manage their homes from afar.

Ms. De Mille visited with her family and friends and spoke with them over the telephone as soon as she was able, thus continuing her close relationships. However, one of her major reknitting acts was to send for the manuscript of the book she was working on before she became ill. She knew at this point that she could not resume her work as a dancer and choreographer (though she did read scripts sent to her) but could attempt another important, and at least currently more feasible, aspect of her Self: writing. Coming through strikingly in her description of this period is her frustration – yet determination as she attempted to manipulate her papers and get her thoughts down – since the papers kept slipping through her fingers, falling to the floor, or getting lost, and even out of reach. She relates how she kept the papers on a special table next to her where they would be constantly visible. When she couldn't work on them, she says she let her eyes rest on them. Why? For her, they represented a means toward the tomorrow future, a goal she could obtain through work in the immediate present. "I thought of them in the night and they were a promise. It is not enough to live in 'now' as we have been told. We have to surmise 'tomorrow'" (p. 76). Again, in this phase she asks the biographical questions in terms of time: when, when, when? "Will I begin lecturing in six months? In eight? Can I take a theater job in a year? How soon?"

Retraining an uncooperative body is not only hard work; it can also be frightening. Normally, the body parts work in unison and we can control them at will. But the failed body often becomes the uncontrollable body and that makes it quite another matter. "I began the real exercises, the exercises that were not boring, like hand therapy, but frightening. I stood between the double bars, one hand on each, which was comfortable and felt safe, except that the right hand was of no use and kept falling off." Also, "I was encouraged to stand and then walk. It was terrifying . . . . Every time I put the right foot out, I trusted the whole of the rest of the mechanism – my head and my breath and my heart and my viscera – to what?" (pp. 83–4).

The past biography is not only something to look back upon with grief: It becomes a crucial resource. Since an individual rarely loses all the various aspects of Self because of illness, those aspects remaining from the past can enter into the present either to enhance or deter development of a better future. The past biography, acting as a positive force in the present, is especially visible in the comeback of Ms. De Mille. She says that people were constantly expressing with sympathy how hard it must be for her, a dancer, to be faced with so many limitations. Her response to that statement was that precisely because she was a dancer and had submitted herself to the physical and mental discipline required to train her body, she was now able to meet the strenuous demands needed in the present for retraining her body (p. 86).

While a straight, steady, uphill climb would certainly be the ideal, life unfortunately is rarely so uncomplicated. Often, a comeback course is marred by setbacks or interruptions which may result in a temporary or permanent standstill or even a reversal in the progress already painfully made. A setback was experienced by Ms. De Mille when she developed an embolism. Limitations stretching and reknitting came to a halt temporarily while the focus of attention was diverted to the physical problem and potential crisis.

It is important to note that during the interim between the development of symptoms and the time that medical action was finally taken, she relates she felt as though she was slowly slipping away. Then she does an accounting review of her biography, and in those moments she gains closure, coming to terms with both her past and her projection of near-future death.

Case 2

Social world/arena: danger and debates

The presentation (by Strauss) of this second case history contrasts sharply with the previous. Rather than one actor and a supporting cast, and a single actor's trajectory development, there are now a series of events occurring within a scientific community, and participated in by a multitude of scientists and nonscientists who are scattered around the country. The events also involve several relevant organizations. What is portrayed in this case history are the events as indicators of subcategories in relation to a core category (danger to the social world). Data were collected with the concepts of social world and arena in mind – both derived from previous research of the author's (Strauss 1978, 1982, 1984) – and were extended by examining issues attending danger to the social world, as well as regulation of and within it.

First, we need a few orienting sentences about social worlds and arenas. A social world is a community, not necessarily spread out or contiguous in space, which has at least one primary activity (along with related clusters of activity); sites where the activities occur; technology for implementing the activity; and organizations to further one or another aspect of the world's activities. Unless very small, there are also subworlds, segments of the larger world. Within each social world, various issues are debated, negotiated, fought out, forced, and manipulated by representatives of implicated subworlds. Arenas involve political activity but not necessarily legislative bodies and courts of law. Issues are also fought out within subworlds by their members. Representatives of other subworlds (the same and other ones) may also enter into the fray.

We begin with an analytic discussion of a specific feature of social worlds: namely, that they and their key foci can be endangered by internal and/or external contingencies. Perceived danger brings about vigorous argument within the social world and sometimes leads its members into external policy arenas relevant to the perceived "danger." Following on that relatively abstract discussion of key concepts, we recount the evolving narrative of the so-called "DNA controversy" (Lear 1978) in case history form, in highly selective fashion and in accordance with the interwoven analysis.

That analysis is designed to bring out the evolving both of interrelated arenas and the development of complex relationships among the social worlds implicated in those arenas. The arenas are policy arenas at several levels of scope: scientific subspecialty, a discipline, science in general, state and federal governmental, and that more amorphous arena called, in common parlance, a "public arena."

Every social world faces one grave problem – if not always and immediately, at least potentially. That problem is how to survive as a world in the face of changing contingencies. The potential danger is to the social world itself, not merely to individual members or to particular organizations within the world. There are always uncertainties and unknown consequences which attend the carrying out of its central activities. These may, but do not necessarily, affect the survival of the world as a functioning unit. So the point at which to examine this important theoretical and research issue lies further back in a basic condition for permitting a social world's existence in the first place. No social world exists in total isolation. However insulated its members may feel themselves whether geographically, socially, or economically, that insulation – protection – is only provisional and subject to threateningly changed circumstances. The danger from outside to the world and its core activities may be only relative – danger lies on a continuum – but the threat is always potential. Only in a platonic environment devoid of all other communities could a social world expect no externally derived constraints that can hamper or kill off its cherished activities. Other communities may resist the presence or expansion of its activities: Also their activities may act as restraints. Those constraints and restraints can be spatial, legal, temporal, financial, political: All this in more or less degree.

That, then, is the underlying general condition for danger to the social world. To this, it is necessary to add those inevitable segmentation processes which produce subworlds within the parent world. These processes set further conditions which not only can be disruptive to, but begin to threaten (or are perceived to) the larger world. Segmentation can lead to feuding, jurisdictional fights, reputational injuries, immigration from the world of either the entire segment or sufficient numbers of members to make a significant difference; indeed, enough to affect its chances of survival. This all being so, each social world must make arrangements that neutralize or forestall and, if necessary, vigorously defend against grave threat. At its most explicit, this amounts to sets of defensive maneuvers. At its most implicit, the arrangements are informal, subtle, and hardly noticed or in fact taken for granted as part of the nature of things.

Social worlds operate under the aegis of implicit agreements that both constrain them from unduly interfering with or otherwise harming other worlds and their members, but permit them, in turn, to pursue their own activities under minimal conditions of being disturbed. These more or less silent agreements – including those made so long in the past that the current membership has forgotten them – are most likely to persist between contiguous worlds. That is, those that frequently or customarily intersect around space, time, money, labor, and other resources: Resources that each world needs to ensure its own purposes and implicated values. In short, to carry on in accustomed, preferred, or at least breathing-space ways.

When contingencies change sufficiently, those agreements are challenged, either partially or more totally. Hence, the need then to defend the world's turf by entering new and more explicit agreements. Those are inevitably accompanied by positional maneuvering which in its more extreme versions can entail force, violence, covert strategies, deception, and the like. In fact, even implicit or taken-for-granted agreements do shift ground imperceptibly over time, with hardly any awareness by the contracting parties that this has occurred.

External constraints on resources and activities entail what commonly is referred to as regulation. It is easy to see why. Less easy to grasp is that every world requires not only social control of its constituent organizations and members, but also a considerable degree of self-regulation. There are internal regulatory rules – whether in the form of explicit codes, with clear negative sanctions, or informal agreements or merely implicit understandings. When these begin to break down, a degree of social world danger can arise. (That, too, can lead to further, potentially harmful segmentation.) If, as it sometimes does, the disintegration of internal regulation has discernible consequences for neighboring worlds, or if they think it does, then of course they will take countervailing steps, including involving the name and institutional machinery of the larger society and its legal and policing processes. One usually thinks of self-regulation in terms of professions like medicine (Freidson 1970) and law – mandated to police themselves because of their societally esteemed work – but self-regulation also exists, if more implicit a phenomena, in less professional worlds. That is true also for the communities of "pure science." Even worlds that seem so free, like the hobby or collectors' worlds, have implicit inner regulation built into their structures: If not, or if ineffective, they run into public troubles.

Indeed, when the "more general public" – meaning important communities – are aroused by threats to themselves, their values, or their members, then they will take steps to curtail the offending world's activities, perhaps even to the extent of immobilizing or completely destroying that world if possible. Threat of legal recourse to compel harsh negotiative terms is perhaps the radical first step, then following through on the threat if necessary. The constraints are not merely on some particular organization within the world – though that may be the obvious target – but on the contextual world which it stands for; for it is that world's values and activities which are basically called into question rather than one or two of its organizations. Enforcing laws against or getting injunctions against particular organizations (viz., demonstrations against specific nuclear-reactor plants) are only immediate aims: The entire industry, social movement, or other type of social world is the villain.

But defensive strategies and defensive maneuvers are not the whole story, albeit for the actors these seem, and often are, crucial to their winning and losing the important battles. For the researcher, it is just as important – probably more so – to focus on defensive *processes.* By this we mean sequences of events by which the community begins to mobilize against, and attempts to manage, perceived threats to its activities or to its very existence. *Perceived* is a key term here, for not only may there be a gap between real and presumed danger (indeed, it can be more, it can be less), but in order for social worlds and their representatives to act they must first define the threat.

To define is not necessarily as simple a process as might appear at first glance: It can be a very complex and drawn-out affair. And to define with any clarity, somebody first has to discern, or recognize, potential danger. Moreover, that cannot just be any somebody – for he or she or they must be able to convince important others in the social world that indeed *this* definition is accurate. Since worlds (and subworlds) are often quite complex in structure, there are likely to be quite diverse views of whether there is danger, how great is its potential, from what direction it is coming, and what is to be done about warding it off. So internal arenas, and their debates, and the maneuvering of participants are constant features of social world defensive stands and actions. Getting a complex world to act with some semblance of unity is difficult, and such consensus as arises among disparate segments – including those that begin to form around the debates themselves – is fragile. Maintaining as well as achieving some measure of consensus involves persistent work. Even within older worlds with their well-established organs for defense (like the American Farm Bureau, the Anti-Defamation League, the NAACP) the in-world fighting for control over either the organizations themselves or for pressure to utilize alternative organizations, can be persistent and ofttimes ferocious in intensity.

To amplify and clarify the points touched on in foregoing pages, as well as to add some important new ones, we turn next to an examination of the so-called DNA controversy, but only in its defense-process aspects. The story in brief is as follows. During the 1950s and 1960s, a number of biologists became interested in "phage" research. This would later lead to contemporary molecular biology. The phage people came from different subcommunities within biology, eventually forming an international community of like-minded researchers: So much so that eventually they decided to negotiate among themselves a "phage treaty." This agreement was necessary to get some order into studies that would otherwise not be readily comparable. By the early 1970s, molecular biologists were making rapid strides in their work, had institutionalized settings in which to meet and work, regularized channels of funding, and so on. In 1975 the wider world was startled to read or hear that this scientific world had declared a "moratorium" on certain kinds of possibly unsafe experimentation, until either a safer technology could be developed or further research proved there was little or no danger from the prohibited experiments. The general interpretation of the unfolding of the DNA story – by the wider public, various social scientists, and most of the scientists themselves – was that the molecular biologists were reacting not so much, or at least not only, to the safety issue as to fears of external regulation that would occur unless they themselves took steps to handle both the safety and external regulation issues. Considerable public gaze became focused on both issues by public interest groups, responsive politicians, and by scientists themselves (reflected in many articles and news reports in *Science* which followed the evolving events over many months). For a short period of time there also was a certain amount of public furor at various localities around major universities where the presumed unsafe research was or might be going on. Eventually, as the usual interpretive story has it, this pressure died down by virtue of the biologists' collaboration with the National Institutes of Health (NIH) to set up guidelines to safe gene-splicing research – which could be changed over time as the possible danger seemed to lessen.

A more complex and subtle interpretation turns around the idea of defensive *processes* rather than merely strategies. Think of the matter in this way and consider the following well-known events. The first person to discern possible danger was Pollack, a virologist, who happened to hear during a set of annual summer workshops for biologists about some projected experiments by Berg, a microbiologist. Pollack was a specialist in biological safety techniques and he worked on the East Coast; Berg knew virtually nothing about the virological aspects of the techniques he was using, and he lived on the West Coast. This virologist placed a telephone call to the geneticist which resulted in the latter's "thinking it over" and then calling a temporary halt to those particular experiments. He was exerting self-regulation in a rudimentary form. But note that the representatives of two biological subworlds were now intersecting on this issue of safety (and implicit regulation).

A relatively short time later, two other geneticists, Boyer and Cohen, made a breakthrough with a revolutionary DNA technique. Berg quickly realized (Cohen was at the same university, and a friend) that its availability would not only bring a speedy invasion of outsiders from other branches of genetics and from biology in general, but put a potentially hazardous technique in their hands. He realized that DNA could then be put into a great variety of organisms, in "shotgun" fashion with absolutely unpredictable results for many organisms – which might get from the labs into the surrounding environment. Berg slowly began to discern that if the microbiologists could not quickly agree on this matter (the classic science way, as with the phage agreement) then – especially if anything untoward happened – the outside world would move with its legal and administrative restraints. Inevitably, these would be very detrimental to the pure-research enterprise. About this time, Boyer unexpectedly revealed the revolutionary technique at the Gorden conference, an annual affair attended by many different kinds of biologists and covering a great range of topics – not including either Berg's projected experiments or the Boyer–Cohen work. His revelation startled and dismayed many conferees aroused to potential hazards; so they immediately held ad hoc meetings about what to do about it, with much disagreement about this. But they finally reached agreement that an open letter be sent to the journal, *Science*, about this current state of affairs. From then on, the wider science community and the wider public community were participants in the definitional process.

But what *was* the definition? Which social world or subworld would develop the definition – after internal debate over it – that would weigh most, or weigh at all, in what would be done about this potentially grave safety issue? Indeed, would the larger issue be phrased in terms of safety or regulation, or both, and in what relationship? The subworld of microbiology was hardly of one mind on all of this, let alone the larger field of biology in general. Virologists (for instance, those experts on bacteria and bacterial transmission) thought the microbiologists relatively ignorant about the dangers they might be courting. So we see the virologists soon working hard to convince the innocents – just as in the first instance a cross-country telephone call by one of them alerted a previously unthinking geneticist.

Biologists came to use traditional institutionalized forms for grappling with these issues – committees, especially; but it is notable that other institutionalized forms – like the Gorden conference and two conferences held at Assilomar – were utilized to address safety (and implicit regulation) issues in tandem with the scientists' usual business.

So far, we have emphasized the slow, gradual process of discovering or recognizing potential threat (external regulation) to this biological specialty and its central activities. There was also internal threat because some segments within the specialty deeply resented and fought against aspects of the NIH guidelines – rules laid down by this government agency in cooperation with the microbiologists themselves. These guidelines were a combination of self-regulation and a bowing to the real or perceived threats of external regulation. But there was also a mobilization process: What to do about the threats. If we only look at strategies, we can see a certain portion of the iceberg, for it is easy to see that both the scientists themselves and the agency officials were consciously working in tandem to forestall crippling constraints and to develop a somewhat more formalized inner-world control. Yet, for us only to see strategy and not process is to miss a central feature of this evolving DNA story.

Relevant questions here are: How did this scientific world of biology mobilize itself? What were the stages of its mobilization? What did its members have to go through psychologically, intellectually, interactionally, and organizationally in order to mobilize? What were the barriers they overcame, or failed to overcome? Which segments allied with what other inner and outer social world segments in getting their views of what was to be done actually operationalized?

So as not to make this account drawn out and too complex (and though the story really is very complex), we shall only point next to two other processes. The first is *monitoring*, which consists of at least two subprocesses: monitoring for efficacy of mobilized defense efforts, and monitoring for potential new dangers. Such assessment of whether defense strategies were working was partly political and partly technical in character. Politically, the greatest outside pressures for regulation came from the legislative branches, mostly federal, but also from some state governments. Researchers helped to allay governmental fears about potential hazards from their experimentation through vigorous educative or lobbying efforts. Early in the DNA debate story, a powerful member of the U.S. Senate, Kennedy, seemed to favor considerable regulation because he was, anyhow, openly espousing more governmental oversight over science in general; so, in this instance, biologists needed additional allies from the larger scientific world. A particularly sensitive assessment agent (as well as effective negotiator) was the director of NIH, who balanced his governmental representative–outsider duties against his science representative–insider commitments. Presumably, he shuttled in Kissinger-like fashion, but more silently, among the various implicated communities. He was not only negotiating and assessing, but also busy at convincing influential people from all these worlds. As for the public interest groups – some of whom were represented on the important NIH committee (RAC) which debated and voted on each of the guidelines covering gene-splicing research – these public interest groups continue, even today, to act as watchdogs both over the committee and, to some extent through their scientist members, over the research itself.

But how did the microbiologists assess the efficacy of their defense measures taken against these outside groups? Though we don't really know, probably they made this assessment by simply noting the dying down of public outcry, the falling off of media attention, and the reflected lessening of pressure on federal and state legislators. All of this political involvement on the part of the scientists – the effort needed plus the assessment – was a completely new venture for them, except for those scientists who previously had been vocal in the public interest groups that were now alarmed by the proposed genetic experimentation. This was unfamiliar territory for the biologists: the kind of outer-world policy arena in which they had never operated.

By contrast, the technological efforts to lessen the possible hazards of their gene splicing was done on comfortably familiar terrain. An outstanding biological researcher temporarily put aside his own pure research to develop a form of *Escherichia coli* bacteria which could not survive contact with air outside of what would be strictly controlled experimental conditions. This weakened strain of bacteria quickly went into general use among the geneticists.

Furthermore, as they continued their collective research – having abstained from the potentially more hazardous experiments forbidden by the NIH guidelines – their own anxiety lessened (among those of them who had anxieties) because of no untoward incidents. Under the virologists' tutelage, the geneticists became more careful about handling their lab materials; but also, the very progress of their collective enterprise gave them more knowledge about DNA phenomena. This progress was very important in terms of the safety issue because earlier, and certainly at the time of Assilomar II (the second conference) and the subsequent hammering out of formal guidelines that led to the NIH ones, the biologists were, as one of them said, "flying by the seat of our pants." They did not really know much about the potential hazards. To sum up this complex assessment process, then: It involved both technological and political work, and somewhat different, though not entirely different, talents and scientists were involved in each kind of work. A few got into both types.

Another important feature of assessing is reassessing. In the instance of the DNA drama, this meant first of all that, from most biologists' viewpoints, ideas and therefore guidelines concerning hazard could now be greatly revised downward. But this kind of reassessment fed imperceptibly into reassessment of the "external" regulatory constraints also. To some, and possibly many, gene-splicers eager to get on with their prohibited experiments, the NIH guidelines represented external regulation quite as much as they did a jointly negotiated insider–outsider production. From their inception, skeptical voices had been raised against them, albeit silenced publicly with the counterarguments that worse might descend on microbiology if the guidelines were too quickly relaxed. So there was constant tension in the larger biological community and even within the smaller microbiology one over the issue. This tension flowed into more organized revolt around 1979–80 when revisionists among the gene-splicers began attacking the too-stringent guidelines, publishing their urgent messages in microbiology journals. They tended not only to dismiss the need for much, if any, restrictions on the gene-splicing experiments, but went further to accuse leaders (and followers) implicated in the original imposition and persistence of guidelines as entirely too alarmist – to the detriment of the pace of scientific progress. But they were not merely trying to revise the microbiological view of danger; they were also unwittingly revising history, since the accused had based their earlier judgments on the very shaky scientific knowledge of that earlier period.

Very recently this in-world debate was reflected within the NIH committee which oversees the guidelines and has responsibility for revising them after considered reasoning. At one of its periodic meetings, in the spring of 1983, the committee finally decided not to relax the guidelines much further – to continue to go slowly in doing that – because although potential danger was no longer as great, nevertheless a public reaction against further loosening of constraints might still be possible. The assessment process was still at work!

In closing off this portion of our account of the DNA drama, we wish to emphasize two additional points. First of all, note that the various danger processes are overlapping in time, in their chronological appearance. Even discerning and defining the danger continue to the end, except later everyone is more experienced, more people are involved, as are more subworlds; also more organizational machinery has evolved for discerning new dangers, although the debate over danger signs still continues. A second point is this: as should now be evident, the safety and the regulatory issues were often compounded – participants not distinguishing between the two. And we have seen why. In terms of our own interest here, in danger to the social world, it is the regulatory issue which is central to this particular case. The danger lay in the degree of constraint that might be more or less permanently placed on the pure research of the geneticists.

As we know, danger to a world's core activities can come from diverse sources, and arousing other worlds to restraining acts because of physical or biological or even symbolic danger is only one such source. So the DNA case is special; but then again, it is not so special since safety issues (think of novels and art labeled as pornographic, or of airlines, of the flareups of concern over deaths from boxing) sometimes are major precipitants in those counterreactions. About this particular DNA case, it is simply necessary for analytic purposes to think clearly about how safety and regulatory issues were related to each other. Simply put, they were distinctly different issues, but fed into each other. As we have seen, for instance, the safer the technology began to look, the more drive there was toward lessening both external and internal regulation. After all, it was the safety issue which touched off both forms of the regulatory issue. On the other hand, the latter issue forced the scientists themselves to throw more effort into being more careful in their experimentation and into minimizing danger by developing a technology to decrease the potential hazard.

11 Grounded formal theory: awareness contexts

In 1979 the author of this book read a paper at the annual meetings of the Society for the Study of Symbolic Interaction in San Francisco. It is reproduced here as a chapter to illustrate two points, neither of which has been emphasized in the previous chapters. First, how one goes about developing a formal theory; while most sociologists seem not to be personally interested in creating these higher-level theories, being content either to develop substantive theories about particular topical areas or just to describe ethnographically behavior in those areas, nevertheless the writing of formal theories is, from the grounded theory perspective, viewed as being ultimately of the greatest importance. Second, the chapter will illustrate how one can use written materials – technical or otherwise – for developing formal theory.

The emphasis will be on the use of theoretical sampling and the associated comparative analysis done right from the beginning of the research project, consequently the focus is more on open than on selective coding, nor are the special issues involved with the integration and writing up of formal theory discussed here. Getting off the ground with open coding is probably the most difficult step in developing a formal theory, as with substantive theory, so this chapter should prove useful to those who wish to learn this necessary skill, as applied now to the development of formal theories. Selective coding for formal theory seems not to present special issues except that the analysis is more abstract and based on more diverse kinds of data than for substantive theories. This chapter does touch, however, on selective coding insofar as developing a formal theory involves relating subcategories to the core category (the phenomenon chosen for study itself).

The *rules of thumb* implicit in the account given below are these:

1. Choose a phenomenon, and give it a name, for this will be your core category, to which all your codes will relate.
2. Select and examine some data in which your phenomenon, named as the core category, appears. This data may be drawn from an interview, fieldnote,

newspaper account, article in a popular magazine, paper in a technical magazine, novel – in fact, any document – or from your own or someone else's experience.

3. Begin to code these data in the usual fashion: dimensions, subcategories, etc., and in accordance with the coding paradigm.
4. Begin to write theoretical memos incorporating your initial ideas and the results of your coding.
5. Employ theoretical sampling, seeking your next data in a different substantive area. This will yield new subcategories and begin to give variance to the previous analyses.
6. Continue to do that, theoretically sampling within the same substantive areas but also in widely differing ones. This tends to greatly extend the similarities and differences brought into the analysis, while continuing to densify the analysis itself.
7. At every step of the analysis, think comparatively – not merely to suggest new theoretical samples (sources, events, actors, organizations, processes) but to enrich your specific codes and theoretical memos.
8. As you do all this, you will notice both the usual development of conceptual density and find yourself, almost from the outset, not merely doing open coding but beginning to do selective coding. This is because you will be relating subcategories to the core category under study. So the final rule of thumb illustrated in this chapter is: Be very aware of how all codes that you develop bear on the core phenomenon, and make the connections as *specific as possible.*

Before presenting the paper that forms the content of this chapter, it is necessary to contrast briefly the nature of formal (sometimes termed *general*) and substantive theories. Comparative analysis can be used to generate both. The latter is theory developed for a substantive, or empirical, area of inquiry, such as patient care, professional education, or industrial relations. Formal theory is developed for a formal, or conceptual, area of inquiry such as stigma, formal organization, or socialization. Both types of theory exist on distinguishable levels of generality, which differ only in degree. Therefore in any one study each type of theory can shade at points into the other. The analyst, however, needs to focus clearly on one level or the other, or on a specific combination, because the strategies vary for arriving at each one. Thus, if the focus is on the higher level of generality, then the comparative analysis should be made among different kinds of substantive cases and their theories, which fall within the formal area. This is done without relating theory to any one particular substantive area. In the paper reproduced below, the analyst will be seen using that research strategy.

The reasons for developing formal theories, as well as how and how *not* to write them, are subjects about which Barney Glaser and I have written many pages together (1967 and 1970; see also Glaser 1978). But we have never offered a set of concrete images for how one might develop a particular formal theory. I shall try to do that today, perhaps in an overly personalized way, giving a few of the steps I am now following in developing a theory of awareness contexts. My talk is meant to show a theorist at work, to offer a prescriptive set of generalized steps in the formulation process. My hope is that you will neither take my illustration as the only mode of doing this necessary job in social science nor dismiss my style of working as idiosyncratic, or as feasible only for someone who already has considerable experience in developing theory.

When listening to this sketch of my procedures, you will notice that some are exactly like those recommended for developing substantive theory. The analytic work begins immediately with the collection of data – it does not await the piling up of data. Analytic memos are written continually. The first phases of the analytic work – which may take several months – are focused conspicuously on open coding. They also are focused on densification, or the building of relationships among those categories – including noting relevant conditions, strategies, tactics, interactions, agents, consequences. Theoretical integration begins through that densification, but is not yet at the forefront of the enterprise. Theoretical sampling begins almost at once and largely directs the collection of data. During these early phases, the differences between developing a substantive theory and developing a formal one is that, for the latter, *theoretical sampling is done across many substantive areas*, and the open coding and densifying is done at distinctly more abstract levels than in substantive theorizing.

The prevalent mode of formulating formal theory is to move directly from a substantive to a formal theory, without grounding the latter in any additional data. The theorist, for instance, suggests that his or her substantive findings and perhaps theory about, say, physician–patient relationships, have implications for a general theory of professional–client relationships, but does not do the further work of studying the latter relationships comparatively. As we have noted, this kind of rewriting technique produces:

> only an adequate start toward formal theory, not an adequate formal theory itself. The researcher has raised the conceptual level of his work mechanically; he has not raised it through comparative understanding. He has done nothing to broaden the scope of his theory on the formal level by comparative investigation of different substantive areas. He has not escaped the time and place of his substantive research. Moreover, the formal theory cannot fit or work very well when written from only one substantive area (and often only one case of the area), because in reality it cannot be developed sufficiently to take into account all the contingencies and qualifications that will be met in the diverse substantive areas to which it will be applied (Glaser and Strauss 1967, p. 179).

In contrast, the general strategy we advocate involves the comparative analysis of data drawn from *many* substantive areas, this analysis directed if possible along its full course by *theoretical* sampling. A good substantive theory can provide an excellent stepping stone for attaining a powerful formal theory; but of course even a good substantive theory only provides the initial stimulus that moves the theorist toward his or her necessary comparative work.

Three months ago I began the comparative work intended to lead to a grounded formal theory of *awareness contexts*. Why awareness contexts, other than that they represent an old substantive interest of mine? Awareness-context behavior is universal, occurring everywhere and in very many areas of life. What comes immediately to mind are secrets of all kinds: their protection, penetration, and disclosure; also illicit and illegal behavior of various kinds (fraud, corruption, plea bargaining, confidence games). Like negotiation, awareness contexts constitute quite possibly what my friend Fritz Schuetze of the University of Kassel calls "mesostructures of social order." These are neither the macrostructures nor the microstructures that we are familiar with but something between, linking both, and very significant for our understanding of social order.

So, three months ago, I began my investigation by comparing several kinds of "data in the head" – data remembered from the *Awareness of Dying* study, and data drawn from what I remembered reading about spies, about the gay world, about "passing," about the handling of stigmatized diseases like leprosy. My aim was to abstract from those data some major categories that might possibly pertain to all awareness phenomena. Then, these would suggest theoretical samples to be looked at soon. My initial dimensional analysis resulted in a typed memo, which listed and briefly discussed the following: (1) the kind of *informational object* at stake (for instance, identity, activity, object); (2) the kinds of *information* involved; (3) the *visibility* of the information; (4) the *accessibility* of the information; (5) the *interpretation* of the information. Later I would add to this paradigmatic scheme two additional categories: *evaluation* of the information, and the *convincing* of relevant others about the interpretation of the information. I also guessed from these initial data that various structural properties might contribute to variation in the awareness phenomena: the number of participants involved the context, the degree of their knowledgeability, the social worlds they came from, the stakes in obtaining the knowledge, and so on.

So, my first scrutiny of new data involved looking at a book titled *The Vindicators*, full of true accounts about how various persons had been mistaken for criminals, been punished, but eventually had been vindicated in their protestations of noncriminal identities. Mostly their vindications came about not through their own efforts – since their resources for discovering why they had been misidentified were slim – but through the efforts of skilled lawyer–detectives who had many resources for searching out and interpreting the new and often relatively inaccessible information (evidence) needed to rectify the record. The vindicators need to convince not only themselves but the authorities – this was how I realized that "convincing" was also an important category. I called the whole process *rectification*, and wrote notes with such headings as misidentification, rectifying actions, conditions preventing rectification, and the convincing of relevant audiences. There were other notes bearing on previously recognized categories, especially the major ones.

The next memos touched spottily on similar topics, as I thumbed through an old copy of Richard Wright's *Black Boy*. This book also brought home, although I did not need the reminder, how the management of information is a matter of positioning and control, as well as how awareness management cannot be understood except in terms of larger macrostructural conditions: the relationships extant between blacks and whites in the United States.

Next, I recollected that Orrin Klapp's *Symbolic Leaders* had data bearing on celebrities, whose public and private identities were sometimes discrepant. If so, then a form of public misrepresentation was occurring. Most of Klapp's data on celebrities bear on the manufacture of a false public identity by the celebrity and his or her agents, and its maintenance in the face of a discrepant private life; also the breakdown of hidden private secrets when the public learns of the private life; also the consequences for the celebrity and for others when that happens. But the private and public identities of the celebrity can also be consonant: Under what conditions and with what consequences then, we can ask. Another new concept appears in this same memo: *discrediting* – either in the face of unwitting or deliberate disclosure of identity information. I called the latter, *public rectification.*

Shortly after, I scanned another book in my library, *Disappointed Guests*, essays by Commonwealth students about their disillusioning experiences while living in London. I chose this volume next because I suspected another rectification phenomenon might appear. It did: I called it *rectification by cumulative incident or event* – for the students gradually realized the true British interpretations of their own identities: outsiders, and of low value. The attendant conditions, consequences, and some of the interactions and tactics are noted in this same memo. Cumulative retification is contrasted with sudden-disclosure rectification, as it appears, say, in the shockingly harsh revelatory incidents in Wright's autobiography – when the child was brutally shown by whites what it meant to be black in the American South. Cumulative rectification was linked in my notes especially with the major category of *interpretation of information.*

Thus far I have given you only major categories and headings in the memos: These are mostly instances of open coding. You should imagine also that some of my notes helped to densify the analysis – that is, to put analytic meat on the analytic bones, in the form of relevant conditions, consequences, strategies, interactions, agents, etc, as specifically noted when coding the data.

My next theoretical sampling – one I had early thought of but delayed getting to – centered around the following contrast: an awareness context kept closed by an experienced and resourceful team, even by an entire organization (for profitable stakes) against a relatively inexperienced and lacking-in-resource opponent. The actors in this drama were the con and the mark: The source of data was Edwin Sutherland's *Professional Thief.* I was especially interested in the division of labor within the confidence-game team. The category of *betrayal* comes immediately into the foreground, since betrayal is potential wherever more than one person is in on a secret. Later I would begin to explore betrayal in relation to awareness, moving toward a scrutiny of materials on spy organizations. At this point, however, I found myself writing memo notes about the requirements of certain social worlds for keeping vital secrets hidden, and asking questions about who was more likely to betray these secrets. "It is at the intersections between worlds that we would expect to find the betrayals, the informers, the giveaways, etc." I told myself that it was worth looking closely at such worlds, since they should have special mechanisms for protecting their secrets, and possibly for discovering the secrets of those who would endanger their own. The thief's description of the criminal "fix," and how it operated, also reminded me of what I had known so well from research on hospitalized dying: Namely, the management of awareness is part and parcel of people getting their work done – work as they perceive it. Of course my memos continued to flower with minor notes on such matters as fake and true settings, courtrooms as disclosure sites, and so on.

Picking up now the thread of social world secrets, I analyzed data found in Carol Warren's *Gay World*, especially that bearing on visibility and accessibility of the signs of homosexuality, to outsiders and insiders to the gay world. And I linked the recognition or nonrecognition of those signs to space: the sites where they were displayed or muffled. Data on space and movement in space, as they relate to maintaining social world secrets, seemed likely also to be salient to a group like the Gypsies, so I looked at a book on them and found a number of their specific and masterful tactics for keeping Gypsy activities secret from outsiders. There were good data, too, on conditions under which their tactics occasionally fail, and what consequences then ensue.

You may have been wondering about the sequence of my theoretical sampling. Leaving aside fortuitous circumstances like just happening to own certain books, the sampling really is being directed by the continuing analysis. In general, the sampling steps have been directed by one or more of four considerations. First: The further exploration of a major dimension, such as accessibility (relatively easy or difficult), which more or less characterizes the secrecy of the gay world; second: a structural condition that might saliently affect the awareness context: for instance, the organized character of confidence games, or the organized control of information by the endangered Gypsies as they drift through enemy spaces; third: the deliberate strategy of building maximum structural variation into the theory; this is combined sometimes with, fourth: theoretical sampling in terms of other formal theories, such as theory of status passage or of negotiation, as in the next instance that will be described. (It is very important of course to link one's formal theory with others, provided they are grounded, not speculative.)

For it had occurred to me that private secrets and intimate relations – seemingly the polar opposite of social world secrets and relationships – might yield theoretical gold. Murray Davis's *Intimate Relations* provided the test ground for this hunch, giving a wealth of data on a passage (from "familiar" to "confidant") manipulated by one or both persons involved in that passage. My memo touches on that passage itself; on the management of both public and private identities; on the characteristics of crucial and mundane secrets; on the dangers of disclosure; on testing tactics; on levels of dangerous information; on psychological hostageship; on mutual hostages; on failure of disclosure efforts, or at convincing; on the interpretation of cues and its timing; on unwitting and deliberate betrayal, on mutual betrayal, on sequential betrayal. In short, I did further open coding, but also further densification. I wrote several methodological directives, telling myself to "look into" this and that, soon.

The Davis monograph raises an important issue: How can the formal theorist utilize the substantive theory which he or she finds in such a publication? In general, my experiences can be summarized as follows. First: Sometimes the substantive theory contains concepts potentially useful for one's formal theory; for example, Davis's "psychological hostages" and "crucial secrets." Second: If the concepts seem useful, then the substantive theorist probably has also offered some analysis of conditions, consequences, and so forth, that are associated with the referents pointed to by the concepts. For instance, Davis has a good analytic discussion of mutual hostages, of conditions and consequences of this, and of phases in moving into those statuses. Likewise, in the Olesen and Whittaker study of nursing-student socialization (*Silent Dialogue*), we are given useful analytic information on public–private identities as they related to students' presentations of selves. And in Fred Davis's paper on "Deviance Disavowal," there is a very useful substantive analysis of phases and tactics in preventive rectification by the visibly handicapped. When using such materials, the formal theorist must be careful to bend the original analysis to his or her formal–theoretical purposes. That is done not only by linking the former, in one's memo writing, with one's categories – especially with the major ones – but by asking further questions that lead to "next steps" in the investigation. For instance, and I quote from a memo here: "Identity disavowal, however, can be false, unlike that in the Davis article. And this misrepresentation can be accepted, be found suspicious, or rejected by other persons. What are the conditions for all of that?"

Now, if I am leaving you with the imagery of a fairly deliberate, step-by-step choice of theoretical samples, that imagery would not be accurate. The procedural picture needs to be filled in with at least two additional images of the theoretical craftsman or -woman at work. This theorist does not merely work when at the desk or in the library: Work goes on subliminally, and while other activities are taking place. So, I have found myself thinking about awareness contexts while walking, driving, even during the duller moments of concert listening. Then, quite fortuitously, one also comes across sparking and even confirming data in various forms: in newspaper articles, friends' stories, books read for entertainment, in events occurring in one's own life. *Everything is grist for the formal theorist's mill*, especially perhaps in the earliest phases of the investigation. An anecdote from Malcolm X's autobiography suggests how public identity may be disclosed to the privileged powerless but not to those perceived as powerful and therefore dangerous. A reseeing of that great old film, *Bridge on the River Kwai*, gives rise to a memo note on the failure to distinguish proper priorities of private identity, and the reading or misreading of its signs by other actors in the drama. A chance bit of thinking about the *Taming of the Shrew* leads to a note about misrepresentation of another's "good." A happenstance reading of a sociological article turns up interesting material on conditions and tactics of awareness management in casual conversations.

It has been my experience that this fortuitous and continuous contribution to the theorist's enterprise is highly useful, both on logical and psychological grounds. Logical, because it helps to build that systematic theoretical structure at which one is aiming. Psychological, because it often gives that "Aha!" feeling of delight ("Of course!") or surprise ("Damn, of course I should have thought of that; it's expectable"). When the theorist is also a teacher in a graduate program, or a consultant to other people's research, then he finds that these other researches unwittingly add to his cumulative data, forcing further analysis and memo writing.

One of the more unexpected and exciting dividends, at least for me, is the gradually enlargening substantive scope of the investigation. One begins to think of, and then actually to scrutinize, substantive areas undreamed of earlier in the research. I always knew that I would look intensively at spy organizations, at the making of artistic fakes, and at Goffman's book on stigma. I did not realize that the work of creating a believable drama on the stage would be a target for my inquiry, along with such interesting topics or categories as "believable or credible performances."

Perhaps I should close this performance of my own – hoping it too is believable – with an open request for any substantive areas *you* would think would be profitably explored for building this particular formal theory: For, as we said specifically in *Discovery* (and later in Barney Glaser's *Theoretical Sensitivity*), if one wishes to develop a *grounded* formal theory, then it should be a "multi-area" theory, based on the comparative analysis of diverse substantive areas, and of numerous incidents drawn from those areas. While I shall not repeat or elaborate further, here, on why we so urgently need grounded formal theories, I certainly hope this talk will encourage you to bend your research efforts in that direction, rather than settling merely for substantive theorizing and even for doing ethnography, however accurate and useful these may be. A good formal theory ought to be at least the equivalent weight of a ton of ethnographies, and perhaps half a gross of substantive theories?

Perhaps that last sentence is too strong: The overemphasis on formal theory was made only to counterbalance the overwhelming development of substantive theories against the paucity of formal ones, since social researchers have been primarily concerned with the former. Indeed, as pointed out years ago (1967), the preponderance of general theories, at least in sociology, were then relatively ungrounded, and they still are. So, we recommend the rules of thumb given at the outset of this chapter, if one wishes to develop a grounded theory at a higher level of generality than the usual substantive theory.

12 Reading and writing research publications

Reading for analytic logic

An important preliminary skill to be learned, if only to improve one's writing, is the acquired ability to read research publications for their underlying analytic logics (see, especially, Glaser 1978, pp. 129–30). Of course, everyone reads publications for their ideas, substantive findings, and perhaps, for useful data. But not everyone knows how to examine them for the analytic structures embedded in them. What is meant by "underlying analytic logics" is whether the publication is organized around proof or causality or concern for consequences, or a setting out of strategies or of topologies, or . . . . Some researchers are quite explicit about the foci of their publications, others are not; and sometimes they themselves are not clear about their analytic purposes.

So, it is very good practice for fledgling analysts to be able to read and think in terms of the logic of analysis. Having learned this skill, it helps them to think more clearly about their own writing, to organize it with more facility, and to give critical attention to the presentation of its underlying analysis. It also contributes to the novice researcher's understanding of how grounded theory is developed, and how its presentation often looks quite different from others written in different "qualitative" styles. So the students are presented very early in the seminar with examples of underlying analytic logics; then must practice finding these by themselves.

We shall give several examples of such analytic logics, briefly discussed, since once the idea is grasped it only takes a bit of practice to carry out these examinations. The examples chosen are either from literature which is very well known and/or easily available for subsidiary study, if necessary.

Before entering into this discussion of analytic logic, consideration of four additional points may be useful. First: Readers may well ask –

"But how do we learn to make these kinds of diagrams, and how shall we know if any given one accurately reflects the analytic logic of what we have read?" The answer to the first question is that the diagram is entirely subsidiary to careful scrutiny of the publication's emphases and then extraction of its analytic structure. You don't really need the diagram, but it can be very helpful; it helps fix in memory the revealed analytic logic. This visual aid can also help in presenting your conclusions to someone else. And how do you know whether your diagram is "really accurate?" You don't. You only know that it is an accurate, or approximate, visual rendition of what it takes verbally to characterize the publication's elicited analytic structure. The burden is on you, not the diagram.

The second point is this: Although five publications – and all from interactionist literature – have been scrutinized, readers who hitherto have not done this kind of scrutinizing should surely now be able to do so for any social science research publication. Thus reports of experiments tend to blackbox the interaction and concentrate on cause(s) and effect(s). Quantitative evaluation studies do the same but this research concentrates especially, from its very nature, on determining which consequences are positive or negative and why. Survey panel studies are concerned primarily with measuring the change of something (attitudes, opinions, etc.) over time (phases) with why (conditions).

Our third point: In taking apart any research publication, the aim is not primarily to judge the analysis, but to grasp it. Then and only then, knowing the analyst's intention, should an assessment be made, based on clarity of the analysis, appropriateness and sufficiency of data, explanatory power of the concepts, plausibility of the findings, etc. You may then note what has been omitted from the analysis – from your point of view – because it failed, by its very nature, to do what you would like to see done. That last represents a legitimate judgment, but only when it is done after previous careful scrutiny of the analytic logic, so as not to shortchange its author's accomplishment. We have found that unless students are told this, they tend to denigrate some publications quickly and often harshly, quite out of hand. In consequence, they learn from reading these publications not even what *not* to do.

Finally, a fourth point: The ability to read for analytic logic should become so habitual that, even without undue concentration on it and without diagramming its logic, the skill becomes integral to one's reading of research publications. A side dividend is that the scanning of longer publications for their logic can save many an hour of subsidiary reading time, unless one really wishes to study a particular report or book with unusual care.

We close with a cautionary fifth point. The foregoing discussion has been addressed to the reading of research publications. Of course, social scientists publish many different kinds of writing: Among them are topical reviews, descriptions of procedural methods, critiques of other writings, commentary on the writings of great men, presidential addresses, rhetorical blasts, speculative theory, speculations about the theories of speculative theorists, and so on. All of those types of writing have their own "functions," and of course some can be exceedingly useful or at least fateful for the evolution of a discipline, more so indeed than many a research report. However, they do not usually embody an analytic logic (in the sense discussed above) as do research publications, or "grounded theories" which refer to bodies of data. So they should be read for quite other reasons, including even regarding them as raw data spread out before the eager analyst who can now go to work on any given publication, analyzing it for its embodied data. In that regard, they are no different than, say, the yellow pages of a telephone book. On the other hand, neither should they be mistaken for research publications any more than are telephone books. It is well to recognize the difference – some readers (and authors) seem not to.

The readings

Opiate Addiction (1947) by Alfred Lindesmith: This is a recognized classic in the literature of sociological social psychology and of drug addiction. Written some years ago, this excellent theoretical formulation of the genesis of addiction to opiates, though controversial ever since first presented, is still very much part of the arena of debate over addiction. Briefly, Lindesmith's interpretation of the data is that a person does not become addicted merely by taking sufficient amounts of morphine, but only through a combined physiological and social psychological process. Once entered on the path of taking increasing quantities of the drug and then taken off of it – or taking oneself off of it – there is the well-known phenomenon of withdrawal distress: of terrific and terrifying physiological reactions. If – and only if – the individual takes or is given the opiate again in order to relieve those physiological reactions, *and also recognizes* that the ensuing relief was due to the resumed intake of the drug, *then* he or she is forever addicted. Even later efforts to become unaddicted will lead, sooner or later, to relapse, since the victim is irretrievably "hooked."

Leaving aside Lindesmith's extraordinarily careful efforts to find even a single exception to this generalization (he was unable to) by combing the entire literature of addiction, let us look at the logic of his analysis. A diagram (Figure

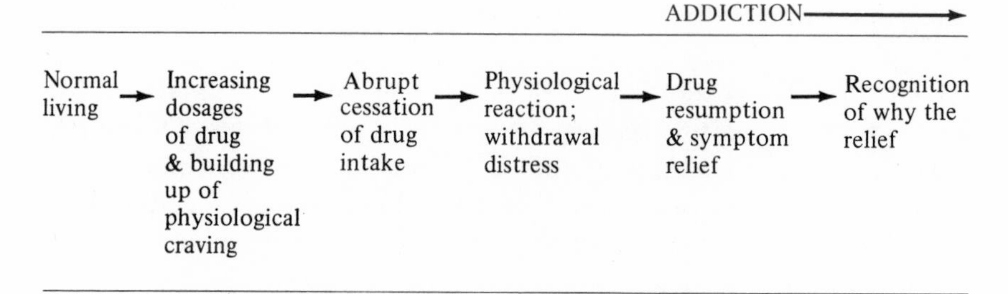

Figure 17. Diagramming Alfred Lindesmith's *Opiate Addiction*.

17) may help. Having inspected the diagram, think next of this if–then proposition: *If* the symptoms are relieved by drug resumptions (and they always are) *and* the reason for that disappearance is recognized, *then* addiction. Clearly this is a causal proposition: a set of conjoined conditions and a dire and persistent consequence. That the proposition is argued by Lindesmith as universal – that only one negative case would destroy it – is not our main issue here. Nor are we concerned with its truth or falsity. Lindesmith himself, of course, is concerned with those issues and also is perfectly explicit about the guiding logic of his study.

Stylistically, one can observe from the organization of the monograph how this researcher has put forth his causal argument. He begins with presentation of his method and purpose of his monograph, having begun by immediately noting the "central theoretical problem" of how people become addicted.

He notes, too, how alternative theories have failed to explain addiction. The next chapters detail the effects of opiates, the habituation and addiction phenomena, and the nature of addiction. In what is the central chapter, titled "The Process of Addiction," he presents and gives evidence for his theory of addiction; followed by another chapter dealing with "cure and relapse." Then follows a detailed criticism of alternative theories of addiction in light of the same data on which his own interpretation has been based. G. H. Mead's framework of self-reflection and self-indication is drawn upon in formulating his central theory of addiction, but since the effort is neither to dot the *i*'s of Mead's conceptions nor to check them out, rather to use them in the service of the causal proposition about addiction, this is exactly how Mead's conceptions function in the analysis.[1]

In one of his earliest papers, Erving Goffman (1975) addressed the phenomenon of what he termed *facework*, giving both a description and theoretical interpretation of it. When, during the course of an ordinary interaction, a participant to it commits a gaffe of some kind – whether verbal, gestural, or

[1] Interestingly enough there is really a second, though minor, less well-worked out part of Lindesmith's analysis: Namely, how addicts attempt to kick the habit, may actually seem to for quite some time, but eventually relapse – that is, they never genuinely get unaddicted. It is important to recognize in some of the debate over Lindesmith's theory that the two segments of analysis have not always been kept separate, so the arguments are correspondingly unclear.

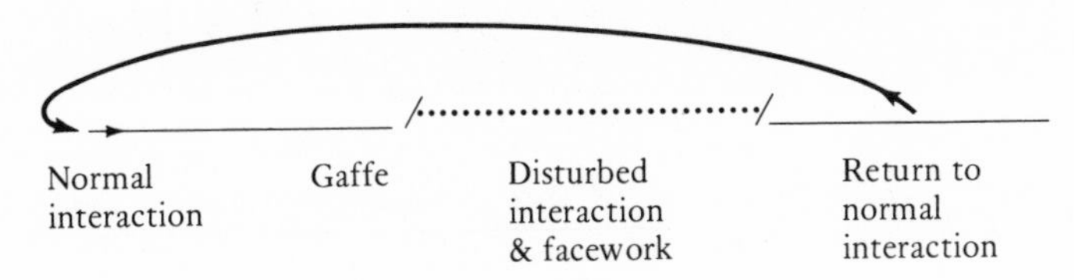

Figure 18. Diagramming Erving Goffman's "Facework."

situational – then the normal smoothness of interaction is radically disturbed. The parties are also disturbed by the untoward event, including the perpetrator if he or she notices it. There then follow observable efforts by the other actors to move quickly beyond the event, to get the interaction back to its ordinary, undisturbed flow. They accomplish this by a series of strategic actions (facework): such as studiously ignoring the gaffe or deliberately misconstruing it so as to neutralize its meaning and impact. By dint of such strategic actions, the interaction moves quickly back in mood and style to its previous state.

We can diagram this as Figure 18. The diagram, as well as the format of the article, speedily make evident the focus of Goffman's analysis. Facework is a universally found interpretation. A critical event (whose cause or causes do not concern the analyst) precipitates a period of disturbed interaction. Goffman is not so much concerned with this consequence as in the *strategic consequences* which ensure that this phase is quickly terminated and that the interaction returns to its original status. So the bulk of his article is devoted both to the nature of the disturbed interaction and to repairing strategies whereby, not incidentally, the perpetrator's "face" is saved: a necessity, since his or her restored poise is part of the salvaging. Goffman is not concerned, either, with tracing out the next phases of interaction – that is, other processual consequences of the disturbed phase – any more than in the conditions that precede the gaffe.

Understandably, there is an informal listing and discussion of kinds of strategies, though not really a formally presented typology, since that does not seem necessary to the analyst for his main analytic job. Perhaps if he were interested in why one class of strategy was used more than another, or when precisely it was used instead of another time, or how sometimes it succeeded and at other times failed, or which class of actor used it most or most successfully – that is, if Goffman were interested in variance – then he would have been more apt to have developed an explicit and elaborate typology.

Howard Becker, et al., on *Boys in White* (1961): (The monograph is mainly the work of Howard Becker and Blanche Geer.) This is a study, based on intensive fieldwork, of medical students undergoing training at the University of Kansas Medical School in the late 1950s. The researchers began by focusing on what happens to medical students as they move through medical school, and then shifted over increasingly to the issue of what happens in medical school as an organization that affects the students and the level and direction of their academic work. Chapter headings reflect some of the major concerns of the researchers and are relatively self-explanatory. Here are some of them:

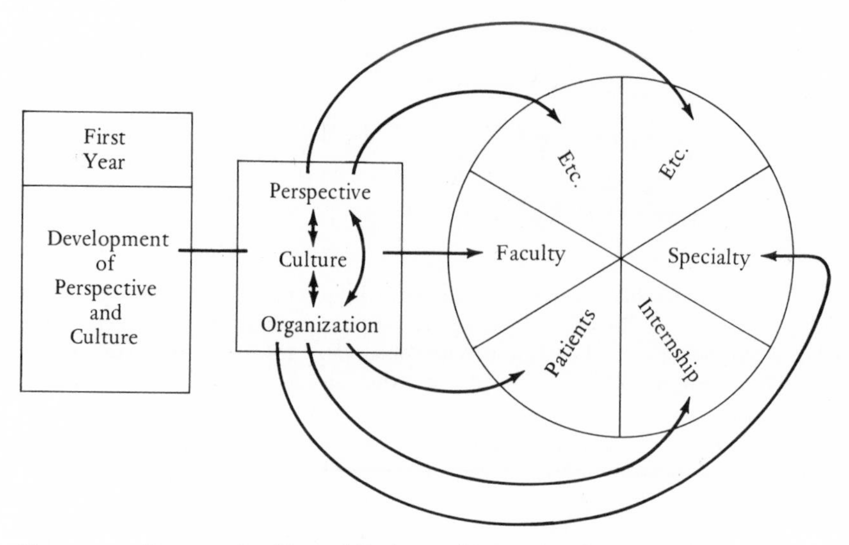

Figure 19. Diagramming Howard Becker et al.'s *Boys in White*.

Perspective, Culture, and Organization
The Long-range Perspective: "The Best of All Professions"
The Initial Perspective: An Effort to "Learn It All"
The Provisional Perspective: "You Can't Do It All"
The Final Perspective: "What They Want Us to Know"
The Work of the Clinical Years
The Responsibility and Experience Perspectives
The Assimilation of Medical Values by Students: The Responsibility and Experience Perspectives
The Academic Perspective: Dealing with the Faculty
Student Cooperation
Students and Patients
Student Perspectives on Internships
Student Views of Specialties

The book is divided into three sections. The first deals with student culture in the first year of study; the second deals with the next two years of clinical study; and the third deals with perspectives on the future. Despite this rough chronology, the analytic framework is not basically chronological beyond what is suggested by the first chapters above about the gradual development of perspectives. Even the perspectives on future choices (internships, specializations) are not pinpointed to specific months during the clinical years. The chapters include ample and often extended quotations which give a vivid sense of the students (and teachers and house staff) interacting, of their working (including students' levels and direction of work), and their perspectives on those activities and interactions (especially, but not exclusively, the students' perspectives).

At the risk of somewhat simplifying the scope of the analysis, Figure 19 is an approximately accurate diagram of the underlying logic. In each chapter –

embracing each topic – of the last two parts of the book, the analysis shows relationships of the three central conditions of: (1) student perspectives, (2) student culture, (3) organization (i.e., medical school and hospital). In addition, for the students, the development as a collective process of student culture and student perspective is shown in Part 1 of the book. However, that development is analyzed closely neither in terms of stages nor phases, but as reflected in titles of the initial chapters listed above ("initial," "provisional," etc.), and phrases such as "but are beginning to" and "at this time."

The diagram does three important things. First, it captures well the essential analytic thrust of the book; and second, it reflects clearly the basic organization of the book. Third, it does something additional, which can be quickly verified by scanning the index. The central categories discovered during the research, and treated in the book, are listed in the index; but few others are isolated and labeled; for instance, the many student strategies. In short, the diagram suggests in this regard that subsidiary additional categories may possibly be utilized in the body of each chapter – but, maybe not. They are really not. Student culture and perspective are the core categories, as is evident in the writing, chapter headings, and the index. In short, this study is carefully ordered around core categories – integrated well – but is not especially dense in its conceptualization, not even in dimensionalizing and subdimensionalizing its core categories. That part of the conceptualization tends, as the diagram hints, to be more implicit than explicit; and in that regard is not very different than many other excellent qualitative analyses. The implicit character of much of the analysis, but then again its underdevelopment, was once underlined for me by two readers of this monograph when it was first published. A physician friend said that reading the book made him uncomfortable because "these outsiders got to know so much!" By contrast, a sociologist, highly sympathetic to this kind of research, nevertheless finished it with a sense of disappointment: "It has so few *ideas*, really, in it."

Fred Davis on "Deviance Disavowal" (1961): This well-known paper reports conclusions drawn from interviews with persons who suffer from various kinds of visible handicaps (see also Appendix). The researcher is principally interested, as one of the subheads reflects, in "the handicap as a threat to social interaction," and in the "process of deviance disavowal and normalization." The movement of interaction portrayed is from very strained to less, and even to "managed" normal interaction. The paper's title totally gives the major thrust of all that interaction per phase: "Deviance Disavowal: The Management of Strained Interaction."

The handicap as a threat actually breaks down into three conditions; namely, as: (1) a focal point of interaction; (2) an inundating potential; and (3) a contradiction of attributes. Also the phases of deviance disavowal and normalization are labeled as: (1) fictional acceptance; (2) the facilitation of reciprocal role-taking around a normalized projection of self; and (3) the institutionalization in the relationship of a definition of self "that is normal in its moral dimension, however qualified it may be with respect to its situational contexts."

Various responses in each phase are noted as consequences and conditions of interaction. Also, a number of interactional tactics used by the handicapped

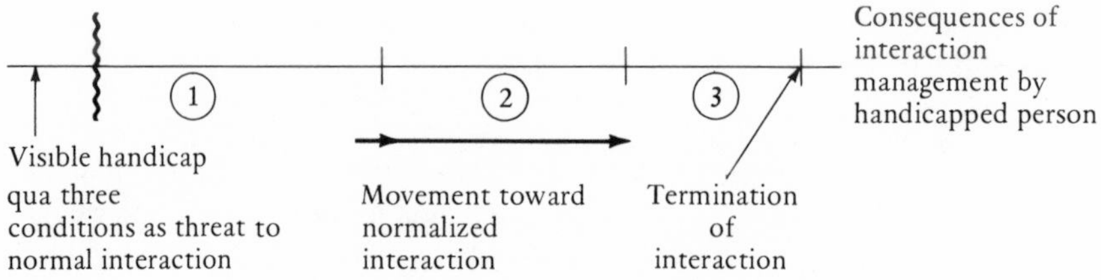

Consequences of interaction management by handicapped person

1 2 3

Visible handicap qua three conditions as threat to normal interaction

Movement toward normalized interaction

Termination of interaction

Figure 20. Diagramming Fred Davis's "Strained Interaction": three phases.

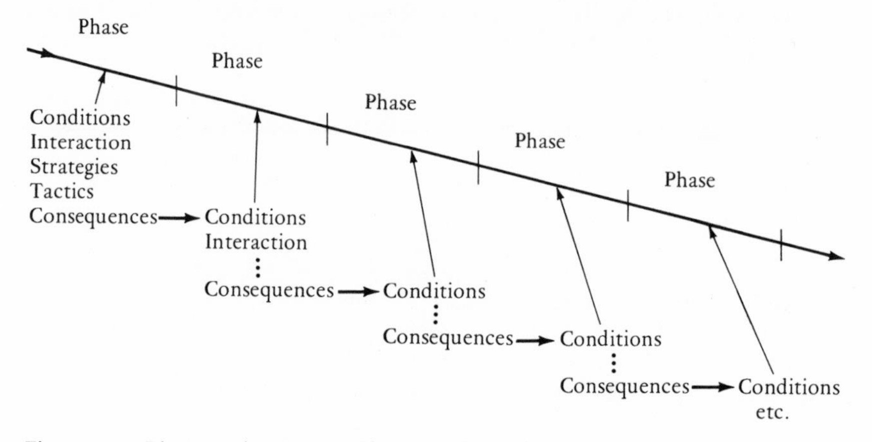

Figure 21. Diagramming Barney Glaser's and Anselm Strauss's "Trajectory."

are noted. Mentioned, in passing, also are a couple of long-range consequences. Close friends frequently "overlook the fact of the handicap and the restrictions it imposes"; and the normal person has to surrender some of his or her ordinary "normalcy, by joining the handicapped in a marginal, half-alienated, half-tolerant outsider's orientation to 'the Philistine world of normals.'"

The focus of this paper, then, is on process (phases), interactions by phase, conditions for precipitating the strained interaction, and the need of the handicapped to manage it so as to move it toward nonstrained, normal interaction. Getting interaction normalized is a handicapped person's main strategy. Specific tactics are mentioned but are not at the center of the analysis. A diagram that will capture most of that analytic structure might look like Figure 20.

Barney Glaser and Anselm Strauss on *Time for Dying* (1968): This is the second major monograph done in the grounded theory mode of analysis, and it will illustrate many of the points made in earlier pages. The monograph was organized around the core category of *trajectory*, and each chapter follows its phasing. Substantively, the phases portray the movement of hospitalized sick persons toward their deaths and the complex interactions that go on around them, which involve their actions also. For each phase, crucial conditions are

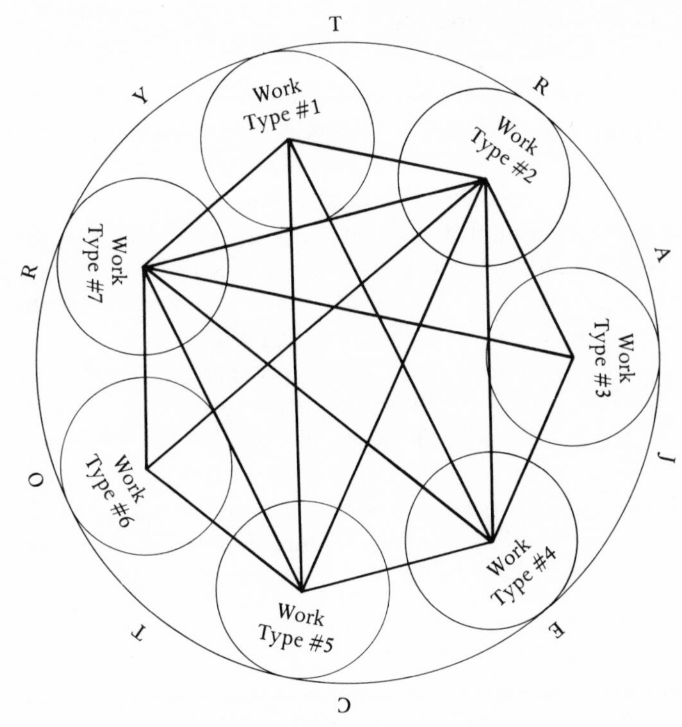

Figure 22. Diagramming Anselm Strauss et al.'s *The Social Organization of Medical Work.*

spelled out; interaction is densely analyzed in terms of a number of categories; strategies and associated tactics are explicitly noted and related to the conditions under which actors are acting; and consequences are listed for patient, staff, ward, and hospital. *In addition,* these consequences are converted into conditions for the *next* phase. (We underline these words to emphasize that the trajectory phases evolve progressively and the researchers have sought to present that movement analytically, rather than just descriptively or as composed of bricks laid down progressively without much connection except their sequence.) The diagram is Figure 21.

We caution readers against thinking that this is the only analytic logic that should guide or underlie a grounded theory publication! Of course, the logic must depend on the study – or which aspect of the study – presented. A good antidote to any such mistaken belief about "only one logic is possible" is now to analyze *Awareness of Dying,* the companion monograph, for its logic is far different, albeit the "rules" of coding, integration, and so on are identical. As a perceptive reader once said: "Reading the latter monograph is like walking slowly around a statue, while reading *Time for Dying* is like walking along a time line where a great deal is going on at every step." A virtually self-explanatory diagram for a recent monograph by Strauss, et al., *The Social Organization of Medical Work* (1985), would look something like Figure 22.

Writing from grounded theory analysis[2]

From the perspective of grounded theory endeavor, writing is the reverse, the mirror image, of reading. That is why this section on writing follows a section on reading. The point is underscored by the fact that our students in research seminars, or so it seems to us as teachers, experience the most difficulty in writing either because of incomplete carrying through with their analyses or a lack of confidence in their analyses, even when their analyses are actually more than sufficient bases for effective writing. We would tend to say, indeed, that the latter impediment stems for the most part from the former – and so major – barrier to writing. A third contributor to hesitation, slowness, even blockage, especially of papers rather than of theses or books, derives from uncertainty about or how to write for specific audiences whom the author wishes to address. Let us look at each of these hindrances to getting into, moving along, and finishing one's writing.

Clarity of analysis

Any professional writer knows that in explanatory forms of writing, except when following a well-worn format, one has to get the story line clear in one's own head in order to present it clearly, so as to get it then clearly into the reader's head. In that regard, any presentation of a grounded theory product is not a whit different. The materials of our presentation, in fact, may not be any more complicated or complexly interpreted than presentation in another mode. The main difference lies in the preliminary type of analysis, which presumably ought to affect the ensuing compositional process.

And it does. But how? To begin with, any writing whether for social scientists or laypersons, whether a paper or a book, will or at least should rest on preceding months of coding, memoing, sorting, and then reexamination of memos and reflection on that reexamination. We have also recommended the use of integrative diagrams. As Barney Glaser has written (1978), if this analysis has been well done, then the researcher should not find inordinately difficult the task of converting it into the first draft of a manuscript.

[2] Readers should study Chapter 8, "Theoretical Writing," in Barney Glaser's *Theoretical Sensitivity* (1978) as a supplement to this section.

Note the qualifying word, "inordinately," for certainly the task still may not be easy, especially under the conditions we shall discuss later of audience and psychological blocking. But here we are directly concerned with the point that getting the basic analytic logic clear is a prerequisite to ease the pain of making preliminary outlines for sketching the organizational skeleton of the manuscript itself, which thereafter facilitates the writing.

Of course, as we have repeatedly said, when writing you must leave yourself completely open to expanding, filling in, and even somewhat altering your analytic structure, including the outline which reflects it. You will also find that good memos, or sections of them, may fit nicely into sections of your outline, and can be used, with appropriate polishing up of language and reworking for clarity of expression. A good summary memo can even provide some of the headings in a subsection of the larger outline. And if you discover that a valuable memo was never really followed through, analytically speaking, you may find yourself doing that now – even returning to codes or data – and so, densifying that section of the manuscript on which you are now working.

Yet, when all is said and done, after the first draft is finished your writing tasks are most assuredly not completed. There are at least three remaining tasks that pertain to the analytic story line. The first is that you live with it for some time: How long will vary by person, experience with this kind of writing, or this kind of topic, etc. If then you are not satisfied with its details, you change them. Presumably, you should not have to alter the analytic skeleton radically. The second task is to think about the clarity with which the analysis has been presented. Is it really clear, unambiguous? A third task involves the question not so much of clarity as of effectiveness of presentation: Has it been said in its most effective form (for a given expected audience)? Can it be said more effectively in this place or that in the manuscript, and how? (But that last task brings us to the issue of audience, which will be discussed after the next section, on confidence in analysis.)

Confidence in analysis

We noted earlier that our students may find themselves blocked in beginning to write, let alone during the writing itself, if they lack confidence in their analysis. Do I really have it right? Have I left out something essential? Do I really have the core category? And if yes, still, do I have all this is enough detail (conceptual density)? The answers may be, yes, no, or maybe! But the issue here is not whether the analysis has been adequately and sufficiently done, but confidence that one really knows the answers to those questions. Even experienced researchers may not always be certain before they have chewed on their suspended pencils long enough to know where precisely are the holes – or be certain that, after review, they know there are no important holes – in their analyses. Whether experienced or inexperienced, a common tactic for reducing uncertainty is "the trial" – try it out on other people, individuals, or groups, informally or formally. Seminars and informal student-research groups meeting regularly or convened can give presenters confidence in their analyses, whether in preliminary or almost final form, as well as confidence in the analyses embodied in their writing. Speeches given at conventions, if favorably received, can add further validation of an analysis and its effective reflection in readable prose.

Nonetheless, when approaching or even during the writing period, there is almost invariably a considerable amount of anxiety about whether this can be, or is being, accomplished effectively. After all, some people are perfectionists and cannot seem to settle for less than an ideal performance. That can mean, of course, no performance at all or a greatly delayed one. Others lack some measure of confidence in themselves generally, and this spills over into questions about ability to accomplish this particular kind of task. Most have styles and habits of writing shaped by years of turning out school compositions and term papers, where writing is done quickly and often facilely – and with no careful rewriting of the draft – under the pressure of tight scheduling. Now, though there may be a planned termination date for the writing, the researchers must do their own scheduling of phases and, anyhow, discover that the dashed-off, often facile style of writing is quite antithetical to the rather complicated, dense, tightly organized style appropriate to the grounded theory methodology, especially when writing theses and monographs.

This anxiety and anguish, for grounded theorists as for any social scientist, can be further mitigated not only by preliminary or concurrent oral presentation, but by writing a paper or two before embarking – at least, seriously – on the long and major writing task. For students, besides, it is a good idea to get into that mode of production as a preparation for future research enterprises and publication. Also, getting a paper or two accepted for publication can give a considerable boost to flagging confidence or lingering doubts about one's ability at research.

Audience

Writing a monograph should pose no great problem of "which audience?" since the researcher usually is addressing colleagues in his or her specialty or discipline, and perhaps also in neighboring disciplines. How to address them, in the sense of choice of style, should present no great problems. Doctoral theses do present problems of presentation to the committee, since doctoral candidates the world over worry about the specific faculty members on their thesis committees and how to manage "getting by" all of them simultaneously. None of that is unique to grounded theory writing. However, there are a few issues that perhaps are directly relevant to the latter.

First of all, writing for journals can encounter the same negative reading as does writing based on other modes of qualitative analysis when read by quantitatively trained reviewers. Perhaps grounded theorists run an additional danger, even from reviewers trained in qualitative analysis, of not having their methods of analysis understood. This lack of understanding will not be helped by the use of jargon, which assumes too much familiarity with these terms by the reviewers, or runs a risk of annoying them. Neither will citing grounded theory publications for purposes of legitimation or information necessarily help. What is more to the point is a clear spelling out of precise procedures used, quite possibly with a few examples – if only in the footnote – so that both the reviewers and later the readers will understand better what has been accomplished in the paper. If ultimately the reviewers remain unconvinced and the journal editor rejects the paper, then it can always be sent to another journal. (That should be the procedure followed by any scientist whose paper has been rejected, a procedure quite customary among professional trade writers.) Also, when turned down by a journal, you should immediately request the reviews, if not already received: That is how you learn about your ineffectiveness in writing for the particular audiences the reviewers reflect, and indeed, learn how those audiences tend to respond to this particular paper. Having learned this may be of help in rewriting the manuscript for essentially the same kinds of readers of another professional journal. The specific negative reviews in hand also allow you to clarify your purposes to the rejecting editor: Sometimes grounded theorists have been able quite effectively to present them and get another reading and eventual acceptance of their papers.

Perhaps an especially difficult barrier is faced by nurses or educationists or others who have used grounded theory methodology, and who are now sending a paper to a journal in their own fields. As in nursing, which is hellbent (in the United States, at least) on becoming properly scientific, any qualitative analysis may be quickly rejected – and certainly this version termed "grounded theory." What can be done other than to read the negative comments of reviewers and try again, if possible, or try another journal? We suggest a sharp, if politely worded, point-by-point critique of the reviewers' comments, showing their misunderstandings or misreadings.

A second issue that faces would be authors – and again the problem is shared with other social scientists – is how to write either for non-social scientists or for a combination of social scientists and other audiences. But first let us look at the simpler job of writing for a single nonscientific audience, say for health professionals. One should suit the style to the particular readership (and its journal), prune away as much jargon as possible, and so on – again, procedures shared with other social scientists. In this regard, there really are no special problems faced by grounded theorists. The best advice that can be given is simple: Study the target journal, do not write in a style inappropriate for it (length of article, lengthy fieldnote quotes, etc.); but also be empathic with the thought processes of your anticipated readers and the trends in their field or events in their lives.

When writing for a combination of collegial audience and, say, other professionals (government officials, health professionals, social workers, lawyers, engineers – whatever) the trick for grounded theorists is to write theory, *but* in such a fashion that other expected audiences will read the paper or monograph as presenting "findings" or reflecting "reality." For instance, health workers read *Awareness of Dying* and *Time for Dying* and often agreed or exclaimed, "That's the way it is!" (perhaps saying "but," and then adding a variation to the unrecognized embedded theory). The chief thing to remember when writing is that both collegial and other audiences must be able to read the publication and understand it in their *own* terms.[3]

While it is not necessary that they understand it in the researcher's terms – they cannot really do that anyhow – what is essential is that

[3] Other than the points covered in the above subsections on clarity of analysis, confidence in analysis, and audience, there are also the bread-and-butter issues of how to write effective prose and to do that with a minimum of hesitation or psychological blocking. Since these issues are faced by any writer – social scientist or not – we shall not address them here. Very good advice is offered by Howard Becker (1985) on the writing problems encountered by his students, including some attention to standard problems of writing clearly and reasons why some authors do not or cannot. He gives some pointers also on getting over the usual blocks to writing.

writings do something more than get them to nod "yes," or function as a mere fortification of their beliefs. Researchers should aim to have them reflect more and in new ways about their own experiences and beliefs, as portrayed in the publications. "I never thought of it in that way," is the metaphysical gong that we should try to ring. In John Dewey's terminology (albeit he was referring to lay people's responses to art), we should not seek for "recognition": "Bare recognition is satisfied when a 'proper' tag or label is attached . . . . It involves no stir . . . no inner commotion" (1934, p. 53). We want active, thoughtful, deeper responses – or else why write for these audiences?

Writing theory

When writing monographs and long articles, *and* where you wish to present your theory and its embedded findings rather extensively, then you face an issue shared by all social scientists who write theory. That is, they (and you) confront it in actuality, but may not know that and so resolve it in traditional ways. Thus, some authors present formal propositions (for instance, "if *x*, then *y*") combined with explanatory discussion, though not very many qualitative researchers do this, for reasons discussed in a moment. Or they may present propositions in less formal terms, and nevertheless be self-conscious about the science-propositions imperative. Though others would call that an ideology rather than an imperative, or more likely simply regard the strict propositional mode as not feasible for their own writing – and often dull to read, when actually done. Qualitative researchers generally write their theories discursively and, as discussed earlier (in Chapters 1 and 12, for instance), often quote many lines, paragraphs, or even pages from their data collection. This is true, whether they have description as their aim or wish to maximally convey their respondents' perspectives, or whether they wish to convince us that their interpretations of data ring true. A great many genuine propositions may actually be written into the monograph or paper, but rarely presented as formal propositions – rarely indexed as such in the index either – and they are always embedded in a more or less rich discussional and interpretive context.

As for people who write in the grounded theory mode – with their emphasis on complex theory, along with the usual concerns for conveying actors' perspectives, the credibility of explanations, and the like – they also abjure the formal propositional form and make presentations in which implicit propositions are embedded in a discussional and descriptive context. In addition (and I quote here from a recent monograph, Strauss, et al. 1985, p. 296):

> In our monographs . . . we attempt to analyze data closely . . . so as to construct an integrated and dense theory. So the interview and fieldnote quotations tend to be brief, and often are woven in with the analysis within the same or closely related sentences. Longer quotes (especially from fieldnotes) are used for case illustrations . . . or when the events and actions described in the fieldnotes might help the reader visualize them better in tandem with the analytic points being made, especially when the events or actions might otherwise be difficult to grasp . . . . And, since many readers may be quite unfamiliar with what goes on in hospitals, the illustrative material is used sometimes to fill that gap, though generally our own words should handle that problem.

Our emphasis on formulating and writing *complex* theory, and understandable, readable, usable, and evolving theory, led us to write some years earlier (Glaser and Strauss 1968, p. 32) that:

> The discussional form of formulating theory gives a feeling of "ever-developing" to the theory, allows it to become quite rich, complex, and dense, and makes its fit and relevance easy to comprehend. On the other hand, to state a theory in propositional form, except perhaps for a few scattered core propositions, would make it less complex, dense, and rich, and more laborious to read. It would also tend by implications to "freeze" the theory instead of giving the feeling of a need for continued development. If necessary for [further] verificational studies, parts of the theoretical discussion can at any point be rephrased as a set of propositions. This rephrasing is simply a formal exercise, though, since the concepts are already related in the discussion. Also, with either a propositional or discussional grounded theory . . . [one] can then logically deduce further hypotheses. Indeed, deductions from grounded theory, as it develops, are the method by which the research directs . . . theoretical sampling.

It is important that you, as a qualitative analyst, grounded theorist or not, understand those points. For that understanding will give you confidence in at least *this* aspect of your writing, protect you from feelings of not being sufficiently "scientific," enhance your ability to resist standard styles of presentation, and thus lead you to explore partially or radically new forms of presentation of complex theory.[4]

[4] For a nice example, see Fisher and Galler, forthcoming. Fisher was trained in grounded theory style; their article is on how disability affects friendship among women, and their presentational mode is suited to a feminist-movement audience. Yet it is as rigorously discussional–propositional, as integrated and dense, as some of the best qualitative analyses published in the more usual social science publications.

13 Questions and answers

Over the years, students in our research seminars have repeatedly asked some of the same questions when puzzled, confused, not yet grasping, or psychologically thrown by certain features or procedures in learning the grounded theory style of analysis. Most questions arise because a student's stage of development does not yet allow him or her to answer those questions: For instance, "How do I know when I have a category?" and later, "a core category?" Other questions arise because students are plagued by lingering doubts about the validity or efficacy of grounded theory (and qualitative analysis generally), having been schooled or partly sold on quantitative methods. Hence, questions are raised about sampling, reliability, and so on. Associated questions pertain to how to convince others who are committed to quantitative methods about the validity or reliability of the qualitative methodology that the students are learning. Other questions stem from puzzlement about some procedures, like converting questions of psychological motivation into sociological questions and analysis. And understandably, there are still other sets of queries about how to write up the results of analysis for publication or speech making.

That certainly does not exhaust the list of sources for questions, but it does help to explain why the bulk of queries are raised – often more than once by the same student, who either remains unconvinced or cannot yet grasp the answer given to the gnawing question. Most questions have already been addressed in this book, but since readers will probably raise some of the same questions, it should be useful to them also to have at least brief additional answers given to the most-frequently raised ones.

"Is it necessary that the analyst be skilled in interviewing or doing field observations?"

This question does not especially apply to our own students, who take a semester or two of training in those techniques. However, there is considerable danger that people who have heard of "grounded theory," or read *Discovery* or *Theoretical Sensitivity*, will attempt to ground their analyses on quite inadequate interviews or fieldnotes. Good analysis is predicted on good data: not so much in the initial generating of possibly relevant categories and relationships, as in the later grounding and verifying of the emergent theory. In this respect, grounded theory methodology is no different than any other mode of analysis. However, it should be realized that not all qualitative analysis leading to useful theory needs to be based on interviews or field observations, and as long as the alternative selections of data (documents, tape-recorded sessions, etc.) are adequate, seem trustworthy, then there is no burden on the analyst to be skilled in actual generation of this data. Yet, most sociologists who do qualitative research, and comparable researchers in fields like anthropology, education, social welfare, health affairs, rely on their own gathering of data. Hence, the above question of data adequacy is very pertinent to their research results.

"Is it necessary to transcribe all your interviews or fieldnotes?"

The general rule of thumb here is to transcribe only as much as is needed. But that is not necessarily an easy decision to make, nor can it be made sensibly, either, immediately – perhaps not until well into the course of the study itself. "Only as much as is needed," and some of the advice given below, should *not* be read as giving license to transcribe just a few of your first interviews or your taped fieldnotes. And indeed, if either this is your first study and so you are still quite inexperienced in this kind of research, or you are doing such a small-scale study as to have relatively few interviews or fieldnotes, then it is wiser just to transcribe *all* of those materials. With more experience and with larger amounts of material, the considerations below will obtain.

There may or may not be the need – for your particular research purposes – to transcribe all of your taped materials, or indeed, every paragraph or line of each interview or taped fieldnote. The actual transcribing (which can involve considerable time, energy, and money) should be *selective*. The mode of operation in the grounded theory style of analysis is generally as follows. The very first interviews or fieldnotes, as you know by now of course, are provisionally analyzed, if feasible, before going on to the next interviews or field observations. This preliminary coding gives further guidance to the next field observations and/or interviews. Let's say you have done four interviews, transcribing each as quickly as possible for easier coding: You *may* then decide to transcribe the next, and the next, and the next – or you may *not*. Why not? What alternative do we have? One possible, sensible answer is to listen to the tapes, take notes on them, then do your analyses. Another possibility is to transcribe only selected passages (sentences, paragraphs, entire sections) for further intense analysis in accordance with your evolving theory. Transcribing those selected portions of the material will not necessarily relieve you later of further transcribing, but it certainly may. That last phrase refers explicitly to the injunction that theory should guide not only what you look for and where you look for it in the field, but for what and where you look in your data, too. This even leaves open the possibility that you may collect data, as you know, near the very close of your study when your analysis then tells you there is a hole in your theoretical formulations that needs closing – and that only further data collection will close it. But, in terms of the transcription issue, this means you might want to transcribe especially those portions of the interview or spoken fieldnote that are especially pertinent to the theoretical gap that sent you back to the field anyhow.

In the end, it is you who must decide – unless a thesis committee or advisor insists otherwise – for the decision is yours. You must decide what the purpose of your study is, what kind of additional analytic (theoretically sensitive as well as "psychologically" sensitive) contribution given portions of transcribed versus untranscribed materials are making to the total study. You may need nearly full transcription in order to obtain the density of theory you desire. You may want full transcription also if you have the money to have the tapes transcribed – but then you *must* (notice the italics!) listen to the tapes intensely, and more than once anyhow. Why? To remind yourself of things that you observed that you didn't fully record; or to remind yourself of things you noticed during an interview and have now forgotten – or never noticed at the time. Listening as well as transcribing is essential for full and varied analysis. Of course, if your study calls for intense analysis of intonation, and other subtle verbal aspects of the interviewee's talk, then you opt for transcribing also, but certainly you must listen repeatedly to your tapes. A final note: The full rule of thumb is, then, *better more than less* – but in the end both the responsibility and the judgment are yours.

In line-by-line analysis: "What does this line, or word, really mean?"

This query is one of the first hurdles to be gotten over in learning to do qualitative analysis. As remarked earlier (Chapters 1 and 3), the line-by-line analysis allows the researcher to fracture the data; to get analytic distance from them; thus escaping the seductiveness of their intrinsic interest by generating the coding, and raising the initial questions, stimulated by the scrutinized lines and words, as well as stimulating the formulation of provisional answers to the questions. Of course, the microscopic analysis also helps to suggest useful comparative analyses and theoretical samplings. Nonetheless, novice researchers – despite explanations about these functions of microscopic analytic procedures – worry about the "true" or "real" meaning of the examined line or word. The answer to this is: *Don't* worry! The point is not at all to settle that question (which anyhow is not resolvable without further interviews or other data), but for the analysis to function as indicated above. The propensity to dig for so-called real meanings, which often translates into "the actor's motives," is especially marked in those who have had a rich background in some branch of psychology, where "motivation," of course, can be a major issue. Even these students can master their trained incapacity, soon learning not to ask the true-meaning questions; for they, too, begin quite literally to see what can be gotten from close scrutiny of just a few opening lines of a document. The excitement of this analytic game, combined with "results," gives them confidence in this procedure, so much so that students characteristically look forward to trying it again in the next class sessions. Many make a personal challenge of mastering the procedure, understanding very well that it is one of the most important initial steps in becoming generally competent at analysis. As one of them (Nan Chico) said, after reading this section: "This is probably the key question, which when mastered (usually after an 'Aha!' experience) makes the rest more accessible, because your perspective, or mind-set is now different."

Perhaps it is not necessary to add at this point in the book, but we shall anyhow, that grounded theorists *are* concerned with meaning in general. Not to be concerned would make their analyses – to make a double entendre – meaningless, since many analyses do rest on including the actors' viewpoints about events, organizations, other actors, and so on. It is just that line-by-line analysis should not be burdened with digging for actors' motives, or with worrying prematurely about the meanings that, at this point, are only provisionally analyzed. We are interested, after all, not in the viewpoints of specific individuals but in the general patterns evinced by classes of individuals.

"What is a category, an indicator, a dimension, a code? How do I recognize one?"

These are questions that novices also sometimes worry over inordinately, and they may do that for some weeks after the instructor's explanations, as well as after having read both *Discovery* and *Theoretical Sensitivity* and even monographs written in the grounded theory mode. Definitions of each of the above-mentioned concepts do not necessarily satisfy those who raise the questions.

Why do they worry about them? The answer seems to be that students must actually see how categories are discovered and named; how indicators are pounced on in the data by the instructor and by others in the seminar; and so on. And until they themselves have done that, they cannot recognize categories and other codes purely by definition or explanation. Furthermore, some students remain uncertain even after they have done those activities and had their efforts approved of by the seminar participants or the instructor. They remain anxious until they have done it "by themselves," and can do it repeatedly and with some ease. The instructor, meanwhile, has been advising them to "relax, you will learn to know by doing and watching others do." But the instructor knows that ultimately only self-accomplishment will allay the anxiety. Fortunately, other students quickly achieve personal satisfaction through the class activity, and so escape this species of personal worry.

"Core categories: How do I know when I have one, or more?"

This question was addressed at some length in Chapter 1. Nevertheless, until young analysts have gone through the experience in analytic discussions – seminars or conferences – they remain uncertain about their abilities to recognize a core category. ("Of course, the instructor, or other experienced researcher, can do it, but can I?") This uncertainty and often extreme anxiety can be avoided – put off until later – during the early sessions of the research seminar because nobody is yet concerned with determining which category is core. Sometimes, however, analysis proceeds so rapidly that even the instructor may indicate that maybe this *might* turn out to be core, though one will not know until more data collection and analysis are done. If a student has collected too much data before trying any analysis or bringing the data to the seminar, then he or she is more likely to be eager to close in on the final product: that is, the main themes cohering around the core category(ies). Also, as the weeks march by and the inexperienced researcher is deep into discovering categories and their relationships – lots of them! – the question keeps nagging: *Which* of these is (are) going to be *the* really important one(s)? The answer to that question is, simply: Do the necessary work and be patient, while following the triadic procedures (data collecting, analyzing, memoing) discussed earlier in this book. Also, as one of our students (Nan Chico) has quite properly learned and then suggested: "Also, if you have several, play around analytically – choose any one to be core and see how much of the data get explained by it." (Though, if "any" one is not advisable, choose two or three that appear most frequently and try them out, with no urgency about the final choice or choices.)

"How many core categories are permissible, advisable?"

One core category, if properly handled, will – of course – give both conceptual richness and integration. In *Theoretical Sensitivity* (pp. 93–4), Barney Glaser has suggested that it is best if two categories are discovered – then one can be chosen, once sure of its relevance, and the other demoted by filtering it into the theory as another relevant, near-core, but not core category. Thus, in *Time for Dying*, we included ideas about awareness, but only *insofar* as they affected trajectory. And in *Awareness of Dying*, we did the reverse. The analyst can be sure that the other core does not disappear by using this method. The latter can still take a central focus in another writing. Many studies yield two or (sometimes) three core categories. "To try to write about them all at once with no relative emphasis is to denude each of its powerful theoretical functions." Probably one core category is all that analysts who are still inexperienced at grounded theory method can manage effectively. More experienced people can perhaps manage two in one monograph (not usually in a single paper). We do not suggest more: Not only because of the difficulty in integrating but because more expenditure of time and effort are needed to do the job effectively. However, it is undoubtedly easier for a researcher to integrate two categories when he or she has had several years of continuity in the line of research, one project leading to the next, so that a mass of data has accumulated.

"I don't understand how to convert psychological language into sociological language and perspective. How do I do that?"

The young researcher who asks such a question does grasp the necessity that, as a sociologist (and we are speaking of sociologists here, though the same is true for anthropologists, political scientists, etc.), his or her job is *not* to do a psychological analysis. True, such psychological terms as hope, denial, panic, hate, affection, horror, may certainly point to important phenomena, especially if used by the actors themselves; but ultimately, they must receive sociological treatment. The general answer to this issue is: "Think like a sociologist!"

Yet, that is hardly an operational answer. The additional specific operations consist of the following: first, regard any psychological term as a black box which needs to be opened up by careful analysis. Second, the analysis is to be done through utilization of the categories already developed during the course of the research project(s). Third, the analysis must raise questions about structural conditions, strategies, interactions, and consequences – as always. As an instance of how this takes place, we reproduce below a memo summarizing a conference dealing with sick people, between a junior and senior researcher (Julie Corbin and Anselm Strauss). This conference was precipitated by the younger researcher's persistence in pushing her insight, derived from sensitivity to the important phenomenon of "hope," until there was adequate consideration of it in the conjoint session. (See the related memo sequence, Chapter 9.)

Hope – 12/82
Into the concept of hope is built the conviction that you will get there. It is the trajectory projection you want, not the one you fear. When you have hope, you have goals for the future and belief that you will obtain them. There is hope for both illness and biographical trajectories.

There are certain markers along the way which are looked for and which identify whether the desired action is taking place, giving comeback, winning the fight or battle over illness. In this way, hope is tied into the major process according to the disease trajectory.

There are forecasting reviews of images of the future, which lie behind the work and a total commitment to get there. The individual buys into the view they have of the trajectory, or the doctor gives them (sometimes it is the doctor who destroys the hope). Hope is a condition for continuing the work.

Ties into "coming to terms," especially when trajectory stalls or starts downward; no hope for a cure, must then come to terms with death, level of limitations, etc. It is the biographical aspects which are impacted that one must come to terms with. Not necessarily come to terms in any deep sense – have to come to terms with each phase, or turn, change in conditions.

Problem arises when individuals lose hope, desire to fight, won't stretch limitations; in other words won't do the work like Coleman's husband – just retreated into self.

Individual may lose hope and not come to terms. (Anselm, we have not worked this out completely yet; we still are not there.) I think that is despair. That is when you lost hope, but have not come to terms with death. I'll think more about this.

"How do I learn to ask generative questions?" (*Implicitly the questioner is also saying to the instructor or more-experienced researcher: "You are so good at it!"*)

This is perhaps the most difficult question to attempt answering. At one extreme, one could simply reply: It's a consequence of much experience, so be patient about this, too. At the other extreme, one could say: Well, not everybody has the gift – you will have to discover whether or not you are gifted at this. Both answers are more than a little true: for, without some experience and without some gift . . . . Yet most novices can learn fairly quickly to ask very good analytic questions of their own and other persons' data; but this is not identical with asking questions that open up whole avenues of inquiry – over the next minutes, hours, even days. Some students quickly display the ability to raise questions that precipitate fertile discussion. Some never seem to learn how to do this. As teachers, we do not really yet know how to develop this ability in persons slow to learn, or how to improve the abilities of those who speedily begin to ask generative questions. Sometimes, it is a psychological matter: basically, encouraging the uncertain and the shy to try their wings, and then pointing out how successful their query has been in laying the groundwork for succeeding analytic discussion. Sometimes a good question has been asked but the seminar participants rush quickly past it to another issue or topic. Then the instructor can bring them back to the potentially generative query, asking them to "think about that and its implications." This becomes a lesson for everyone, including the questioner, if the question, indeed, turns out to be a stimulating one. Sometimes the instructor afterward can retrace the discussion – or the students can do this on request – pointing out the role of the initial query and how it was formulated. Sometimes the instructor will wish to rephrase a query to make it more striking or attackable, and later explain why that was so.

When all is said and done, however (and agreeing that much experience at analysis can vastly improve the ability to ask effective questions), nevertheless, we do agree that some persons have an instinct for the analytic jugular. They ask the right questions at the right times; and they usually know both that they can do this, having done it often before, and are doing it right now in all probability. Sociologists like E. C. Hughes think or have thought that this ability is enhanced by wide reading in one's own and neighboring disciplines: Undoubtedly, being bathed in the technical literature does help many fecund analysts, but certainly not everyone who is well read is an exceptional, let alone even a competent analyst.

We are inclined to believe, rather, that facility in thinking *comparatively* (whether with examples drawn from the scholarly literature, from one's data, or from personal experience) is the key element in improving the generation of far-reaching analytic questions. If this is so, then it follows that it is this skill that needs most to be developed, worked at, as a requisite for achieving the highest order of analytic ability. Said another way, it is a skill at thinking in terms of variations, that never settles for one answer, but always presses on with the query of "under what specific (and different) conditions?" In this regard, young analysts should train themselves to think analytically about what they see, hear, experience, read about in the course of their daily lives. This is an indispensable training for raising generative questions when doing explicit analysis – with emphasis on variation – of data.

"How do I choose a research topic?"

This query is no different than that asked by any relatively inexperienced researcher who is being trained in any tradition or in any discipline. Conventionally in the physical or biological sciences, students are assigned or adapt research topics that are offshoots of or connected with their sponsors' projects. Conventionally in the social sciences, students are directed to, or find their way to, topics derived from theories and/or research findings in the technical literature. E. C. Hughes used to say to his classes that a good sociologist could make sociology out of anything which he or she observed – meaning that comprehensive scholarly reading plus confrontation with any kind of data would set the researcher's brain into whirling and effective action. Barney Glaser, with much the same perspective, suggests "study the unstudied."

As we have indicated earlier, analysts in the grounded theory mode can opt for a research topic that is derived from previous grounded theory; but they are more likely to follow Hughes's prescription, beginning with choice – or fortuitous happenstance – of something that interests, intrigues, fascinates, bothers them. So, for our students the question, "How do I choose a research topic?" seems primarily to be translatable into: "How do I know whether my topic is really manageable, or at least manageable for someone at my stage of research development?" Phrasing the question in those terms makes immediately clear that the student is partly addressing reality and partly needing guidance in deciding how specifically to direct the inquiry, or how psychologically to handle accumulated data by settling for an analysis addressed only to a portion of the issues raised by the data.

In our experience most, but not all, of our students in training encounter no great difficulty in moving into a personally satisfying substantive area. So, sooner or later the acquired analytic skills enable them, with relatively little conferencing with the instructor, to focus down on what can reasonably be accomplished for a thesis. The analytic procedures reviewed in this book, if properly learned, will indeed lead to additional skills necessary for formulating specific research problems, delimiting problem boundaries, and avoiding the many potential diversions into intriguing, alternative, and often associated research topics. Of course, most novices also need some guidance, some cautionary comment, like: "Maybe that's too much to do, how about cutting it back this way . . . ."; and certainly, some encouraging remarks to the effect that what they are proposing to do, or are already doing, is feasible and phenomenologically salient. Alas, not everyone pilots safely between the Scylla of self-doubt and the Charybdis of less-than-optimum resources (including research skills), and so – to continue with that particular metaphor – may drift for many months without final, even any genuine, commitment. The students who learn grounded theory procedures best, or at least feel most secure in their learning, do not seem to undergo this travail, or at least not for long and not severely. The added dividend is that this experience contributes to the next step – if they wish to take it – of moving on relatively easily to their next research project after graduation.

"Now that I've finished my thesis, I have several questions about theoretical sampling."

1. There are a lot of problems I want to study that don't come wrapped up in single-site packages, or I don't know if they do or not. The procedures for theoretical sampling between sites are extremely unclear to me. Suppose I want to study a relatively infrequent phenomenon that's spread out over many sites, but still want to do field work to study it? Or, conversely, I want to study a slow-moving, broad-scale phenomenon (like triangulation of results between disciplines that may not exactly appear to the daily observer in the lab).
2. Also, in talking with an experienced field researcher recently, I noted that getting access to field sites takes a lot of time and investment. He laughingly remarked that this is why researchers usually end up working at only one site. Also, the reasons for choosing substantive fields or topics at different stages of the research are especially unclear – does theoretical sampling pertain to that?

Much qualitative research, whether in grounded theory mode or not, consists of what can be loosely termed one-site studies (fieldnotes from one site, interviews around one major topic, a single cache of documents). But of course, qualitative research can entail observations, interviews, and documents drawn from multiple sources. The guidelines for using theoretical sampling are not appreciably different in either case. The key point about theoretical sampling is this: Once you have even the beginnings of a theory (after the first days of data collection and analysis), then you begin to leave selective sampling and move directly to the theoretical sampling. With this new data:

for field sites
  theoretically sample the total site (for instance, a hospital)
  theoretically sample on this particular site (a ward, or department: actors, events, behaviors, procedures, etc.)
for interviews
  theoretically sample the total interviews
  theoretically sample within the interview
for documents
  theoretically sample the total cache
  theoretically sample within each document.

With previously collected data however, whether your own or someone else's, then you theoretically sample in exactly the same way and whether it is one or multiple site(s) or cache(s).

However, note that if you wish to do a multiple-site study, then you can begin with selective sampling at multiple sites, or with informants from multiple sites. Or you can actually begin with a single site, develop initial theory; then utilizing theoretical sampling move to the multiple sites as suggested by your theorizing.

As for studying a phenomenon that is slow moving, the answer is that if documents exist that cover the time span, they can be sampled. If not, then you interview survivors and knowledgeable actors in order to locate probable (possible) times, sites, actors, and behaviors relevant to your phenomenon. When you have some purchase on that, then you can theoretically sample for times, places, people, events, behaviors.

Confining yourself to one site, because of difficulty of access and time and energy, is conventional ethnographic and qualitative research procedure. If, however, access to the sites is available, then one doesn't have to spend a lot of time and energy in the typical fieldworker's fashion because theoretical sampling allows for more efficient, shorter-time observation and interviewing. (Some researchers never have done single-site studies. For examples, see monographs by the author of this method book and his co-authors on other projects – i.e., *Psychiatric Ideologies and Institutions*, 1964; *Awareness of Dying*, 1965; *Politics of Pain Management*, 1977; *The Social Organization of Medical Work*, 1984.)

As for rationales for choosing fields or topics to study: Do you mean substantive issues and problems, or whole specialties like the sociology of science? As remarked earlier, E. C. Hughes used to say that a good researcher could make sociology out of any topic; and alternatively, when people study something because they are compelled personally to study it, then they often do very well indeed; also much good research gets done on the initial basis of its social relevance. However, the theoretical sampling basis for project selection does not eliminate the possibility of its feeding into the other bases for choice. The answer is to choose fields, topics, problems from previous theory on a theoretical sampling basis. For instance, the medical terminology project mentioned followed from some years of previous work by the principal investigator, though he was personally, as well as theoretically, involved with that area and also recognized technology as of considerable social and occupational relevance. But, after you have chosen a topic, on whatever basis, you can actually begin to do theoretical sampling "in the head" on it – providing that you have some data in your head, too.

"But how is theoretical sampling actually DONE – what are the specific operations? How does one THINK of the samples? For instance, if I'm interested in anomalies that appear in scientific work, how do I sample using comparative analysis, since you say that's the basic strategy?"

You take the phenomenon under study and turn it around as if it's a sphere: Look at it from above, below, from many sides. In other words, you think comparatively along any of its dimensions. Think in terms of *variation* along the given dimension, say size, intensity, or flexibility. What is, perhaps, its opposite? Or extremely different? Or somewhat different? Or just a little different? Or what other dimensions might possibly be relevant other than the one you have already thought of, that you may have overlooked? With your instance of scientific anomalies, your research has led to distinctions among anomalies, mistakes, and artifacts. And of course you now know the salient dimensions of each. So you can compare anomalies with mistakes and artifacts along different dimensions. To repeat, train yourself to think easily – almost automatically – in terms of comparison. Some people compare very well in terms of internal comparisons (*close-in* comparisons). But it pays to do that more externally – widely, diversely. Perhaps this is especially so early in the research, but not exclusively since even later this may suggest additional theoretical samples (events, actions, behaviors, etc.). On these you gather data and then analyze. Or you use previous data to raise new questions now in terms of the additional theoretical sampling.

Perhaps we should add that this theory-directed type of sampling also allows for and encourages some measure of chance or fortuitous circumstance. The researcher is alerted to *recognize* relevant comparisons when he or she happens to come across them in a document, an interview, or a fieldnote. Indeed, experienced researchers will attempt to maximize these happy discoveries by positioning themselves at certain sites – in the library or in the field – where they calculate there is some probability that relevant theoretical samples will strike their prepared eyes.

Moreover, on research teams and in research seminars, the activity of theoretically sampling can be made more efficient by having the parties feed their data into a common discussion, and through it develop a shared set of related concepts which will continually direct all their attempts at theoretical sampling. (See Chapter 6 on teamwork.) However, any attempt to build a sizable research program involving many people, particularly if they are at different stages of their research careers, is likely to encounter certain obstacles in maximizing the efficiency of theoretical sampling. (We say "likely" since this has not yet been tried.) Among the obstacles is the following. The researchers are likely to have somewhat different substantive interests. They are fascinated by different substantive materials, and have differential commitments to developing shared substantive or formal theory applicable to those materials. A tightly welded research unit might provide the proper interactional circumstances for developing relatively high-level theory that furthers the research of the participants who are working in different substantive areas. But the pull away into substantive specificity is likely to be great for the reasons noted above; besides, there are many temptations to become experts in given topical areas. Career considerations may also enhance the focus on substantive specialization.

It is anticipatable, in fact, that unless program participants are carefully chosen for their capacities at effective discussional exchange and also for abilities at developing higher-level theory, then quite different talents will be exhibited in the group. Some researchers will be better at and more absorbed with lower-level theory, or with filling in the details of higher-level analysis; others will have a greater bent toward and interest in creating the higher abstractions. So some special organizational problems may be encountered in maximizing the efficacy of theoretical sampling beyond what usually occurs within the smaller research teams.

"How do I rid myself of habits of thinking in terms of quantitative methods which I learned earlier and which now interfere with my learning to do qualitative analysis?"

This is a problem for students who have had considerable experience with using quantitative procedures. What impedes the new learning is not at all the capacity to use the previously learned procedures, but the style of thinking that is associated with their use: For instance, it is no longer appropriate to ask – and this operational question will occur implicitly to the researcher – "What percentage of *x* will do *y*, with what probability?" And it is difficult for you not to reason in terms of standard modes of sampling. The main impediment is, however, a strong commitment to questions derived from quantitatively oriented research, which understandably cannot be dislodged until the learner of qualitative research begins to have comparable confidence in the efficiency of the new style and its associated procedures. Quite like the issues of "What are the actor's motives?" and the associated interference of a psychological perspective despite the desire for a sociological one, this problem of an overly exclusive attachment to quantitative analysis can be overcome principally by just asking the typical qualitative analytic questions of one's data: What *are* the strategies, what *are* the differential conditions, what kinds of theoretically derived comparisons would be useful here, or what theoretical samplings are implied by the analysis so far? Perhaps the steepest barrier to overcome psychologically for the novice is, however, the injunction to allow the main theoretical concepts to "emerge" during the research itself, rather than have them before the research. From the quantitatively trained student, that demands an act of faith: If it cannot be made, then the game is over before it begins.

"When am I really going to get competent at this kind of qualitative analysis?"

This cry of impatience is not just characteristic of grounded theory learners but, of course, for learners of any kind of complex skill. Some of our students learn – or think they learn – much too slowly for their own peace of mind, evincing all the signs of wishing instant, or at least faster, command of skills. Probably the instructor's best tactic here is to set up one or two personal conferences, and to work on whatever analytic weakness is most salient currently. If it is an inability to rise very far above a descriptive level, then work on getting the student to dimensionalize systematically and to find categories in the data and name them. If a difficulty is encountered with sensing integration, then demonstrate how much integration has indeed already taken place in the student's memos, and set tasks involving more integration through sorting or diagramming. A helping hand over such actual or perceived barriers can do wonders for the impatience, or even despair, of anxious learners.

"What do I do when I have done a whole bunch of integrative diagrams, all rather different?"

The answer to this is: There should not be an aggregate of multiple diagrams but *successive* ones. Each later one should incorporate elements of earlier ones; or alternatively, the larger, more summarizing diagrams should encompass most of what is sketched in the earlier diagrams. (So then to choose among them is not an issue.) In short, the diagrams should *cumulate* in snowball fashion. Like the sorting of memos, past diagrams should either be integrated immediately into the next ones, or reexamined periodically for inclusion into the next "big" one. Furthermore, the making of such diagrams should become a regular feature of the research. Beginners especially need to force themselves to diagram – as one testified in class: "When in doubt, force yourself to do it, even if the first diagrams you produce seem silly." Although some researchers are more visually minded than others, it is a good idea, anyhow, to get into the routine of trying to cumulate successive memos early.

"After you decide on a core category and are well along in integration, what do you do with all those other ideas that don't fit well or not at all – that seem not to belong to the main analytic story?"

This is the important problem of "letting go" of many bright ideas and categories that still seem tantalizing, exciting – but our methodology commands: "Drop them!" What is entailed in doing that?

As we know, early in the study a researcher thinks of many intriguing questions, discovers and names many interesting categories, follows many leads; but eventually one or two categories are decided upon as core and their relationships with other categories are systematically traced. Meanwhile, a lot of seemingly unrelated or not clearly related categories, along with not clearly related categories and relationships have accumulated; so have many associated but topically scattered memos. Now, since the point of the research is to produce well-integrated, conceptually dense theory, the novice researcher is likely to be faced with two problems.

The first is purely psychological but double-edged in nature: How to discard all those bright ideas, even when one is quite certain they don't really belong? If not certain, the sensible choice is to wait longer, continue to code, memo, and collect data until quite sure of what is the core category. If certain, then the nonfitting analytic items must be pruned away, however wedded to them is the researcher. "But they are such good ideas and so much has been expended on them!" Discard them – the aim is not to be bright, and the work has not been in vain. "But they might add to the richness of my analysis." Discard them – the aim is certainly to get conceptual richness, but you have that, or will have it, *plus* integration. Otherwise you have a lot of totally or relatively unintegrated ideas floating around on the margins of your central analysis. So, alas, the novice researcher has somehow to endure the pain of severance. Later, when more experienced, the pruning operation will be easier to perform.

"But can't I do *something* with those severed parts?" This is the second problem: What, if anything, can be done with the discarded bits or segments of analysis? The answer is: "It depends." There is always the possibility of writing a paper, quite aside from a thesis or monograph, which will not be focused around the main analysis. (In fact, on sizable research projects, this is done often, regardless of the discipline or tradition.) If the topic is sufficiently engrossing or important, if the researcher has something to say about it, and if he or she estimates that there is an audience for this paper, then it should be written. For lack of time and because it is in competition with the main publication, the paper may not get written until later. Alternatively, because its composition takes less time, energy, and commitment, perhaps it will be written first. Even if published, the pruned material may be found useful for teaching, immediately or later. The third alternative for dealing with pruned analyses is to let them lie fallow and then later follow up the leads provided by them. That may result in a full-fledged research project, but again it may at least lead to completing enough details so that a good – if sometimes exploratory but nonetheless useful – paper can be published.

All of those alternatives require calculation of personal resources, personal pacing, and commitment. After the researcher becomes more experienced, he or she may well juggle, following the main storyline through to the end. Or, much earlier, choose to write up for publication the most compelling one(s) of the pruned analyses. On team projects, the likelihood of side-issue publication is even more probable. That is because more people are involved; there are more divergent interests represented; and more ideas are likely to be generated, albeit there is still a main analysis which all team members are committed to and working on.

"I still don't quite know how to relate my efforts to the technical literature. Aside from the point you've made about sometimes taking off from someone else's grounded theory, can you say again how one fits the analysis together with the literature?"

There are roughly five modes of relating one's work to the literature, aside from taking off from a nonspeculative theory. First, researchers' general knowledge of the literature in their discipline and related ones gives a basic substratum of "the" discipline's perspective, which furthers thinking in characteristic disciplinary modes. This perspectival view provides a sensitivity (psychologists used to call this an *apperception mass*) to features of the phenomenon under study – or leads initially to study of it because you sense its relevance to the discipline itself. It also leads you to raise some of the kinds of questions that you do about your data. However, this generalized knowledge does not necessarily supply a specific theory from which you make specific deducations in order to depart from that theory, as illustrated in the Discovering New Theory section of the Appendix.

A second mode of relating to literature is to read papers and monographs that directly deal with the phenomenon of your study, doing that primarily for the raw data they contain, like reproduced documents, segments of interviews, or fieldnotes: That is, those data can supplement one's own data base.

This is especially true of phenomena that are somewhat different but closely related to your own. For instance, if doing research on patients ill from a given disease, you would systematically look at studies (and popular articles too) dealing with other diseases. Also, you should examine these early in your research, for they provide comparative data which can greatly enrich the conceptualization of your primary data. For instance, people with hypertension usually suffer from no symptoms whatever, so often tend not to believe the professional's diagnosis and quickly stray from adherence to the prescribed regimen, or reject it. All of that is quite different than what happens, say, when someone's angina is diagnosed as indicating a severe cardiac condition. This recommended procedure should be followed regardless of the phenomenon under study, and some can even be examined comparatively in terms of occurrence overseas, as when an American industry is compared with others both here and abroad. This procedure does not call for exhaustive comparative work, but only enough to further the conceptual densification in the study of your specific phenomenon.

Third, if the publications deal with other phenomena than you are studying, then the raw data as well as the "ideas" expressed there may still stimulate comparative analyses and thus precipitate additional ways of approaching, coding, and memoing your own data.

Fourth (and equally important): Publications about the same or related phenomena under study may contain theories, views, or analyses different than you are pursuing in your own research study. Reading of these materials is probably best avoided until your own main analytic story (core category) has emerged and stabilized. Otherwise the reading can reduce your creativity, or at the very least take energy out of, or the edge off, your own drive. But after your theory has begun to integrate and densify to a considerable degree, then supplementary or complementary or conflicting analyses should be grappled with. They should be integrated into your theory if possible (including some of their categories, conditions, etc.); or criticized in terms of what you are finding; or if their approaches to the phenomena are so different as to lead to quite different places (as when a sociologist reads a study by a political scientist), then that might be discussed also. And, of course, it makes sense eventually, in terms of the necessary accumulation of knowledge, anyhow to fit your own study into the larger framework of preceding studies, providing there are any, though quite often you will find that yours is the first exploration of the phenomena. Your findings can, perhaps, also be related to the more general disciplinary theory (cf. theories of social order); but again, providing this is done without undue reaching out for faint possible connections or just dotting an I ("This is an instance of . . . .").

The advice about delaying the scrutiny of related technical literature applies full force to inexperienced researchers, but perhaps less so to the experienced. Why? Because the latter are more practiced at immediately subjecting any theoretical statement to a comparative analysis. "Yes, that seems true, but what about if the phenomenon under question occurs under other conditions, such as . . . ?" or, "This theoretical statement indicates only this one consequence, but what about consequences for these other actors or for the organization itself or its different parts?" Yet even experienced researchers may need to be wary when entering a substantive area new to them. They are less likely to be captured by others' seemingly informed theoretical formulations if first they feel more secure about substantive details and their own initial interpretive formulations.

Finally, we should add that when someone takes off from *your* published theory, it is requisite that his or her findings be related not just generally but specifically to your preceding study.[1] (And so should yours, if you were in his or her position.) When the relationship is not discussed with specificity, which is common, there is no cumulative integration.

"How do I negate received theories, already in my head from past education and reading, which block me from seeing afresh?"

[1] For a quite good example of how this can be done, see the series of research papers addressed to negotiated order in *Urban Life*, 1982, vol. 11.

This question reflects the trained incapacity (to use Veblen's old but accurate term) of people who have had good educations derived mainly from reading and listening in classrooms. When they begin trying to learn new modes of thinking about phenomena (data) their learning slows them down, and they wrestle with what to do with their educational acquisitions. In general, there are three answers to this problem.

First, if the received theory is indeed grounded, then you should follow procedures suggested in the section on discovering new theory from old theory, and the preceding discussion in the current chapter on incorporating "the literature." That is, use the theory as a legitimate springboard into qualitative analysis and data collection based on theoretical sampling, comparative analysis, careful and extensive coding, etc.

Second, if the theory is not particularly grounded, rather abstract (for example, Parsonian), then the problem may derive from some of the assumptions behind the theory: for example, functional or neo-Marxist ones. You then have to decide whether you can use functionalism or neo-Marxist perspectives as a useful sensitizing instrument (as discussed earlier re using the technical literature). If the perspective indeed facilitates getting you off the ground in your research and doesn't block you from discovering new categories *specific to the phenomena* under scrutiny, then it can be useful.

Third, the real problem for many literature-educated beginners, however, is that running through their heads are Goffmanesque or Marxist or Garfinkelian, etc., refrains which embody specific concepts. The journals are full of articles that offer readers rich, raw data interpreted by applying such concepts, essentially as exemplifications of the theorist's ideas, rather than as qualifications or extensions of theory carried out by specific pinning down of variable conditions that affect the phenomena under study. Such labeling is useless and should be avoided like the plague.

How to do that? The answer is: Firmly resist temptations to code for those specific concepts *unless* they make sense in terms of line-by-line or paragraph-by-paragraph analysis. If that microscopic analysis draws out those categories, as well as others, then they eventually are to be related to each other like any other sets of categories. They are not to be made core categories unless they earn that status by virtue of using the usual procedures. Furthermore, the literature-derived categories must, like all others, be coded in terms of conditions, consequences, etc., as discoverd through the microscopic coding of both current and new data collected through the direction of your emerging theory.

"Is there a difference in working as an individual researcher and with a partner or as a member of a team?"

Under current conditions of graduate study in social science, most students do not often learn their research skills – and certainly not qualitative analysis – as responsible and full-time members of research teams. Our students mostly do not either, although they are sometimes invited into team meetings, or volunteer for, or are research assistants in, small side projects of a faculty's larger research project. However, team research done by grounded theorists does not call for additional analytic skills; yet as in all collaborative work, whether between partners or with teammates, various additional conditions profoundly affect the work itself (its style, scope, pacing, division of labor, direction). Only actual immersion, or at the very least some experience in, collaborative research can give the lone researcher the necessary imagery of what transpires in that relationship, along with the additional skills necessary for doing good grounded theory research under those collaborative conditions.

"How much can I do a day? How do I pace myself in doing these analyses?"

The answers to these questions are no different than for any other kind of "head work." If working well, having fun, keep going. If tired, don't bother, unless you are meeting a tight schedule. Learn your daily rhythms: including when you work most efficiently at analysis. Try to work regularly, every day or every few days; but if you find you are not working well on a given day, then quit for a while or for the entire day – do something else. Of course, if you are on a kick, a "high," when everything is developing marvelously, creating lots of excitement in you, then you may wish to keep going even if you pay the consequences later in tiredness or need to recuperate. The payoff may be worth the price. (For a fuller, excellent discussion of these issues, see *Theoretical Sensitivity*, Chapter 2, "Theoretical Pacing.")

"Now that the research seminars are finished, and I am alone with my own research, I have difficulty in grappling with or making progress with my own analysis: What can be done about that?"

One of the untoward consequences of learning to do analysis cooperatively is that – with all its positive aspects – students miss the collective stimulation and support when they begin to work quite on their own. Likewise, people who are learning this style of analysis by themselves, as with most readers of this book, are likely to feel "lonely" (their term) since researchers around them will be differently trained. To our students, we suggest attempting to schedule time and effort so as to get another person in their class cohort to work at least occasionally with them; or to call together a group of students who have been through the research seminars, whether the members be from their own cohort or not. In fact, students (and even former students) at work on research projects of their own have with some frequency called together working groups, which met regularly or from time to time. Students may need this focused discussion with others for quite different reasons depending on such matters as their particular research problems, their psychological blocks, and their stages of research. Some become flooded with too many ideas and memos, and are worried about finding or having chosen the main story line – that is, about integration. Some believe that they are progressing well, but need validation for that belief. Others require help when beginning to write, for despite years of writing at school they have never done writing in *this* mode; and besides, organizing a thesis has its own special horrific imagery. So, the voluntary working groups have functioned fairly successfully for many participants. However, for the lone learner we have no really effective counsel to give, although occasionally we have given useful research advice by correspondence or telephone, or face to face if the researcher travels to us, provided that some data or analytic materials can be scrutinized first.

"How do I present grounded theory results to quantitatively trained people? How, for example, do I write an effective proposal for funding, given that most reviewers are not trained in qualitative methods and almost certainly not in grounded theory methodology?"

The first question does not represent much of an issue at first, but later students begin to confront it when talking with people trained differently, who raise skeptical and perhaps horrified eyebrows. How to counter their often blunt criticisms? The appropriate answer to be given to perplexed or bedeviled students is, simply: Insist that science requires a variety of methods (and give them ammunition to use in that regard) and note that their own are especially effective for handling certain kinds of data and certain kinds of issues and probably many more ("process," "subjectivity," "theory derived from qualitative data"). Understandably, many quantitatively trained researchers will never be convinced by that argument, not even when presented with accomplished monographs and effective papers. (Flipping through the pages of *Awareness of Dying* one skeptic scornfully threw it down, saying, "Horse manure, no tables!") Some proportion of those researchers, however, will exhibit tolerance or at least a wait-and-see attitude, and some, having become dissatisfied with the methods taught them, may actually be envious and wish to learn these methods. People trained in other styles of doing qualitative analysis, however, may be just as critical of you, believing there is no difference in their own analytic modes and those of grounded theory. The most convincing argument for them, we believe, is to underline the lack of conceptual density or integration in even the best monographs based on qualitative analysis, while admitting their many virtues.

As for the entirely legitimate question about writing proposals that embody grounded theory procedures: There are two paths to be followed. Neither guarantees success with funding agencies, since that is so dependent on other factors, including the composition of the review board. However, these paths represent the best ones that can be pursued. First, write up the proposal as tightly and effectively as possible, as would any other experienced researcher. Dot all the Is and cross all the Ts. Be as explicit and clear as possible. Second, and more particularly, write the methodology section clearly and straightforwardly (not defensively!), with a minimum of jargon; and when terms like *theoretical sampling* are used, define them carefully and give illustrations and explanations of why that operation is useful or requisite. Also, it is advisable to give specific examples of codes derived from preliminary analyses of already gathered data. (No proposal should be written without preliminary data collection and analysis.) Only in this way can reviewers who are unacquainted with grounded theory procedures get both a sense of how these basic operations work and some confidence that they will work for *this* specific researcher.

With all those precautions, like anybody else's proposal, yours may fall by the funding wayside. Some grounded theorists have reasoned (or been advised) that a mixture of qualitative and quantitative methodology may raise the odds of getting funded by a particular agency. Sometimes that works, sometimes not. Bluntly speaking, it is usually better not to get funded than to settle for a poor or personally unsatisfying project. And since grounded theory proposals have been funded by government agencies, perhaps if yours is well conceived and properly presented, it will be funded, too.

14 Research consultations and teaching: guidelines, strategies, and style

It would be less than honest if we did not signal to our readers some of the guidelines, strategies, and general style that lie behind both our teaching of grounded theory methods to students and our consultations with research associates, colleagues, and others who seek advice on the conduct of their own research. Not everyone who is committed to this particular analytic approach would necessarily concur with what will be written below – for strategies and styles are linked with individual temperaments, personal predilections, and teaching/consulting contingencies. But readers will understand much better the contents of this book if they keep in mind that the teaching and consulting portrayed in it are informed by the points covered in this chapter.

Again, these are presented in the spirit of their being used as guidelines rather than rules. Please do not regard them as dogmatically held prescriptions for teaching and learning. We use these guidelines also in working with research partners and teammates. Presumably they could also help lone researchers working with – and teaching – themselves.

These are guidelines for teaching and consulting, where the aim is not merely to instruct in techniques (though that, too) or to solve technical problems. The aim is to help in enhancing and sometimes in unlocking the creativity of students and consultees. While research has, of course, its routines and its routine stretches of activity nevertheless, the best research – Can anyone seriously doubt it? – involves a creative process by creative minds. The issue here, then, is how to further it and them.

However, since there are different structural conditions that affect consulting with colleagues or teammates, consulting with students, and teaching in student seminars, we shall discuss separately the activity in each of those situations. However, the bulk of the discussion will be addressed to the seminar situation.

Before doing that, here are a few general remarks about these situations. Giving counsel and teaching about qualitative analysis cannot be given by formula, but require advisors and teachers who are highly sensitive and alert to overtones in the interaction between themselves and the others. Moreover, the latter enter the sessions at different phases of their specific research and research experience – and so does the advisor/teacher. Those phases are crucial to the interaction, affecting what will be heard, attended to, reacted to, and said. Both consultation and teaching situations illustrate the point, too, that the most effective, or at least the most efficient, analysis is often done not by a lone researcher but by two or more people working together; for even in consultation sessions, both parties should be doing that. Often, indeed, interaction can turn out to be a voyage of discovery for each interactant, albeit each may discover something quite different during the seminar or consultation session. We should add that until researchers become quite experienced in analyzing data, they will encounter *some* difficulties when handling those mountains of collected data, which as a reader of a draft of this book remarked, "really appear to be chaos, and that can be anguishing." Consultation sessions and teaching sessions are frequently addressed to that endemic problem.

Consultation

The basic assumption about consulting with colleagues or research associates is that there is a reason why the person requiring consultation has come for counsel about his or her research. A corollary assumption is that the person who is doing the counseling should listen carefully for that reason (or reasons) – or elicit it if necessary – and attempt to answer the consultee's problem, issue, question, or handle appropriately the "psychological condition" that is blocking an effective research analysis. The worst tack that can be taken is to disregard the explicit or implicit message carried by the consulting person's words and gestures. So, the first rule is: Listen! And the second rule is: This is not the time to show your own brilliance or to give expression to your own needs. It is a time for meeting the other's requirements, not your own.

So, the initial questions the advisor poses, either silently or openly, are: What is on this person's mind? What does he or she want from this session? Do they know what it is that they want? Where are they in their research – or in their lives, insofar as that affects their research?

Since the reasons why people come for consultation are varied, and are not always immediately or directly revealed, and sometimes are not clearly apprehended by the person him/herself, it follows that the advisor may need to be watchful and patient, and to listen to undercurrents of speech with great sensitivity: in short, to be a receptive listener. Or, the advisor may need skillful tactics to elicit or help the other formulate the basic issue that should be addressed.

Once that is grasped, or guessed at, by the advisor, then the next general rule is: Work within the research framework of the other – unless the latter has declared or indicated dissatisfaction, even despair, with the workability of the framework. The advisor attempts to help elaborate the scheme, deepening it, stretching its boundaries, drawing out its analytic implications.

If, however, the framework seems (sooner or later) inadequate to the other, then the advisor can begin suggesting one or more alternative paths into the data. We say "suggesting" because to press an alternative too strongly will usually not really be effective, or if accepted will, very likely, be rejected later. The suggestion should be given and taken – like all ideas, hunches, hypotheses – as provisional: the motto being, "Let's just try it on for size." If accepted in that spirit, then the next step is for both parties to work with the new path, ideas, concept, diagram, framework – whatever – to see where the discussion then goes. The discussion, of course, can vary from genuine grounded analysis to more basic issues such as the researcher's psychological, marital, or other difficulties which are holding back his or her analysis of the data. The essential questions here are: Does my advice really address the issue or issues; and does he or she believe it does?

If the answer to one or the other question is negative, or dubious, then the advisor may probe further, or cautiously suggest other alternatives. If both parties are satisfied, then the advisor should suggest that the other try out some or all of what's been arrived at during their mutual conversational exchange – and if it seems useful or feasible to come back again, then suggest that too. In any event, the advisor should be careful (this is another general rule) not to flood the other with more than he or she can manage to absorb at one session, even if this is to be the only session, for that could destroy some of the effectiveness of the counsel.

We might add as a kind of footnote that, in our experience, one reason why people come for consultation is that they are dissatisfied – sometimes after much experience at research – with more quantitative and positivistic methodologies, or at least sense or discover that the latter do not work too well with their current research projects. Another frequent reason for requesting consultation is that the usual mountain of qualitative data has been gathered, the researcher now quailing before it, not really knowing how to make sense out of all that data or being discontent with his or her current interpretations of them.

Then again, there is another type of consultation, which takes the explicit form of an intellectual conversation about the other's research, with an exchange of views but with advisory or consultatory features left implicit. The signals of "on target" receptivity to one's suggestions are made, sometimes openly ("This has really been very helpful"), but sometimes they take a more muted form. The "advisor's" tactics will anyhow be much the same as when the consultation is more official.

Consultations with students

All the above guidelines apply to the situation of the student who comes for counseling about research. However, two structural conditions are at play now (both pertain also to consultations with young associates or research teammates): Namely, the advisor knows the other more or less well, and will see her or him again in class, or in another face-to-face session, or at research team meetings. Also, students or inexperienced research associates may require more direct advice, supportive mandates, and even explicit directives or commands. One basic rule here is that the advisor–teacher should leave himself or herself open for another advisory session, "when you are ready" – or, a specific time for the next meeting should be laid down or agreed upon if the other needs or feels the necessity of closer supervision, or is too diffident or shy to initiate the next session.

Sometimes students or research associates signal or ask for evaluation ("How am I doing?"), so the advisor may answer to that implicit or explicit request, using his or her best judgment as to what to say and how to say it; including giving advice and/or mutually working out how to manage a weakness of analytic skill or other barrier to effective research. Again, readers must regard that last general rule as only a guideline, since it cannot cover all situations. Indeed, students may profit from a more direct confrontation as a means of getting them over their weaknesses; but again, that kind of radicial surgery, we believe, is not often called for, and if the surgeon misjudges then the results can be disastrous. Anyhow: Whether addressing weaknesses or pointing out strengths, it is generally preferable to be specific – for instance, "You are gifted at gaining access, and your data are rich, but what you have to break through is just translating sentences on a descriptive level, and get down to learning how to really code. That's the next step. Why then not do . . . now?" An implicit or explicit contract is implied: "Do that and I will respond, will take the next step in advising you; but only when you or/and I think it is the proper time for that."

Teaching analysis in seminars

The same guidelines pertain to the teaching of analysis in research seminars. Nevertheless, as can be seen from the case illustrations, the specific teaching tactics will vary according to how much training in analysis the class has had, and according to what the presenting student appears to need from the particular session. Generally, however, the teaching style to be outlined below will obtain.

There are five major considerations with which we are concerned when teaching grounded theory analysis: (1) the form of the seminar itself; (2) the presenting student's requirements plus where the class is in its development; (3) therefore, the teacher's focus during given sessions; (4) the teacher's control over the seminar discussion's directions; (5) the usefulness, sometimes, of raising the class's self-awareness of "what's been going on" during phases, or at the end, of the seminar discussion. These considerations lead to our guidelines for seminar teaching.

The form of the seminar. The seminars are kept small, at a maximum of ten or twelve participants, since more people makes concentrated analysis difficult and stretches out the time before students can repeatedly present their data for group discussion. Visitors are not allowed since their presence – we have found – even if they do not talk, tends to inhibit the presenting student and sometimes the group itself. Auditors are not allowed either, since all participants should have data to offer, thus have data to be worked on and worked over during the sessions. Once the seminar has begun, no new participants should be admitted (and this *is* important), since it takes time for the group to become fused more or less into an effective working unit. Basically, the group must become and remain a unit which works together: making discoveries, teaching each other, being taught by each other as well as by the instructor, working things through, and becoming increasingly self-aware about their thought processes – and some of the psychological processes, too – in order to attain the skills that are requisite to effective analysis.

For those reasons, the instructor, at the first meetings, quickly sets the style of the forthcoming discussions. These sessions are work as well as training sessions: The work is serious even if there is much latitude for play and playfulness. Showing off or playing the grandstand and other presentation-of-self stances are frowned on. Aside from those forbidden actions, the participants are encouraged and permitted to do many things: Engage in flights of imagery, providing they are tied to some kind of reality or data; blue-sky on ideas; and even take chances on ideas, without fear of derision or ending up with complete failure. "The right to be wrong is vital since [even] wrong tracks [may] lead to right ways" (Glaser 1978, p. 34). Also, students must learn, if they need to, not to be defensive *and* not to be put on the defensive, since this too is inhibiting. That does not mean that students cannot be criticized or their failings pointed out, for explicitness often brings rewards to everybody. Participants can "one-up" each other conceptually – but not psychologically – because "the job of all is to raise the conceptualization" of the analysis (Glaser 1977, p. 34). They can also break apart, fracture, the data in any fashion they believe will advance the collective analysis, providing they take into account, do not violate, what the presenting student has signaled or announced is wanted from the analytic session. (A discussion of that request or signal follows.) Students may wish to remain silent for various reasons, either for some minutes or throughout a complete session; and this right to silence must be respected by everybody. Silence may reflect personal pacing, not merely reluctance to speak or to express dissent at the direction of some phase of the discussion. But if the instructor suspects some variety of psychological holdback, then he or she may attempt to break through a student's silence, particularly if the silence continues through several seminar sessions.

In these seminars, one student (though sometimes two) sequentially presents materials for whatever purpose he or she wishes. Copies of the interview, fieldnote, or other document are distributed either before or at the beginning of the session. Meanwhile, another student may take notes which will soon by typed up, xeroxed, and distributed to the participants. All data are kept confidential, are not to be spread about among nonseminar members. Generally, the presenting students are required to stay with one research project, unless there is a very good reason for switching in midstream, since this breaks the continuity of everyone's training, as well as interferes with the progressive analysis of a given body of data. Students often tape the discussions – presenters, and other participants also – in order to better recollect and think through what transpired. Indeed, many students find themselves too flooded with the analytic discussion, or too afraid of missing analytic connections and steps, to entirely trust their memories or the often inadequate, insufficiently detailed, or belatedly distributed reporter's typed notes.

In sum, the purpose of the seminar is to facilitate and force its members' thinking and analyzing. "The learning of the participants is assumed. Eventually such collaboration will become an internal dialogue and the participant is trained to go it alone" (Glaser 1977, p. 34). Students, of course, learn to do this with varying degrees of accomplishment. But there is always the possibility that the students may wish to reconvene the class or confer with colleagues. In fact, some groups of students are adept at reconvening occasionally, even regularly, in order to keep their analytic momentum rolling or to breach analytic or psychological barriers. Understandably, the "seminar is most productive when it occurs in a context of much other learning, such as classes in the same and different fields" (Glaser 1977, p. 35), since those may sensitize them to the potentials of their data and speed up their analytic operations.

We should add that, throughout the actual seminar sessions, the instructor, and sometimes the various participants too, will vary teaching/consulting tactics, depending on judgments made about the next items to be discussed.

The presenting student's requirements. The very first presenter is asked to tell what he or she expects to get from the ensuing class discussion, so that everyone learns quickly to expect that presentations will be prefaced by such announcements or requests. That minimizes both the possibility that the participants will misjudge what is wanted and curb their quite understandable tendencies to impose on the presenter their own paths into the data. This teaching device quickly becomes part of the opening texture of all the sessions that will follow; so much so that presenters will put their requests/expectations into their opening statements and even into the written prefaces to their distributed materials. Their requirements can vary, depending especially on their experiences with analysis, the stage of their research, and their sense of success or failure with the research. (The data most often are gathered by the presenter during a small-scale investigation, but sometimes are partial data taken from a large-scale research project on which the student may be working.)

Some may wish to "try out" their initial ideas or formulations with the seminar participants. Others admittedly are floundering, and need firm guidance or perhaps only encouragement to move in directions already explored. Others have successfully worked out analyses and now wish to see if classmates can suggest additional approaches to the data; and so on. If the presenting student does not explicitly state his or her expectations, participants will ask them point-blank: "What do you want from this session?" In short, the students are being trained to think of themselves as consultants, or future teachers of analysis, who will work within the frameworks and expectations of their own students or consultees.

The teacher's focus. This focus will vary in accordance with two considerations: the presenting student's development and the development of the class in general. Those considerations, of course, must be sensibly balanced by the instructor. Fortunately, furthering the presenter's development is likely to contribute to the other students' progress also. For instance, on occasion it may seem useful to have the presenter continue "in the hot seat" during the following session, and occasionally even through the next two sessions. The class must be asked whether they would wish that extension or told why that would profit everyone: As when conceptual integration is beginning to become apparent through the day's discussion, and might well be followed through during the next session or two so that everyone can better grasp "how it happens."

Sometimes, too, the teacher may quickly grasp where a presenting student stands in relation to his or her data, and may point that out to the class. The presenter is flooded with too much data. Or, her theoretical framework is diffuse, confused. Or, she has "gone native," accepting too readily the interpretations offered by her informants. Or, she has gone stale because of too long a period of struggling with interpretations of the data. Or, she is too impatient about getting quickly to larger interpretations and needs, first, to buckle down to the hard work – the expressive German expression is *sitz fleisch* – of careful if imaginative microscopic coding.

In short, the teacher's tactics must be flexible, suited to where the presenting student is judged to be. The instructor certainly will not always be the one who points out what seems needed: The presenters themselves or the others may do that.

As for the development of the class as a whole: The instructor also has that very much in mind, and so is balancing more generally – if not at every moment – what the presenter needs from the discussion and what the other participants might be getting from it. Sometimes one and sometimes the other may take precedence, but happily the two goals often can be furthered by the same teaching procedures and sequences of discussion.

For instance, in the early sessions, both presenter and others can eventually become very frustrated because the conversation has wandered "all over the lot" – because everyone bolts off expressing his or her own train of thought, usually precipitated by remarks made by someone else. The instructor does not necessarily intervene until a number of students evince increasing impatience, then enters to tell them what is occurring, and usually then sets a framework for the next phase of the discussion. This tactic ultimately helps everybody, since all must learn the discipline that is needed, individually and collectively, to keep analytic thought from going astray.

Instructor's control of discussional directions. That example leads to the next pedagogical point, which is the instructor's concern for where the discussion is going, and might better go. Again, the instructor must juggle having too much and too little control over the content and direction of the discussion. We find that too much control, exerted too often, will not give the class sufficient freedom to try out its slowly developing and hard-won analytic skills. Too little control, of course, even well along in the students' training, can sometimes lead to a certain amount of wasted class time and energy.

For instance, the participants can get so enthralled by the subject matter being presented that they keep asking the presenter for more and more data – just because those are so interesting and some students are resonating to the data with deep feelings of their own. Meanwhile, no analysis whatever is getting done. If this kind of episode occurs relatively early in the training, then the instructor may choose to cut off the filling in of data rather quickly; if it occurs later, the instructor may purposely sit back for as much as an hour before asking the class: "Just what do you think has been going on – and on – and why?" When the class becomes aware of its transgressions (after all, the presenter has asked for analysis!), then several teaching points can be pressed home: You should never forget what the consultee has requested; never get carried away by the sheer excitement of eliciting more data; and some of you (including the presenter) should have blown the whistle on this descriptive game. Generally, however, seminar participants quickly learn to curb any straying from the analytic focus.

It should be evident that in teaching analysis, as in teaching anything, the instructor will find it necessary to adapt specific pedagogical tactics to the stage of students' development; and the cases given earlier in this book should have illustrated that well. Our general *guidelines* pertaining to the instructor's control of discussion, however, are these: If the discussion is going well – as judged by what you think should be the best ending for today's analysis – then let it flow freely. If the discussion begins to wander, then bring it back to the main track. If no focus has emerged in the discussion, then set a provisional framework (not sharply imposed, but suggested), and enjoin the class to remain within it for at least a while. Eventually, a better framework may emerge and then the instructor can point this out, suggesting that the next phases of the discussion keep within its boundaries.

An essential teaching device for directing the discussion along probably profitable lines is the asking of generative questions. Some of these stimulate immediate discussional analysis. Others set frameworks for longer discussion. Some even open such vistas for students that long after the session one of them may recollect what happened after being confronted with such a question. ("A breakthrough for me occurred when we spent so much time on Pat's data on the sex hotline, the sentence that read: 'You're 23?' You insisted that we stay on that sentence in the interview for a long time, and I had no idea what you were after until you pointed out that this sentence suggests that there are questions that are and are not permissible to ask. How come she asked it? This immediately broke me *away* from and *above* the data to a new conceptual level; nothing else has ever had quite as much impact on me, and I now find it much easier to do the same kind of distancing on my *own* data" (Nan Chico, memo to Anselm Strauss).)

The instructor is able to think of those questions because he or she is stimulated by the class discussion, but also because still another tactic can be resorted to. During the first minutes of the seminar, while everybody is scanning the presenter's materials, the instructor writes down where the analysis probably should be carried, its categories, and so on. Sometimes a provisional diagram can be sketched. To those, additional and frequently important points will usually be added later, because the instructor is sparked further by the remarks of various students, or integrates their concepts with his own. All of that becomes a source for many – and carefully paced – generative questions. The integrative diagrams can be used, as the case illustrations have shown, to summarize class discussion at the end of the session, or can be drawn on the blackboard at some point in order to set a framework for guiding the next phases of discussion. Naturally, some students begin to peer around the instructor's shoulder and eventually emulate this style of approach and also the formulation of potentially generative questions.

Raising awareness and self-awareness. A very important aspect of teaching grounded theory methods is the raising of students' awareness of analytic operations and their own use of them. It is one thing to utilize those more or less appropriately, and another to have a keen awareness of just which ones are being used, why, how, whether effective or not; but also when some should be used but aren't being used, and which ones they themselves use well and others not so well.

Since the utilization of analytic means is closely tied to the flow of class discussion, to the nature of the analytic task set by the presenting student and to the class's level of development, there follows another general *guideline* for the instructor. From time to time he or she must point out what is going on, analytically speaking, in the discussion; as well as whether what is transpiring is effective, ineffective, appropriate, inappropriate, and so on. Sooner or later, the students will do this for themselves, either silently or indicating their observations publicly.

To supplement this tactic, the instructor may occasionally query: What's going on now? Or the class may be asked to summarize what has been happening during the last ten or twenty or thirty minutes. Sometimes it is useful to ask this latter question at the very close of the session (especially if it has been very exciting). Or, to request that everyone figure out, after class, what they thought the main line of analytic evolution was today, tracing it step by step, as best they can. Sometimes, too, it is useful to relate a phase or the totality of today's discussion to the development of a specific student or to the entire class, pointing out what he or she or they have just accomplished by comparison with their ability to do this before. If the session has taken a genuinely different turn than most, the instructor will wish to sum up and emphasize what has happened and why. Eventually, the participants learn to engage in these additional pedagogical tactics too.

The teaching of research planning or design

Because grounded theory methodology emphasizes careful and continual efforts to ground theorizing in the data, students sometimes mistakenly come to believe that research planning or design is unnecessary. They tend to confuse planning and design with *deduction* and speculation as opposed to working with data, which is *induction*. So they conclude that they must just "feel their way" throughout the lives of their research projects. This is an entirely incorrect apprehension, for at least three reasons.

First of all, induction, deduction, and verification are all essential operations (see the discussion in Chapter 1, Part 1). Second, the requirement that the theory eventually be an integrated, conceptually dense one makes requisite a coming to grips with the question: Where is this study really going and how can it next, and ultimately, be furthered? This of course requires, in turn, much more than just drifting or "sensing one's way"; rather, one must think and plan ahead, while taking appropriate operational steps in accordance with the planning. Although theoretical sampling will be essential and its specific details cannot be predicted in advance, the lineaments of that do begin to emerge eventually. When that happens, and when integration becomes clearer too, then the interplay between the two should force attention on longer-range planning. And third: More experienced researchers usually begin with at least some provisional plan (if not "design"), knowing that much of it may become revised or even abandoned under the guidance of evolving theory and its increasing integration. Students have to be taught those points – generally most effectively by coaching them through the integrative steps of their own research, as well as by examples drawn from one's own; also by encouraging them to make specific project plans from time to time and to be influenced by but not unduly committed to the succession of drafts.

It is true however that with this style of qualitative analysis – as in most alternative modes of doing "qualitative research" – researchers find it virtually impossible to plan in great detail even after an exploratory phase. So, they find themselves when applying for funds from agencies either spelling out why they cannot do this, although giving some general plans; or they lay out a design that they know full well will not necessarily be adhered to in detail. Perhaps that is different than is the case with, say, standard survey research or with proposals in physical research; but perhaps not. This does not preclude, we tell our students, the obligation to meet agency requirements for specificity as much as possible, at the point at which they apply for funds: explaining how coding will be done, giving examples of theoretical sampling already done, being quite clear about the major problems being attacked, and so on. (Also, see discussion in Chapter 13.) Funding aside, the major issue is how to reach project goals with the greatest creativity within the inevitable limits of project resources.

Four requisite abilities

Aside from the standard problems students have when learning to become competent practitioners of this style of qualitative analysis, what about those who have, in our judgment, gained the most skill at doing it? There seem to be four major ingredients answering to that issue:

1. ability to draw freely on *experiential data*;
2. *sensitivity to nuances of behavior*, the meanings of behavior to the actors being studied, and to their social relationships; also *sensitivity toward theoretical issues* when looking at social phenomena;
3. *mastery of basic grounded theory techniques*, including persistence in learning to use them in their own work;
4. *analytic ability*, in general, but now utilized in conjunction with the foregoing three items.

Let us look briefly at each.

Experiential data. Since the use of experiential data has been discussed earlier, all that needs to be said here, perhaps, is that most students learn quickly to draw on their experience and, often more slowly, on their technical reading, for data and appropriate ideas. However, only with practice in seminar discussions and on their *own* work does ease and speed come to some. They need to be urged to "keep doing it" – also in conjunction with the conjuring up of stimulating comparisons. Those who are working on a topic within their own field of expertise and especially connected with their own job or professional work face just the opposite problem: being flooded with masses of personal experiences and memories.

Sensitivity. Sensitivity to the nuances of social relationships is not such a directly teachable skill, since it depends more on abilities developed since childhood, and perhaps is associated with such terms as personality and temperament. Perhaps it really is possible to develop interactional sensitivity further by teaching, though we have no particular experience in that. Most of our students display considerable sensitivity to social relationships, but perhaps they are self-selected as recruits into our graduate program. Also, Barney Glaser (1978) has a rather full discussion of *theoretical sensitivity*, by which he means being sensitive to theoretical issues when scrutinizing data. For theoretical sensitivity, wide reading in the literature of one's field and related disciplines is very useful, and probably requisite: not for specific ideas or for a scholarly knowledge, but for authors' perspectives and ways of looking at social phenomena, which can help to sensitize one to theoretical issues. When theoretical sensitivity is linked with sensitivity toward social relationships in general, then the qualitative analyst has distinct advantages over someone who is less gifted or trained in one or the other skill. Indirectly, there probably is, anyhow, considerable marriage of theoretical and interactional sensitivity. Not only is the latter and its associated skills heightened by the experience of doing interviews and observing in the field, but certainly also by learning to think comparatively and gaining experience in grappling with the data themselves.

Mastery of grounded theory techniques. This is a technological problem for students: First, how to grasp the perspective of grounded theory analysis; second, how to understand its basic elements; third, how to use the techniques themselves; fourth, how to use them in carrying one's own work to completion. All of our students seem to accomplish the first two, but experience varying degrees of difficulty with the third, and especially the fourth, as well as learning those at different speeds and with differing degrees of apparent success.

Definitely, the techniques cannot be mastered in just a few months. It takes time to try out the techniques, gain confidence in using them, to keep trying, to get validation that "they work for me" and "I can really do it." And carrying an entire project (a thesis for most students) through to completion, utilizing the techniques, is an absolute necessity before these tools of the trade can be used comfortably and confidently, and before being able to see how conceptual density and integration actually occur. Some students are impatient at how long it takes them to achieve mastery and they long for instant learning, and worry also a great deal about whether they will *ever* learn how to do it.

Analytic ability. People vary greatly in their degree of analytic ability. Understandably, there are different kinds of analytic ability, some people's bent being more abstract, etc.; a possible relevance to doing qualitative analysis we have not thought much about. However, what prompts our listing analytic gifts as a fourth point is that people with great analytic ability – *when combined with* the use of experiential data, social and theoretical sensitivity, and technical mastery – seem to do work that is closest to our own in style. So, of course, their work pleases us most. Since pure analytic ability is not enough, and since technical mastery is a rather slow process, it follows that achieving a really impressive level of qualitative analysis is not done without considerable intellectual struggle and persistent work. Those who cannot scale the heights (again, at least in our judgment) can attain varying degrees of analytic mastery: but, surely, there is ample room for varying degrees of analytic accomplishment. The value of the research, after all, depends not on some platonic measure of worth but on its value for appropriate audiences. Thus, social welfare or educational or health practitioners can profit from the grounded theory research of their colleagues without it being carried to the furthest extent. In saying this, we are being neither charitable nor tolerant, rather being knowingly pragmatic and relativistic. This is in keeping with the original discussions in *Discovery of Grounded Theory* (1967) and *Theoretical Sensitivity* (1978) about the fit and relevance of good substantive theory.

After a seminar discussion

In closing this section on teaching in seminars, with its emphasis on collaborative work, we reproduce here two sets of comments by students. The second is by an advanced student writing generally about what she calls *grounded theory culture*. The first consists of the written remarks of a presenting student a few days after she had proffered the seminar an interview and supplementary oral remarks. (This class had worked together only for about two months.) The previous year, this student had lived on an American Indian reservation, working as a public health worker. She had been struck not only by the stereotypes held about the Native Americans by other (Caucasian) personnel, but had concluded also that the anthropologists who had written about this tribe had in large part failed to grasp the viewpoints of the Native Americans themselves. When presenting her materials she expressed those opinions with considerable passion, but she did not yet know how to handle her mass of controverting data. Her written afterthoughts are reproduced here not as a testimonial to the teaching, but because they so vividly convey what these collegial teaching–research sessions can mean to the presenting student–researcher.

> There are many things which can and need to be written about the phenomenon of cross-cultural research. Of particular interest to me in regard to this analysis session are the notions of *integrity* and *precision* as they emerged relative

to method and analysis. By integrity I mean that within the data lie as many definitions and constructions of reality so as to engender depth to one's understanding and knowledge – that the data be expansive and not constrictive, that it open up the mind of the researcher and analysts to varieties and diversities and not confine the researcher–analyst to limited visions. This seems, at least to me, integral to research that crosses cultures or where the researcher is unfamiliar and the foreigner to established ways and meanings of a particular group. By the idea of precision I mean that there is a high degree of fit between how the researcher–analyst views the worlds of the people and how they themselves say they view it. Simply put, I mean to say that how the researcher analyzes the data and interprets the findings accurately reflects the experience of the participants – that those being researched could read the researcher's interpretation and say, "That fits." Precision is tied to integrity. By incorporating both into the research process one can ensure not only validity and reliability but accuracy, relevance, and increased understanding/knowledge.

Prior to this analysis session I was, quite literally, sitting on boxes of data, my mind filled with memories and highly charged mental images from my two years fieldwork on an American Indian reservation. Almost by definition, reservations are isolated, rural communities that are the most economically depressed hollows of this country. For at least a century Indians have been the mainstay of anthropologists and other researchers who arrive on reservation uninvited, eager to study Indian ways frequently breaching the basics of what is considered social courtesy. There are libraries filled with literature on Indians that resulted from such research. Indians joke how so little is done with research that benefits them and much has been done to disturb their life, to erode their self-esteem, and to damage the national image of Indians – "but the reseachers do not look back to see what comes in their wake."

Having been shown the dark side of research by my hosts during my two-year period, I came to the analysis session somewhat recalcitrant, in that I did not want the data "misused," that is to say that I wanted the data to be allowed to speak for itself and the analysis to stay close to the data. I was fearful of students unfamiliar with reservations and Indians opinionating over the data and skewing their interpretation by their preconceptions of what it is to be Indian, or what it means to live on a reservation, or a host of other a priori hypotheses they might want to have tried out on the data.

These concerns and fears were systematically and carefully dispelled over the course of the two-hour session. I watched very carefully and listened intently to what people said and how they worked their ideas and images through the data, carefully questioning me when more information was needed and not jumping to conclusions in advance of important additions. The students seemed to search carefully for the richness in the data, picking out critical issues and playing them off against one another for more meaning, noting several possible interpretations to many situations. Not only was the inherent integrity of the data emergent it was also maintained. I was quite overjoyed at the degree of fit between what these analysts were identifying and what I had heard and seen while doing the work. Both the integrity and precision aspects of these sessions were sparked by and sustained by the pedagogical style, which is to say (or cannot be separated from) the formulations of interactionist epistemology and the conceptual and analytic framework of grounded theory.

Aside from the concerns regarding cross-cultural research, a second important aspect of the analysis session relates to my position in regards to my data. Generally I would describe myself as being at boiling point, with vast amounts of information inside me about the data and also about my feelings about the data. It is a strange experience to live in a "poverty area" and observe and participate in the health care of underserved people. The deprivation at times was overwhelming. People *died* from lack of medical care. This aspect of the environment was difficult to live with and almost as difficult to understand analytically. Emotions run high and analysis seemed a distant mirage or perhaps a panacea that might make sense of it all. In short, I came to the session very close to the data. I needed to maintain what I consider to be a healthy and genuine sensitivity but simultaneously needed help in obtaining some distance from the personal emotion of my experience in the field. This was achieved early on in the session when the professor framed the "good-bad" dilemma within the arena of work and out of the arena of personalities. The exact question was, "What does it take to do good medicine?" It was a marvelous question for me as it engaged portions of my thinking apparatus so that the entire focus shifted and I was cleared to think more completely about a variety of issues that I had been blocked on because of more emphasis on the personality and morality overtones. It did not clear me to avoid those personal or moral issues but to approach them from a less-vested perspective – one that is potentially more dense.

So, in sum, the session left me hopeful and trusting that accurate, viable, and relevant research can be done; that the data is rich and holds within it the integrity of my field experience and of these people's construction of life; that the analytic tools (both collectively and in time individually) will sustain my efforts to complete my project, which is a better understanding of Indian health care. I am indebted to the group for their sensitivity and their implicit and explicit concern for my work and their willingness to help me through this process. Equally important is the transformation from description to analysis: the experience of knowing – first of experiencing – the levels of understanding within the mind and of learning how to make the shift between the two. At this time the awareness of that process is new and somewhat fragile. It was very cathartic when it occurred in class in response to the question of the work conditions for good medicine. I trust that that experience will not always be in response to someone but will, with training, become a self-generating process.

Grounded theory culture

I'm part of a writing group which has met about once a month for a couple of years. We pass around work in progress and criticize it; sometimes help with analytic rough spots. Recently an old member of the group returned, and described to us her unsuccessful attempt to start a similar group in another location. Participants in her group had followed the same procedures we had, in form, but had gotten very harsh with each other's work and focused more on competitive speeches than genuine collaboration. Our group tried to analyze why we'd been successful, and realized that it had had a lot to do with the fact that four of us had been through the grounded theory. It wasn't just that we shared an analytic focus, though, because in fact we're very different. The striking thing was that we had learned to *work together* in a collaborative and supportive way. Several parts of this might be interesting to bring out in the methods book:

1. We all have learned a spirit of adventure about analysis. Taking "fliers" is an integral part of grounded theory, and the corollary is that *ideas are cheap and plentiful.* You can spin out codes and ideas, and feel very free about doing so. They'll just as quickly get shot down, compared, replaced, supplemented, etc. The experience is just the opposite of much of academia, where you get one precious idea and nurse it along in solitude for several years, finally presenting it like unwrapping a fine present. Along with this goes an implicit assumption about:

2. Robustness. We learned that ideas become robust through getting kicked around by many people and "through" many contexts (the constant comparative emphasis). Grounded theories are constantly being updated; they may contain contradictory data at any time; they always involve contradictory viewpoints which must be negotiated, which mitigates against purely central or purely logical organization. Since this is a different idea of robustness from the ones that are commonly accepted as "scientific" (see Wimsatt 1982), it's often difficult to communicate exactly what a *finding* in the grounded theory method is to those not trained in it. But, on the other hand, it's exactly this definition of robustness that evolves from the cooperative work venture and the emphasis on *grounded* theory. When I explained grounded theory the other day to a dean at the University (a scientist), I used the image of scientists working in a laboratory – checking hunches and partial results with each other, taking fliers, and constantly interacting and checking data sets – in contrast to the humanities model of the lone scholar.

3. There's another, related point to be brought out, about *public* versus. *private* relationships to data. There's a unique (in my experience of academia) emphasis in the grounded theory classes on public sharing and criticism of data. In many other settings there's an embarrassment about data or a feeling that it's extremely private. You would never show anyone your fieldnotes or raw interview material; but only show it after you'd gotten it boiled down a bit. Again, I suspect that the public emphasis in the grounded theory classes is a lot more like in the natural sciences.

4. Finally, the *processual* emphasis in the work of doing grounded theory (as well of course as in the finished analysis itself) is very important. There's a long-term "boiling it down" process in the grounded theory classes that is the opposite of looking for *the* true thing or the significant ($p < .05$) finding. Rather, there are first cuts; next cuts; somewhat-gelled memos; really pulled-together chunks; and finally, reports, papers, and books. There's a real interweaving of the process of doing the work and the process of doing the analysis – that's where attention is focused in these classes – as opposed to applying the analysis.

Epilogue

After the first draft of this book was completed, a most apposite statement by C. W. Mills (1959, p. 28) was brought to our attention. Here is the quotation: It seems to round off our own sequence of presentations so well.

> Of method-and-theory-in-general, I do not here need to say any more. I am nowadays quickly made weary by it; so much discussion of it interrupts our proper studies. I feel the need to say that I should much rather have one account by a working student of how he is going about his work than a dozen "codifications of procedure" by specialists who as often as not have never done much work of consequence. Better still: If sometimes in our professional forum we wish to discuss method and theory rather than the substance of our studies, let us ask each man whom we believe to be doing good or superior work to give us a detailed account of his ways of work.

Appendix

Discovering new theory from previous theory

Throughout this book, the data analyzed have been gathered by researchers "in the field," whether in the form of interviews, fieldnotes, or other documents. For the most part, grounded theorists, when developing substantive theory, have tended not to *begin* their researches by following through the implications of previous theory. Perhaps they have been overcautious in looking at other people's theories, once the lineaments of their own have begun to evolve. Yet, as mentioned in the first chapter, there is no reason not to utilize extant theory from the outset – *providing* only that it too was carefully grounded in research – to direct the collection of new data in the service of discovering a new (and probably more encompassing) theory. Using the familiar techniques of coding, theoretical sampling, comparative analysis, and with the usual emphasis on variations associated with dimensions, conditions, consequences, interactions, the extant theory then acts as a springboard for laying out potential lines of research work. The analyst can thereafter choose which lines to pursue, in which directions to begin, following through with a potentially effective and personally interesting research project.

So, it is entirely feasible to begin with someone else's theory, if entry into the research field follows immediately, or at least before a commitment is made to the research project: in order to see if the project is feasible, if data are really available, *and* the deduced lines of work are really relevant to the substantive area under scrutiny. *Also,* the researcher must be immediately sensitive to the new data and their potentials for new coding, conceptual densification, and integration.

Having noted that, it should be useful to reproduce here an article published some years ago (Strauss 1970), to show how researchers can escape, first of all, a standard hazard of utilizing even a grounded theory. This is simply to dot Is and cross Ts, perhaps extending it a little, but not qualifying or altering it much; second, to show how to follow through the implications of such a theory so as to lead *specifically* to the enrichment of one's own theoretical endeavor. In the materials below, an analyst will again be seen suggesting categories, dimensions, theoretical samples, comparisons, and the like – all the usual procedures discussed in this book. It is worth adding that when researchers have themselves developed theories in the course of previous investigations, they should certainly utilize them. Of course they should follow the same guidelines and rules of thumb that helped them before, quite as did the author of the reprinted article.

In his article, "Deviance Disavowal" (1961), Fred Davis has offered a useful, grounded theory about (as his subtitle reads) "The Management of Strained Interaction by the Visibly Handicapped." . . . this presentation is exceptional because its author tells us exactly what his theory applies to, alerting us to what phenomena it does not apply and to matters over which it glosses. (Thus: "Because of the paper's focus on the visibly handicapped person . . . his interactional work is highlighted to the relative glossing over of that of the normal" person.) One of the most valuable features of this paper is that it stimulates us to think of variables which Davis does not discuss fully or omits entirely, including those quite outside of Davis's focus when he developed his theory. Indeed, when rereading the article I have often found myself aching to know more about all those untreated matters. The reader bears the responsibility to carry on this unfinished business, of course, if the author does not elect to do so.

Davis's theory is about (1) *strained* (2) *sociable interaction* (3) in *face-to-face* contact between (4) *two persons*, one of whom has a (5) *visible handicap* and the other of whom is (6) *normal* (no visible handicap). The theory includes propositions about tactics, especially those of the visibly handicapped person. But the central focus is upon *stages of management*, notably (a) fictional acceptance, (b) the facilitation of reciprocal role taking around a normalized projection of self, and (c) the institutionalization in the relationship of a definition of self that is normal in its moral dimension. Emphasis on stages makes this a distinctly processual theory.

The italicized terms . . . begin to suggest what is explicitly or implicitly omitted from Davis's theoretical formulation. The theory is concerned with the visibly (physically) handicapped, not with people whose handicaps are not immediately visible, if at all, to other interactants. The theory is concerned with interaction between two people (not with more than two); or with combinations of normal and handicapped persons (one interacting with two, two with one, two with two). The interaction occurs in situations termed "sociable"; that is, the relations beween interactants are neither impersonal nor intimate. Sociable also means interaction prolonged enough to permit more than a fleeting exchange but not so prolonged that close familiarity ensues. Sociable interaction does not encompass ritualized interaction.

But the interaction is not merely sociable, it is face to face – not, for instance, carried out by telephone or through correspondence. This interaction represents the first meeting between the interactants, not a later meeting or one based on an interpersonal tradition. This meeting is only the first of a series of episodes that may lead to a more intimate relationship, and is so recognized by the handicapped person. Throughout this interaction the handicapped person attempts to minimize his handicap, rather than to highlight or capitalize on it. The normal person also attempts to minimize the handicap, rather than favorably or unfavorably maximizing it. Also, control of the interaction is vested in the handicapped person, who has a willing accomplice in the normal. Moreover, the normal must agree to the game of normalization rather than resisting or being indifferent or even failing to recognize it. In addition, we should note especially that the interaction is strained – that is, the visible handicap tends to intrude into the interaction, posing a threat to sociability, tending to strain the framework of normative rules and assumptions in which sociability develops.

Visible handicaps which pose no particular threat to sociable interaction are not within the province of this theory. Also, because emphasis is on the handicapped person's management of interaction, the theory covers quite thoroughly the tactics and reactions of the handicapped, although it says relatively little about those of the normal. And finally, the theory pertains to a handicapped person who is already quite experienced in managing strained interaction with a normal – who by contrast is relatively inexperienced in interacting with handicapped persons. If we imagine a simple fourfold table, we can quickly supply three contrasting situations involving such experience.

This filling in of what has been left out of the extant theory is a useful first step toward extending its scope. We have supplemented the original theory. (Supplementation does *not* mean remedying defects of a theory.) Supplementation has led to the generation of additional categories, which in turn leads us – unless we cut short our endeavor – to think about those new categories. Thinking about those categories amounts to building hypotheses which involve them, quite as Davis built hypotheses around the categories generated from his data. We can think about those new categories, one at a time: for instance, the nonvisible handicap. We can do this much more efficiently, however, by comparing the new category with others, whether those are newly generated or inherited from Davis.

Imagine what happens in the first episode of face-to-face, sociable interaction when (1) a person with a relatively invisible (although potentially visible) handicap meets a normal person as against (2) when Davis's visibly handicapped meets a normal . . . . Unlike the latter situation, one of its properties may be "secrecy," because the invisibly handicapped, if he is experienced, probably will be much concerned with keeping his handicap thoroughly hidden. If he is more experienced, he will probably be less anxious about betraying his secret. But immediately it must strike us that this person will indeed be experienced unless the handicap is of recent occurrence; as for instance, a woman recently operated on for mastectomy (cancer of the breast) who now appears in public with a breast prosthesis hidden under her dress.

By reaching out for this case as an example, we have begun (in imagination) to sample theoretically; we could, in fact, now either interview or seek existing data not only about such patients but about others who had newly acquired various nonvisible, but potentially visible, handicaps. We can ask, what other kinds of persons (i.e., comparison groups) might those be? . . . Groups of nonvisibly handicapped persons whom we seek might include those born with stigmata which can be readily covered with clothes . . . .

A moment's reflection about those comparison groups of handicapped – visibly or invisibly – tells us that we have generated additional catgories . . . . Thus, there are invisible handicaps which have been present from birth, others which have been acquired whether recently or some time ago or long ago. There are some, whether visible or invisible, which never grow worse, and others which may grow much worse. Some may be temporary, disappearing over varying amounts of time. Some handicaps are seen by most people as stigmatizing, while other handicaps bring compassion or pity or indifference. They may also cause fear (leprosy) or revulsion (syphilitic noses).

If we pursue this analysis (resisting all temptation to shrug away the new categories, saying "oh, that's all quite obvious"), we can eventually develop testable hypotheses about each class of handicapped person as these people interact in sociable or other situations with normal or other handicapped people. The hypotheses are designed not merely to illustrate what happens to this class of handicapped but to add density of conceptual detail to our evolving theory of interaction engaged in by handicapped persons generally.

Think again about women operated on for mastectomy, with their invisible defects. What is likely to be a dominant consideration for them in sociable interaction? Must they guard their secret because the loss of a breast is stigmatizing if it is known? Is the loss more likely to be a dread and guarded secret for unmarried young women than for young mothers? For young mothers than for elderly mothers? (We shall not even bother here with other obvious comparisons such as what happens in their encounters with normal men versus normal women.) It should be easy enough to imagine the kinds of hypotheses that might be generated about each of those situations, including those involving tactics to keep secret the loss of a breast. For instance, we can hypothesize that a woman who has been operated on recently will be fantastically concerned with the selection and arrangement of her clothing and with her appearance when she leaves the house, and that she will play close, if surreptitious, attention to her bosom during the ensuing sociable interactions. If we turn to the experienced women, who have worn their substitute breasts for many years, we can hypothesize that there will be less concern about betraying the secret – so that their social interactions are more like those between two normals. Under what conditions will anxiety about accidental revelation make the secret salient again for these more experienced women?

Our selection of secrecy as an important probable feature of the above interactions suggests that it is a core category, standing somewhat or exactly in the same relation to the nonvisibly handicapped as does "normalization" to the visibly handicapped. Using the terminology developed in *Awareness of Dying*, we may say that the nonvisibly handicapped attempt to keep the context "closed" while the visibly handicapped attempt – with the tacit cooperation of the normal – to maintain a context of "mutual pretense." Secrets and the possibility of disclosure are characteristic properties of closed contexts, not to mention certain tacitly agreed upon matters characteristic of mutual pretense contexts. Note, however, that in our theoretical sampling we have built considerable variation into the probable management of secret handicaps, just as one might for the management of strained interaction by visibly handicapped persons under similar varied conditions.

At this point in his analysis, the theorist has various options. He can pursue further the case of the patient operated on for mastectomy, turning her around as if she were a complexly cut diamond and examining her many facets. The cues for that analysis have been adumbrated. How do these women act in various types of nonsociable interaction? How do they perform in the successive episodes of social interaction, rather than just in the first episode? What happens when women, each of whom has been operated on for mastectomy, meet each other in various kinds of interactional situations? What occurs in intimate interaction when the woman regards her stricken bosom as ugly but her husband does not? Suppose she regards herself as victimized by fate, but he regards her with compassion? As the theorist answers these questions (in imagination or later with data), he/she builds hypotheses of varying scope and different degrees of abstraction, with the variables crosscutting again and again in the analysis; thus continues to build conceptual density into the theory and simultaneously to integrate it.

Instead of continuing to analyze this same comparison group . . . the analyst has the option of examining other groups, especially those that will maximize the power of his comparative analysis because of the great differences among them. Suppose, for instance, he begins to think about the interactional situation in which Davis's visibly handicapped person is inexperienced while the normal person is exceedingly experienced in handling, say, stigmatized handicaps. Physical therapists are not only experienced – as professionals they are much involved in treating and giving "psychological support" to handicapped clients. We can now hypothesize, either from the professional's or the handicapped person's viewpoint, in sociable, nonprofessional encounters. We might even in imagination (and later, in fact) interview physical therapists about their reactions when they encounter different classes of handicapped. How do they react to those who have handicaps identical with or similar to those of their patients, as against those with dissimilar handicaps (the deaf, the astigmatic, the blind)?

If the theorist wishes to build into his theory the phenomenon of handicapped patients interacting with professionals (normals), he can concentrate on comparisons of that type of interaction with the sociable type. Comparisons can include not only the case of the physical therapist managing varied classes of handicapped clients (stroke, polio, arthritic, auto accident cases), but can include those comparisons with the professionalized interaction of physician and mastectomy (and other physically handicapped) patients.

The theorist can, of course, decide to delimit the theory – indeed must draw limits somewhere, restricting it even to as narrow a scope as sociable interaction. Then he will not focus on the nonsociable encounters (except secondarily, to stimulate his thinking about sociable encounters), but will focus steadily on comparisons that will yield more and more hypotheses about this central phenomenon. Again, he will seek to make comparisons among groups that seem quite dissimilar and among those that seem relatively similar. In each comparison, he will look for similarities as well as differences. These comparison groups will, as before, be suggested by the emerging theory. There is no end to the groups he will think of as long as his theory proves stimulating.

When does one stop this process, so fertile that it seems to have run riot? This issue was addressed in *The Discovery of Grounded Theory* . . . . In brief, the directed collection of data through the theoretical sampling leads eventually to a sense of closure. Core and subsidiary categories emerge. Through data collection there is a "saturation" of those categories. Hypotheses at varying levels of abstraction are developed . . . . Those hypotheses are validated or qualified through directed collection of data. Additional categories and hypotheses which arise later in the research will be linked with the theory . . . .

Once we have developed this theory (whether or not we have jumped off from someone else's theory), there is no reason not to link other grounded theory with ours, providing this extant theory fits well and makes sense of our data. The one example given above was the linkage of "awareness theory" with our emergent theory. Useful linkages with other grounded theories may occur to other readers. In turn, our own theory is subject to extension, best done through theoretical sampling and the associated comparative analysis. This extension, perhaps it needs to be said, represents a specifying of the limits of our theory and thus a qualification of it.

References

Becker, H. 1982. *Art Worlds*. Berkeley, CA: University of California Press.

1985. *Writing for Social Scientists*. Berkeley, CA: University of California Press.

Becker, H. and Geer, B. 1960. "Participant Observation: The Analysis of Qualitative Field Data." *In* R. N. Adams and J. Preiss (eds.), *Human Organization Research*. Homewood, IL: Dorsey Press, pp. 267–89.

Becker, H., Geer, B., Hughes, E., and Strauss, A. L. 1961. *Boys in White*. Chicago: University of Chicago Press.

Biernacki, P. 1986. *Pathways from Heroin Addiction*. Philadelphia, PA: Temple University Press.

Bowers, B. 1983. *Intergenerational Caretaking: Processes and Consequences of Creating Knowledge*. Doctoral dissertation, Department of Social and Behavioral Sciences, University of California, San Francisco.

Broadhead, R. 1983. *Private Lives and Professional Identity of Medical Students*. New Brunswick, NJ: Transaction Books.

Cauhape, E. 1983. *Fresh Starts: Men and Women after Divorce*. New York: Basic Books, Inc.

Charmaz, K. 1983. "The Grounded Theory Method: An Explication and Interpretation." *In* R. Emerson (ed.), *Contemporary Field Research*. Boston: Little, Brown, and Co., pp. 109–26.

Clarke, A. 1984. *Emergence of the Reproductive Enterprise: A Sociology of Biology, Medicine, and Agricultural Science in the United States, 1900–1940*. Doctoral dissertation, Department of Social and Behavioral Sciences, University of California, San Francisco.

Conrad, P. and Reinharz, S. 1985. "Special Issue on Computers and Qualitative Data." *Qualitative Sociology* 7: (Summer).

Corbin, J. and Strauss, A. 1986. "Accompaniments of Chronic Illness: Changes in Body, Self, Biography, and Biographical Time." *In* J. Roth and P. Conrad (eds.), *Research in the Sociology of Health Care*, Vol. 5. Greenwich, CT: JAI.

Forthcoming. *The Social Organization of Unending, Openended Work*.

Davis, F. 1961. "Deviance Disavowal." *Social Problems* 9:120–32.

Dewey, J. 1934. *Art as Experience*. New York: Minton, Balach, and Co., p. 65.

1937. *Logic: The theory of Inquiry*. New York: Wiley.

Diesing, P. R. 1971. *Patterns of Discovery in the Social Sciences*. New York: Aldine Publishing Co., pp. 146–8, 156–60.

Fagerhaugh, S. and Strauss, A. 1977. *The Politics of Pain Management*. Menlo Park, CA: Addison-Wesley.

Fagerhaugh, S., Strauss, A., Wiener, C., and Suczek, B. Forthcoming. *The Social Organization of Clinical Safety*. San Francisco: Jossey-Bass.

Fann, K. T. 1980. *Peirce's Theory of Abduction*. The Hague: Martinus Nyhoff.

Fisher, B. and Galler, R. Forthcoming. "Friendship and Fairness: How Disability Affects Friendship between Women." *In* A. Asch and M. Fine (eds.), *Voices from the Margin: Lives of Disabled Girls and Women*. Philadelphia: Temple University Press.

Freidson, E. 1970. *Professional Dominance*. New York: Atherton.

Gerson, E. 1983. "Scientific Work and Social Worlds." *Knowledge* 4:357–73.

1984. "Qualitative Work and the Computer." *Qualitative Sociology* 7:61–74.

Gerson, E. and Star. S. L. 1986. "Modeling Due Process in the Office." *ACM Transactions on Office Information Systems* 4:257–70.

Glaser, B. 1976. *Experts versus Laymen*. New Brunswick, NJ: Transaction Books.

1978. *Theoretical Sensitivity*. Mill Valley, CA: Sociology Press.

Glaser, B. and Strauss, A. 1965. *Awareness of Dying*. Chicago: Aldine Publishing CO.

1967. *The Discovery of Grounded Theory*. Chicago: Aldine Publishing Co.

1968. *Time for Dying*. Chicago: Aldine Publishing Co.

1971. *Status Passage*. Chicago: Aldine Publishing Co.

Goffman, E. 1975. "On Face-Work: An Analysis of Ritual Elements in Social Interaction." *In* Lindesmith, A., Strauss, A., and Denzin, N. (eds.), *Readings in Social Psychology*, 2nd ed. Hinsdale, IL: Dryden Press, pp. 236–54. (Originally published in 1955.)

Hammersley, M. and Atkinson, P. 1983. *Ethnography: Principles in Practice*. London: Tavistock Publication.

Hartshorne, C., Weiss, P., and Burks, A. (eds.) 1958. *Collected Papers of Charles Sanders Peirce*, Vols. 7 and 8. Cambridge: Harvard University Press.

Hazan, A. 1985. *A Study of Shrinking Institutions: The Transformation of Health and Human Services Work*. Doctoral dissertation, Department of Social and Behavioral Sciences, University of California, San Francisco.

Hoffman-Riem, C. 1984. Das adoptierte Kind. Munich: Fink Verlage.

Hughes, E. C. 1970. *The Sociological Eye*. Chicago: Aldine. (Reprinted 1985, New Brunswick, NJ: Transaction Books.)

Jick, T. 1983. "Mixing Qualitative and Quantitative Methods." *In* J. Van Maanen (ed.), *Qualitative Methodology*. Bevery Hills, CA: Sage, pp. 135–48.

Lakoff, G. and Johnson, M. 1982. *Metaphors We Live By*. Chicago: Chicago University Press.

Lear, J. 1978. *Recombinant DNA*. New York: Crown Publishers.

Lifchez, R. and Winslow, B. 1979. *Design for Independent Living*. Berkeley, CA: University of California Press.

Lindesmith, A. 1947. *Opiate Addiction*. Bloomington, IN: Principia Press.

Louis, K. S. 1982. "Multisite/Multimethod Studies." *American Behavioral Scientist* 26:6–22.

McClintock, C., et al. 1983. "Applying the Logic of Sample Surveys to Qualitative Case Studies: The Case-Cluster Method." *In* J. Van Maanen (ed.), *Qualitative Methodology*. Beverly Hills, CA: Sage, pp. 149–78.

Mead, G. 1934. *Mind, Self, and Society*. Chicago: University of Chicago Press.

Miles, M. 1979. "Qualitative Data as an Attractive Nuisance: The Problems of Analysis." *Administrative Science Quarterly* 24:590–601.

Miles, M. and Huberman, M. 1983. *Qualitative Data Analysis*. Beverly Hills, CA: Sage.

Mills, C. 1959. "On Intellectual Craftsmanship." In L. Gross (ed.), *Symposium on Sociological Theory*. Evanston: Row Peterson.

Polsky, N. 1984. *Political Innovation in America*. New Haven, CT: Yale University Press.

Riemann, G. (forthcoming). *Biographieverlaufe Psychiatrischer Patienten aus Soziologischer Sicht*. Munich; Fink Verlage.

Rosenbaum, M. 1981. *Women on Heroin*. New Brunswick, NJ: Rutgers University Press.

Roth, J. 1963. *Timetables*. Indianapolis, IN: Bobbs-Merrill.

Roth, J. and Conrad, P. (eds.). 1985. *Research in the Sociology of Health Care*, vol. 5. Greenwich, CT: JAI.

Schatzman, L. and Strauss, A. 1973. *Field Research*. Englewood Cliffs, NJ: Prentice-Hall, Inc.

Schuetze, F. 1981. "Prozess-Strukturen des Lebenslaufes." In J. Mattes, et al., *Biographie in Handlungswissen Schaftliche Perspektive*. Nuremberg.

Sieber, S. D. 1976. *Synopsis and Critique of Guidelines for Qualitative Analysis Contained in Selected Textbooks*. New York: Project on Social Architecture in Education, Center for Policy Research.

Smith, A. G. and Robins, A. E. 1982. "Structured Ethnography." *American Behavioral Scientist* 26:45–61.

Star, S. 1983. "Simplification and Scientific Work: An Example from Neuroscience Research." *Social Studies of Science*. 13:208–26.

1985. "Scientific Work and Uncertainty." *Social Studies of Science* 15:391–417.

1986. "Triangulating Clinical and Basic Research: British Localizationists, 1870–1906." *History of Science* 24:29–45.

Star, S. and Gerson, E. Forthcoming. "The Management and Dynamics of Anomalies in Scientific Work." *Sociological Quarterly*.

Strauss, A. 1970. "Discovering New Theory from Previous Theory." *In* T. Shibutani (ed.), *Human Nature and Collective Theory*. Englewood Cliffs, NJ: Prentice-Hall Publishing Co.

Strauss, A., Fagerhaugh, S., Suczek, B., and Wiener, C. 1985. *The Social Organization of Medical Work*. Chicago: University of Chicago Press.

Strauss, A., Schatzman, L., Bucher, R., Ehrlich, D., and Sabshin, M. *Psychiatric Ideologies and Institutions*, 2nd ed. New Brunswick, NJ: Transaction Books.

Volberg, R. 1983. *Commitments and Constraints: The Development of Ecology in the United States, 1900–1940*. Doctoral dissertation, Department of Social and Behavioral Sciences, University of California, San Francisco.

Wiener, R. 1983. *The Politics of Alcoholism*. New Brunswick, NJ: Transaction Books.

Wimsatt, W. 1980. "Reductionist Research Strategies." *In* T. Nickles (ed.), *The Scientific Discovery: Case Studies*. Holland: Dordrecht, pp. 213–59.

1981. "Robustness, Reliability, and Overdetermination." *In* M. Brewer and B. Collins (eds.), *Scientific Inquiry and the Social Sciences*. San Francisco, CA: Jossey-Bass, pp. 124–62.

Author index

Subject index